72

Fortschritte der Chemie
organischer Naturstoffe

Progress in the
Chemistry of Organic
Natural Products

Founded by
L. Zechmeister

Edited by
W. Herz, G. W. Kirby,
R. E. Moore, W. Steglich,
and Ch. Tamm

Authors:
E. Fattorusso, R. J. Highet,
D. L. Klayman, A. Mangoni,
R. D. H. Murray, H. Ziffer

SpringerWienNewYork

Prof. W. HERZ, Department of Chemistry,
The Florida State University, Tallahassee, Florida, U.S.A.

Prof. G. W. KIRBY, Chemistry Department,
The University of Glasgow, Glasgow, Scotland

Prof. R. E. MOORE, Department of Chemistry,
University of Hawaii at Manoa, Honolulu, Hawaii, U.S.A.

Prof. Dr. W. STEGLICH, Institut für Organische Chemie der Universität München,
München, Federal Republic of Germany

Prof. Dr. CH. TAMM, Institut für Organische Chemie der Universität Basel,
Basel, Switzerland

This work is subject to copyright.
All rights are reserved, whether the whole or part of the material is concerned, specifically those of translation, reprinting, re-use of illustrations, broadcasting, reproduction by photocopying machines or similar means, and storage in data banks.

© 1997 by Springer-Verlag/Wien
Printed in Austria

Library of Congress Catalog Card Number AC 39-1015

Typesetting: Thomson Press (India) Ltd., New Delhi
Printing: Novographic, Ing. W. Schmid, A-1238 Wien
Graphic design: Ecke Bonk
Printed on acid-free and chlorine-free bleached paper

With 2 Figures

ISSN 0071-7886
ISBN 3-211-82879-6 Springer-Verlag Wien New York

Contents

List of Contributors X

Naturally Occurring Plant Coumarins
By R. D. H. MURRAY 1

  I. Scope of the Review 2

 II. Progress in the Past Six Years 3

III. Introduction to Tables 4
    Table 1.   7-Oxygenated Coumarins 6
               1.1   6-Substituted-7-Oxygenated Coumarins 9
               1.2   8-Substituted-7-Oxygenated Coumarins 15
               1.3   5,6-Disubstituted-7-Oxygenated Coumarins 28
               1.4   6,8-Disubstituted-7-Oxygenated Coumarins 28
    Table 2.   5,7-Dioxygenated Coumarins 29
    Table 3.   6,7-Dioxygenated Coumarins 40
    Table 4.   7,8-Dioxygenated Coumarins 45
    Table 5.   5,6,7-Trioxygenated Coumarins 48
    Table 6.   5,7,8-Trioxygenated Coumarins 50
    Table 7.   6,7,8-Trioxygenated Coumarins 52
    Table 8.   5,6,7,8-Tetraoxygenated Coumarins 54
    Table 9.   3-Substituted Coumarins 56
               9.1   3-Aryl-Substituted Coumarins 58
    Table 10. 4-Substituted Coumarins 59
               10.1 4-Aryl-Substituted Coumarins 61
    Table 11. Miscellaneous Coumarins 63
               11.1 3-Aryl-Oxygenated Coumarins 68
               11.2 Coumestans 69
    Table 12. Biscoumarins 70
    Table 13. Triscoumarins 78

  Amendments/Additions to Entries in Reference (*171*) and/or Reference [*172*] .. 79
    Table 1.   Coumarin and 7-Oxygenated Coumarins 79
               1.1 6-Substituted-7-Oxygenated Coumarins 80
               1.2 8-Substituted-7-Oxygenated Coumarins 82
    Table 2.   5,7-Dioxygenated Coumarins 85
    Table 3.   6,7-Dioxygenated Coumarins 88
    Table 4.   7,8-Dioxygenated Coumarins 89

Table 5.	5,6,7-Trioxygenated Coumarins	90
Table 6.	5,7,8-Trioxygenated Coumarins	91
Table 7.	6,7,8-Trioxygenated Coumarins	91
Table 8.	5,6,7,8-Tetraoxygenated Coumarins	91
Table 9.	3-Substituted Coumarins	92
	9.1 3-Aryl-Substituted Coumarins	93
Table 10.	4-Substituted Coumarins	94
	10.1 4-Aryl-Substituted Coumarins	95
Table 11.	Miscellaneous Coumarins	96
	11.1 3-Aryl-Oxygenated Coumarins	97
	11.2 Coumestans	97
Table 12.	Biscoumarins	98

Formula Index 99

Trivial Name Index 102

References 105

Artemisinin: An Endoperoxidic Antimalarial from *Artemisia annua* L.
By H. ZIFFER, R. J. HIGHET, and D. L. KLAYMAN 121

1. Introduction 123
    1.1. Historical 123
    1.2. History of the Qinghao Plant 124
    1.3. Modern History of *Artemisia annua* L. in Chinese Medicine 125
2. *Artemisia annua* L. and its Constituents 126
    2.1. Taxonomy 126
    2.2. Geographic Distribution 126
    2.3. Cultivation 126
    2.4. Cell Culture 127
    2.5. Isolation of Artemisinin from *A. annua* 127
    2.6. Other Constituents of *A. annua* 127
3. Artemisinin Structure Determination 128
4. Artemisinin Syntheses 129
    4.1. From ($-$)-Isopulegol 129
    4.2. From (R)-($+$)-Hydroxymenthol 131
    4.3. From 3(R)-Methyl-6-phenylsulfinyl-cyclohexanone 133
    4.4. From ($+$)-Isolimonene 133
    4.5. From Artemisinic Acid 133
    4.6. From Arteannunin B 136
5. Physical Measurements and Analyses 137
    5.1. NMR 137
        5.1.1. ^{1}H 137
        5.1.2. ^{13}C 139
    5.2. Circular Dichroism 139
    5.3. Infrared 139
    5.4. Mass Spectroscopy 140
    5.5. X-Ray Crystallography 141

5.6.	Quantitative TLC		141
5.7.	Titrimetric		141
5.8.	HPLC		141
	5.8.1.	Electrochemical Detection	141
	5.8.2.	UV Detection Methods	142
	5.8.3.	Capillary Gas Chromatography	143
	5.8.4.	Diverse Analytical Methods	143
	5.8.5.	Radiolabelling	143
	5.8.6.	Radioimmuno Assay	143

6. Reactions of Artemisinin and its Derivatives 144
   6.1. Thermolysis 144
   6.2. Chemical 145
      6.2.1. Reactions with Alkali 145
      6.2.2. Reactions with Ammonia and Amines 146
      6.2.3. Reactions with Acid 148
         6.2.3.1. Arteether 148
         6.2.3.2. Dihydroartemisinin 149
         6.2.3.3. Acid-Catalyzed Additions to Anhydro-
                  dihydroartemisinin 150
            6.2.3.3.1. Triphenylphosphine Hydrobromide 150
            6.2.3.3.2. p-Toluenesulfonic Acid 150
         6.2.3.4. Acid-Catalyzed Rearrangements of Artemisinin
                  Derivatives 150
            6.2.3.4.1. Lewis Acids 150
            6.2.3.4.2. Silica Gel-Catalyzed Rearrangements 150
      6.2.4. Reaction of Artemisinin with Reducing Agents 153
         6.2.4.1. Lithium Aluminum Hydride 153
         6.2.4.2. Sodium Borohydride 153
         6.2.4.3. A Mixture of Sodium Borohydride and
                  Boron Trifluoride 153
         6.2.4.4. Hydrogenation 154
      6.2.5. Bromination 155
      6.2.6. Fluorinated Artemisinin Derivatives 155
      6.2.7. Epoxidation of Anhydrodihydroartemisinin 156
      6.2.8. Osmium Tetroxide Oxidation 157

7. Dihydroartemisinin Derivatives 158
   7.1. Derivatives with Enhanced Oil Solubility 158
      7.1.1. Ethers 158
      7.1.2. Esters 159
      7.1.3. Carbonates 159
   7.2. Derivatives with Enhanced Water Solubility 160
      7.2.1. Sodium Artesunate 160
      7.2.2. Sodium Artelinate and Related Derivatives 160
   7.3. Artemisinin Derivatives 161
      7.3.1. (+)-Deoxoartemisinin 161
      7.3.2. (+)-Homodeoxoartemisinin 162
      7.3.3. (+)-10-Alkyldeoxoartemisinin 163
      7.3.4. (+)-10β-Allyldeoxoartemisinin 164

7.3.5. C-3 and C-9 Substituted 10-Deoxoartemisinins 164
7.3.6. C-14 Modified Deoxoartemisinins 165

8. Simplified Artemisinin Derivatives 166
   8.1. 9-Desmethylartemisinin 166
   8.2. 6,9-Bisnorartemisinin 166
   8.3. (+)-8a,9-Secoartemisinin 167
   8.4. (+)-4,5-Secoartemisinin 168
   8.5. (+)-Hexahydroisochroman-3-one 170
   8.6. 4,5-Desethanoartemisinin 170
   8.7. 9-Alkyl-9-desmethylartemisinin 170
   8.8. C-3 and C-9 Modified Artemisinin Derivatives 171
   8.9. Carba-Analogs of Artemisinin 172

9. Quantitative Structure-Activity Analyses 173

10. Tricyclic 1,2,4-Trioxane Analogs 174

11. Metabolism 175
    11.1. Microbial Metabolites of Artemisinin and its Derivatives 175
    11.2. Mammalian Metabolites 177

12. Test Data of Artemisinin Derivatives 178
    12.1. *In Vitro* 178
    12.2. *In Vivo* 188

13. Toxicity 193

14. Pharmacology and Pharmacokinetics 194

15. Clinical Evaluation of Artemisinin and Derivatives 195
    15.1. Dihydroartemisinin 196

16. Mechanisms of Action 196

17. Other Peroxides 198
    17.1. Naturally Occurring Peroxides 198
    17.2. Synthetic Peroxides 199

18. Conclusion 201

References 202

Marine Glycolipids
By E. Fattorusso and A. Mangoni 215

1. Introduction 215

2. Isolation Procedures 218

3. Determination of the Structure of Glycolipids 219
   3.1. Determination of the Structure of the Sugar Portion 220
   3.2. Determination of the Structure of the Lipid Portion 222

4. Glycoglycerolipids 223

5. Glycosphingolipids 228
   5.1. Neutral Glycosphingolipids 229

	5.2.	Phosphorus-Containing Glycosphingolipids	247
	5.3.	Gangliosides	260
6.	Other Glycolipids		281
7.	Biological Activities		287
	7.1.	Immunological Activity	287
	7.2.	Pharmacological Activity	288

References 291

Author Index 303

Subject Index 319

List of Contributors

FATTORUSSO, Prof. E., Dipartimento di Chimica delle Sostanze Naturali, Facoltà di Farmacia, Università degli Studi di Napoli Federico II, Via Domenico Montesano 49, I-80131 Napoli, Italy

HIGHET, Prof. R. J., Laboratory of Biophysical Chemistry, National Heart, Lung and Blood Institute, National Institutes of Health, Bethesda, MD 20892, U.S.A.

KLAYMAN, Dr. D. L., Walter Reed Army Institute of Research, Division of Experimental Therapeutics, Washington, DC 20307, U.S.A.

MANGONI, Dr. A., Dipartimento di Chimica delle Sostanze Naturali, Facoltà di Farmacia, Università degli Studi Napoli Federico II, Via Domenico Montesano 49, I-80131 Napoli, Italy

MURRAY, Prof. R. D. H., "Selsey", 25 Haugh Road, Dalbeattie DG5 4AR, Scotland

ZIFFER, Prof. H., Laboratory of Chemical Physics, National Institute of Diabetes, Digestive and Kidney Diseases, National Institutes of Health, Bldg. 5, Room B1-31, Bethesda, MD 20892-0510, U.S.A.

Naturally Occurring Plant Coumarins

R. D. H. MURRAY

Chemistry Department, University of Glasgow,
Glasgow G12 8QQ, Scotland

Contents

I. Scope of the Review	2
II. Progress in the Past Six Years	3
III. Introduction to Tables	4
Table 1. 7-Oxygenated Coumarins	6
1.1 6-Substituted-7-Oxygenated Coumarins	9
1.2 8-Substituted-7-Oxygenated Coumarins	15
1.3 5,6-Disubstituted-7-Oxygenated Coumarins	28
1.4 6,8-Disubstituted-7-Oxygenated Coumarins	28
Table 2. 5,7-Dioxygenated Coumarins	29
Table 3. 6,7-Dioxygenated Coumarins	40
Table 4. 7,8-Dioxygenated Coumarins	45
Table 5. 5,6,7-Trioxygenated Coumarins	48
Table 6. 5,7,8-Trioxygenated Coumarins	50
Table 7. 6,7,8-Trioxygenated Coumarins	52
Table 8. 5,6,7,8-Tetraoxygenated Coumarins	54
Table 9. 3-Substituted Coumarins	56
9.1 3-Aryl-Substituted Coumarins	58
Table 10. 4-Substituted Coumarins	59
10.1 4-Aryl-Substituted Coumarins	61
Table 11. Miscellaneous Coumarins	63
11.1 3-Aryl-Oxygenated Coumarins	68
11.2 Coumestans	69
Table 12. Biscoumarins	70
Table 13. Triscoumarins	78
Amendments/Additions to Entries in Reference (171) and/or Reference [172]	79
Table 1. Coumarin and 7-Oxygenated Coumarins	79
1.1 6-Substituted-7-Oxygenated Coumarins	80
1.2 8-Substituted-7-Oxygenated Coumarins	82
Table 2. 5,7-Dioxygenated Coumarins	85

Table 3.	6,7-Dioxygenated Coumarins	88
Table 4.	7,8-Dioxygenated Coumarins	89
Table 5.	5,6,7-Trioxygenated Coumarins	90
Table 6.	5,7,8-Trioxygenated Coumarins	91
Table 7.	6,7,8-Trioxygenated Coumarins	91
Table 8.	5,6,7,8-Tetraoxygenated Coumarins	91
Table 9.	3-Substituted Coumarins	92
	9.1 3-Aryl-Substituted Coumarins	93
Table 10.	4-Substituted Coumarins	94
	10.1 4-Aryl-Substituted Coumarins	95
Table 11.	Miscellaneous Coumarins	96
	11.1 3-Aryl-Oxygenated Coumarins	97
	11.2 Coumestans	97
Table 12.	Biscoumarins	98
Formula Index		99
Trivial Name Index		102
References		105

I. Scope of the Review

This review of plant coumarins discovered between 1989 and early 1996 has been compiled on the premise that the reader has access to the two previous reviews in this series by the reviewer, those of 1978 in Vol. 35 (*171*) and 1991 in Vol. 58 (*172*). With these other two reviews to hand, the reader will have immediate access to every plant coumarin ever known and should readily be able to determine, for example, if a coumarin just isolated is indeed a new natural plant product.

In the 1978 review, the 502 naturally occurring monomeric plant coumarins known at that time were tabulated principally according to the number and orientation of oxygen atoms on the benzenoid ring and then by the oxidation level of the substituent. For every entry, leading references were given to the isolation, structure determination, stereochemistry assignment where relevant and synthesis, where effected, of the coumarin. A similar format was adopted in 1991 and for comparability the reviewer has here again presented the data in the style used earlier.

The 1978 review specifically excluded aryl-substituted and biscoumarins even though examples were known. This was rectified in the 1991 review with the inclusion of a further 750 monomeric coumarins which included 111 aryl-substituted coumarins and coumestans. Data on 34 biscoumarins and one triscoumarin were also given. A similar format to that of the 1991 review has been adopted here but aflatoxins, benzocoumarins and ellagic acid derivatives have again been excluded.

References, pp. 105–119

II. Progress in the Past Six Years

Comparison of the entries in each of Tables 1–8 of the 1991 review with those of the 1978 review revealed that the numbers of each type of coumarin discovered during 1978–1989 were almost identical with those found during the previous 158 years since the first isolation of coumarin in 1820. Now, some six years later, comparison of the 1996 entries in Tables 1–8 with those of 1991 shows that the number of new coumarins has approximately halved indicating a similar rate of isolation. In the opinion of the reviewer, however, the numbers of new coumarins reported are, not surprisingly, beginning to fall. Compared with the period 1978–1989, there has been a relative increase in 5,7- and 6,7-dioxygenated coumarins (Tables 2 and 3) with a corresponding decrease in ethers of 7-hydroxycoumarin (Table 1) while there has been a marked increase in 5,6,7,8-tetraoxygenated coumarins (Table 8). Thirty biscoumarins, many from *Citrus* plants and hybrids have been newly reported (Table 12) and three triscoumarins (Table 13).

The power of modern spectrometric techniques, especially nuclear magnetic resonance, has been elegantly displayed in many of the structural elucidations, some on extremely small amounts of material. Apart from many isolation procedures to be found in the leading references on specific coumarins, the reader's attention is directed towards additional publications on solvent extraction studies (*31*) and ultrasound-assisted furanocoumarin extraction (*37*) and separation methods using capillary electrophoresis (*182*), micellar electrokinetic capillary chromatography (*40*), centrifugal partition chromatography (*157*) and two-dimensional planar chromatography (*81*). A review has appeared on high-performance liquid chromatography of coumarins (*231*) with additional publications on analysis by high-performance liquid chromatography (*21, 29, 75, 83, 181, 226, 229*) micro high-performance liquid chromatography-mass spectrometry (*34*) and capillary gas chromatography-mass spectrometry (*23*). An easy and absolute diagnosis for coumarin/chromone discrimination used ^{17}O n.m.r. spectroscopy at natural abundance (*173*).

Many of the new coumarins isolated in the past six years could well have been anticipated as natural products being, for example, glycosides of a known aglycone with a different sugar, an alcohol acylated with a different acid or a coumarin with a known side chain but at a higher oxidation level. However, some structures would have been much less easy to predict such as microminutin (**83**), the initially proposed structure (*197*) being shown to be incorrect and later elegantly found to contain a five-carbon bicyclic moiety (*198*), and the unstable dihydroquinone (**250**), a supposed intermediate in the biosynthesis of miroestrol (*45*).

While esters of (−)-*cis*-khellactone (**90**) and (+)-*trans*-khellactone (**91**) are known, bioactivity-guided fractionation has led to the isolation of the cytotoxic parent diols (*62,63*). Anti-HIV bioassay-guided fractionation has also been instrumental in the isolation of calanolide A (**264**) and B (**262**) and related 4,6,8-trisubstituted-5,7-dioxygenated coumarins from *Calophyllum lanigerum* (*122*). The structures assigned to calanolide C and D were subsequently shown to be incorrect (*163,191*) and the two compounds have been renamed pseudocalanolide C (**260**) and D (**261**).

There has been a marked increase in the number of acrimarines, naturally-occurring acridone-coumarin dimers, from three to 13, isolated in Japan mainly from *Citrus* hybrids (Table 1.1) along with five neoacrimarines. The structure originally proposed (*242*) for acrimarine-L is not given in the Tables since the compound was later shown to be identical with acrimarine-C (**360**) and the name acrimarine-L removed (*246*). Two further coumarin-naphthoquinone dimers, pummeloquinone (**80**) (*109*) and toddacoumaquinone (**158**) (*98, 102*) have been reported.

The simple, albeit unexpected, 4-allyloxycoumarin structure assigned to setarin (**277**) has been confirmed by its thermal rearrangement to 3-allyl-4-hydroxycoumarin (*112*). On the other hand, the structure assigned to yuehgesin-A (**65**) (*143*) is most surprising in that it would appear to be unambiguously confirmed by a variety of ^{1}H n.m.r. techniques yet it is an acyclic hemiketal of a tertiary alcohol and acetone and should not be stable.

In the Amendments/Additions Tables which update data in the two earlier reviews many of the entries refer to synthesis. However, supposedly new coumarins continue to be reported but later have their structures revised to those of known coumarins, such as asacoumarin B = galbanic acid (**350**) (*13*) and bakuchicin (**124**) = angelicin (**367**) (*164*).

In an elegant piece of work using supercritical carbon dioxide extraction of *Toddalia asiatica*, the diol toddalolactone (**379**) has been shown to be a genuine natural coumarin but in the original 1933 studies it had been isolated principally as an artifact derived from aculeatin (**377**) during extraction (*101*). The structure of the 6,7,8-trioxygenated coumarin obtusifol (**407**) has been revised for a second and, it is hoped, final time (*27*).

III. Introduction to Tables

The arbitrary but biogenetically-related classification used for tabulating coumarins in the two earlier reviews (*171, 172*) has once again been employed in order to assist the reader. It is based primarily on the number, and then position, of oxygen atoms attached directly to the coumarin

References, pp. 105–119

nucleus. Thereafter, within each table, entries are presented in the following order: (i) coumarins with acyclic substituents, (ii) dihydrofuranocoumarins, (iii) furanocoumarins, (iv) dihydropyranocoumarins, (v) pyranocoumarins. The coumarins of each subclass are listed in order of increasing number of carbon atoms in the substituent and in increasing oxidation level within that group. Phenols are considered before their ethers and glycosides while alcohols precede their glycosides and esters.

For each entry, the trivial name if one has been coined, is given first followed by the year of isolation, the structure and molecular formula. The melting point of crystalline coumarins is quoted; alternatively, the physical form as an oil or gum or amorphous (amorph.) is reported. The $[\alpha]_\lambda^t$ and solvent columns refer to the specific rotation at t°C in the given solvent at a given wavelength, λ (nm). Where no wavelength is quoted, as in most cases, the rotation has been measured at 589 nm. The plant source from which the coumarin was first isolated is then given. Where more than one plant source is quoted, the later reference has provided additional information such as another trivial name and/or different physical constants.

The naturally occurring aryl-substituted coumarins which were not discussed in the 1978 review (*171*), even though a number were then known, but were incorporated in the 1991 review (*172*) are again to be found in Tables 9.1, 10.1 and 11.1, with data on six new coumestans presented in Table 11.2.

An asterisk (*) in the top right of the structure column indicates that some aspect of the stereochemistry remains to be defined. In cases where the relative stereochemistry is shown the asterisk implies that the absolute stereochemistry has not yet been defined; *racemic* substances are so indicated.

In the later tables giving amendments/additions to data in the two earlier reviews, the compound numbers given in parenthesis below numbers (**344–434**) inclusive are those which appeared in the 1978 review while those numbers in square brackets refer to the compound entry numbers in the 1991 review.

Table 1. 7-Oxygenated Coumarins

	Trivial name(s)	Year isolated	Structure	Formula	M.p.	$[\alpha]_\lambda^t$	Solvent	Plant sources	Leading references
1	Crellisin-B	1995	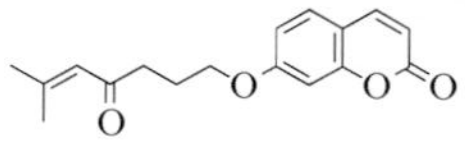	$C_{17}H_{18}O_4$				Cremanthodium ellisi	(265)
2	Anisocoumarin H	1989	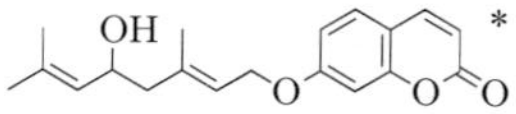	$C_{19}H_{22}O_4$	oil	-20.5^{25}	CHCl$_3$	Clausena anisata	(176)
3	Acetoxyaurapten	1995	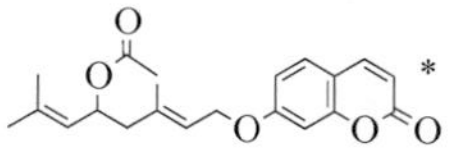	$C_{21}H_{24}O_5$	53–55	-30.0^{23}	CHCl$_3$	Zanthoxylum schinifolium	(42)
4		1992	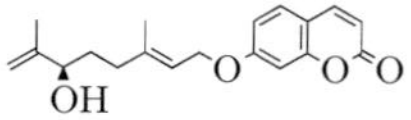	$C_{19}H_{22}O_4$	75	10^{26}	CHCl$_3$	Citrus hassaku	(158)
5		1992	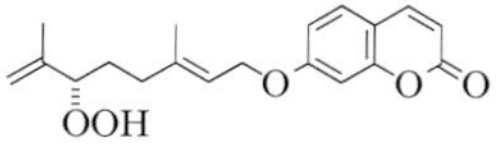	$C_{19}H_{22}O_5$	gum	-11	CHCl$_3$	Phebalium filifolium	(210)
6		1992	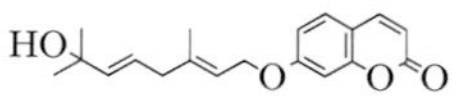	$C_{19}H_{22}O_4$	oil	24	CHCl$_3$	Phebalium anceps	(196)
7	(−)-(S)-trans-Marmin	1995	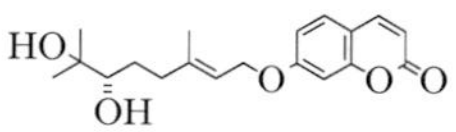	$C_{19}H_{24}O_5$	117–118	-11^{26}	MeOH	Pituranthos triradiatus	(78)
8	Pituranthoside	1995	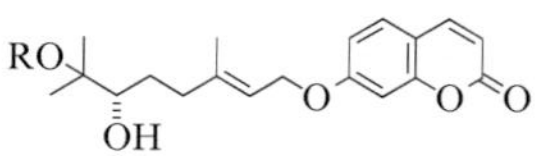 R = β-D-glucosyl	$C_{25}H_{34}O_{10}$	185–186	12^{26}	MeOH	Pituranthos triradiatus	(78)
9		1992	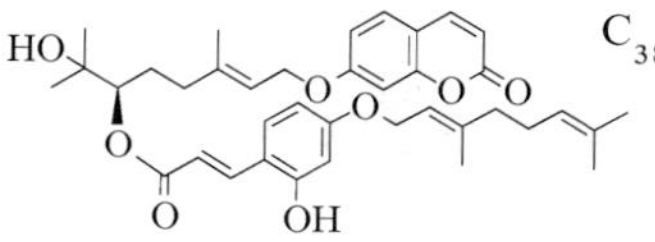	$C_{38}H_{46}O_8$	oil	72^{26}	EtOH	Citrus hassaku	(158)

10	Chloromarmin	1995	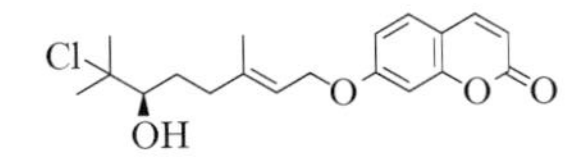 [structure]	$C_{19}H_{23}ClO_4$	oil	27.3^{23}	EtOH	*Aegle marmelos*	(183)
11	Aeglin	1995	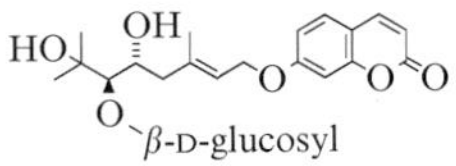 [structure with β-D-glucosyl]	$C_{25}H_{34}O_{11}$	218–219	25.9^{24}	EtOH	*Aegle marmelos*	(183)
12		1994	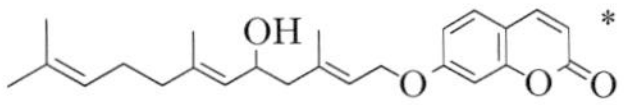 [structure]	*$C_{24}H_{30}O_4$	53	-3.8^{25}	CH_2Cl_2	*Ferula assafoetida*	(13)
13		1994	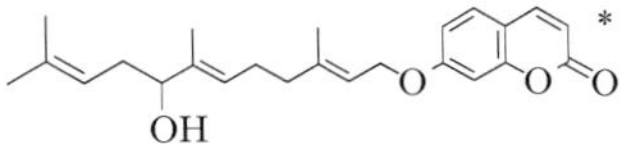 [structure]	*$C_{24}H_{30}O_4$	oil	-2.4^{25}	CH_2Cl_2	*Ferula assafoetida*	(13)
14		1994	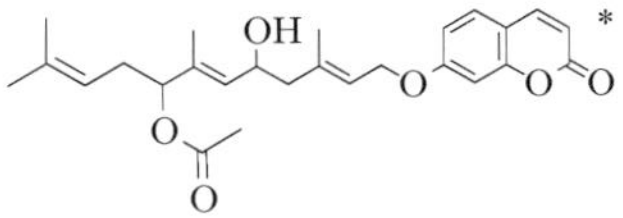 [structure]	*$C_{26}H_{32}O_6$	oil	-2.8^{25}	CH_2Cl_2	*Ferula assafoetida*	(13)
15		1992	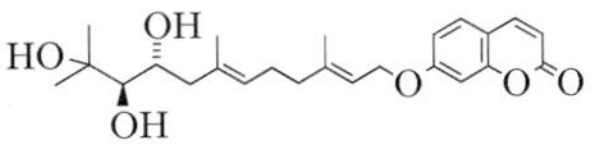 [structure]	$C_{24}H_{32}O_6$	oil	-14.3^{25}	CH_2Cl_2	*Heptaptera anisoptera*	(11)
16	Ligupersin B	1991	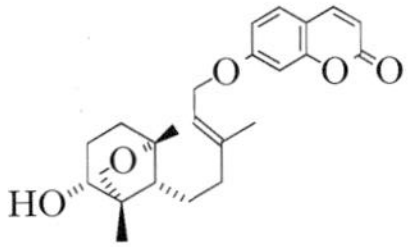 [structure]	$C_{24}H_{30}O_5$	oil	-8^{24}	$CHCl_3$	*Ligularia persica*	(156)
17		1992	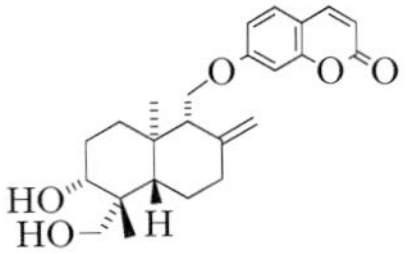 [structure]	$C_{24}H_{30}O_5$	220	-37.4^{25}	CH_2Cl_2	*Heptaptera anisoptera*	(12)

Table 1 (continued)

	Trivial name(s)	Year isolated	Structure	Formula	M.p.	$[\alpha]_\lambda^t$	Solvent	Plant sources	Leading references
18		1992	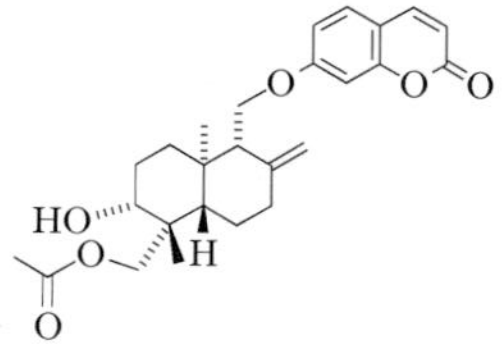	$C_{26}H_{32}O_6$	120	-31.5^{25}	CH_2Cl_2	Heptaptera anisoptera	(12)
19		1992	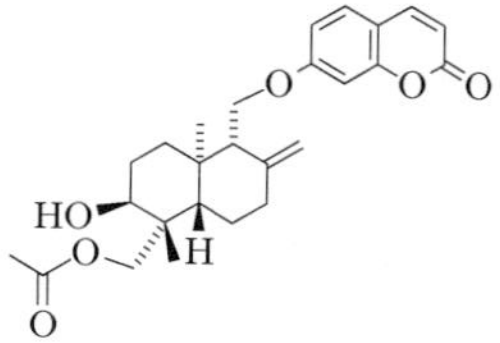	$C_{26}H_{32}O_6$	oil	-22.0^{25}	CH_2Cl_2	Heptaptera anisoptera	(12)
20	Ligupersin A	1991	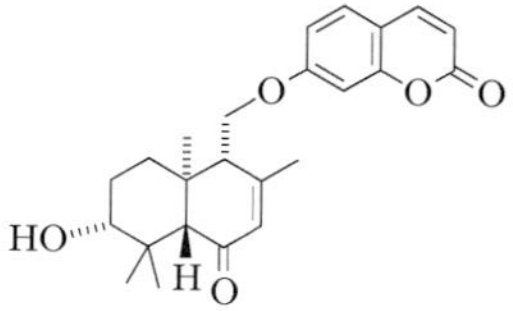	$C_{24}H_{28}O_5$	oil	-70^{24}	$CHCl_3$	Ligularia persica	(156)

Table 1.1. 6-Substituted-7-Oxygenated Coumarins

	Trivial name(s)	Year isolated	Structure	Formula	M.p.	$[\alpha]_\lambda^t$	Solvent	Plant sources	Leading references
21	Buntansin B	1994	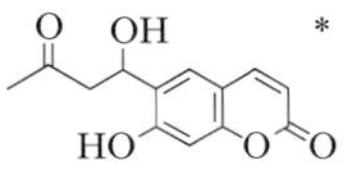 *	$C_{14}H_{14}O_5$	162–163	−3.76	MeOH	*Citrus grandis*	(262)
22		1994	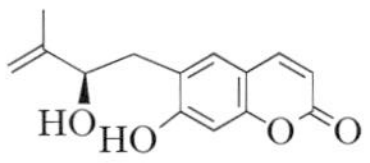	$C_{14}H_{14}O_4$	165–170	8	CHCl$_3$	*Boronia lanceolata*	(6)
23	Peroxytamarin	1991	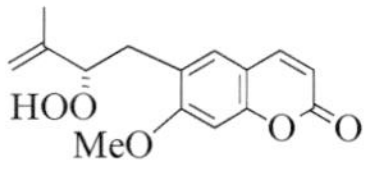	$C_{15}H_{16}O_5$	oil	−5.3	CHCl$_3$	*Citrus sulcata*	(103)
24	Peujaponiside	1994	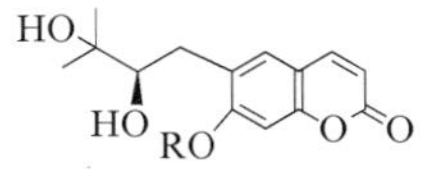 R = 6′-β-D-apiosyl-β-D-glucosyl	$C_{25}H_{34}O_{14}$	powder	−18.7^{22}	MeOH	*Peucedanum japonicum*	(90)
25	Albiflorin-1	1991	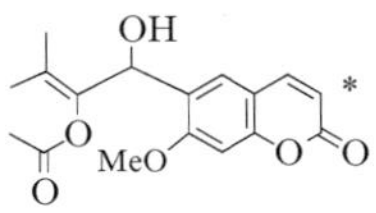 *	$C_{17}H_{18}O_6$	128	−1.0	CHCl$_3$	*Boenninghausenia albiflora*	(115)
26	Angelitriol	1995	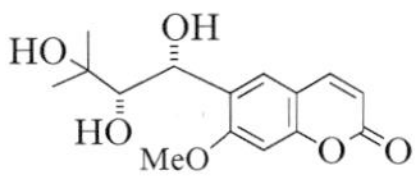	$C_{15}H_{18}O_6$	167–169	−120.4	EtOH	*Angelica pubescens* f. *biserrata*	(147)

Table 1.1 (continued)

	Trivial name(s)	Year isolated	Structure	Formula	M.p.	$[\alpha]_\lambda^t$	Solvent	Plant sources	Leading references
27	Buntansin C	1994	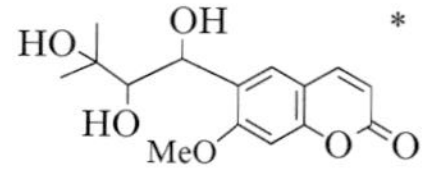	$C_{15}H_{18}O_6$	164–166	−210.1	MeOH	*Citrus grandis*	(262)
28	Angelol J	1995	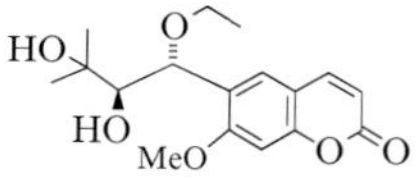	$C_{17}H_{22}O_6$	oil	−15.4	EtOH	*Angelica pubescens* f. *biserrata*	(147)
29	Angelol K	1995	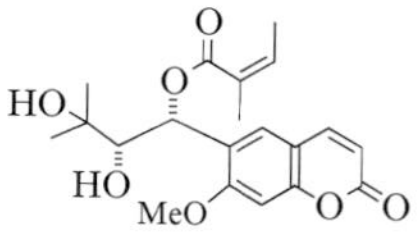	$C_{20}H_{24}O_7$	116–118	−122.4	EtOH	*Angelica pubescens* f. *biserrata*	(146)
30	Angelol L	1995	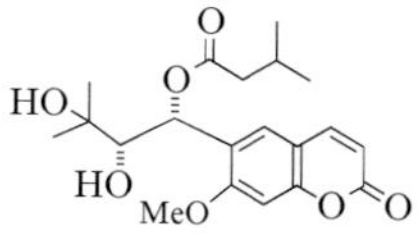	$C_{20}H_{26}O_7$	oil	−79.0	EtOH	*Angelica pubescens* f. *biserrata*	(146)
31		1992	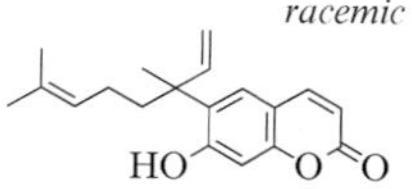	$C_{19}H_{22}O_3$	65			*Citrus hassaku*	(158)

#	Name	Year	Structure	Formula	State	mp/bp	Solvent	Source	Ref
32		1992	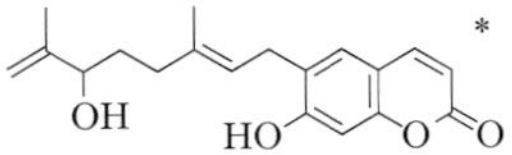	$C_{19}H_{22}O_4$	gum	4	$CHCl_3$	*Eriostemon tomentellus*	(208)
33		1992	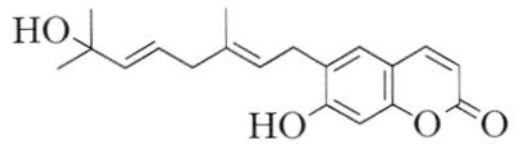	$C_{19}H_{22}O_4$	gum			*Eriostemon tomentellus*	(208)
34		1992	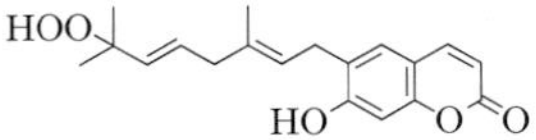	$C_{19}H_{22}O_5$	gum			*Eriostemon tomentellus*	(208)
35	Acrimarine-M	1992	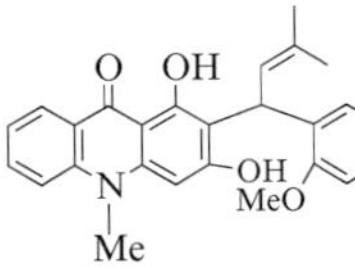 racemic	$C_{29}H_{25}NO_6$	oil			*Citrus paradisi × C. tangerina*	(214, 242)
36	Acrimarine-G	1990	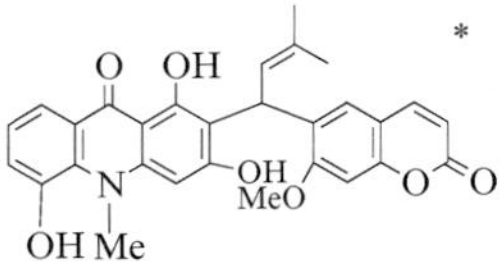	$C_{29}H_{25}NO_7$	oil	8.0	$CHCl_3$	*Citrus funadoko*	(72)
37	Acrimarine-H	1990	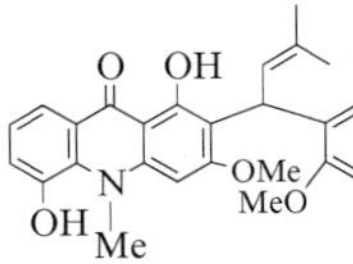 racemic	$C_{30}H_{27}NO_7$	oil			*Citrus unshiu × C. sinensis*	(110)
38	Acrimarine-K	1992	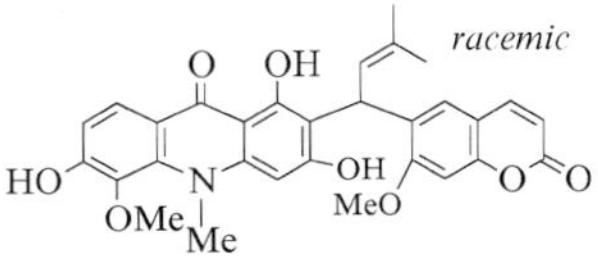 racemic	$C_{30}H_{27}NO_8$	oil			*Citrus paradisi × C. tangerina*	(242)

Table 1.1 (*continued*)

	Trivial name(s)	Year isolated	Structure	Formula	M.p.	$[\alpha]_\lambda^t$	Solvent	Plant sources	Leading references
39	Acrimarine-E	1990	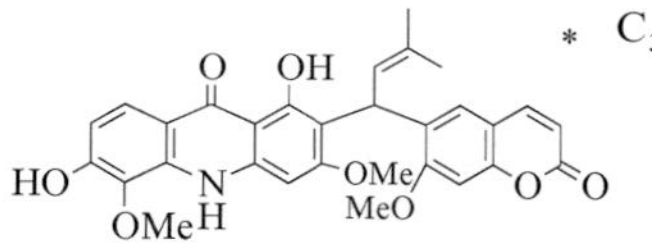 *	$C_{30}H_{27}NO_8$	274–276	20.1	Me_2CO	*Citrus funadoko*	(72)
40	Acrimarine-F	1990	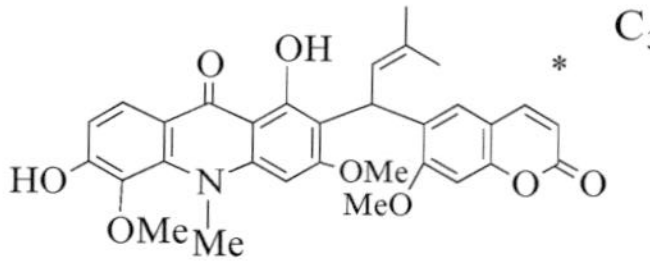 *	$C_{31}H_{29}NO_8$	powder			*Citrus funadoko*	(72)
41	Acrimarine-N	1994	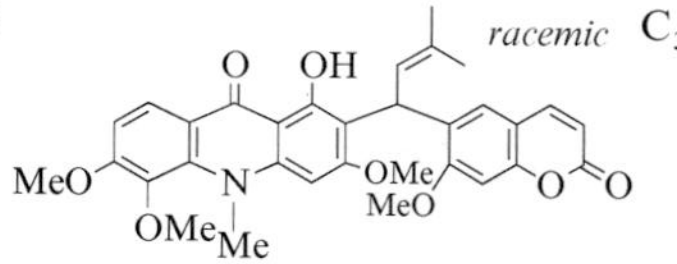 *racemic*	$C_{32}H_{31}NO_8$	oil			*Citrus paradisi* × *C. tangerina*	(246)
42	Acrimarine-I	1992	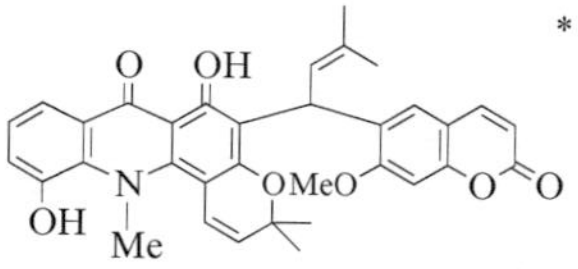 *	$C_{34}H_{31}NO_7$	oil	27.8^{26}	$CHCl_3$	*Citrus paradisi* × *C. tangerina*	(242)
43	Acrimarine-J	1992	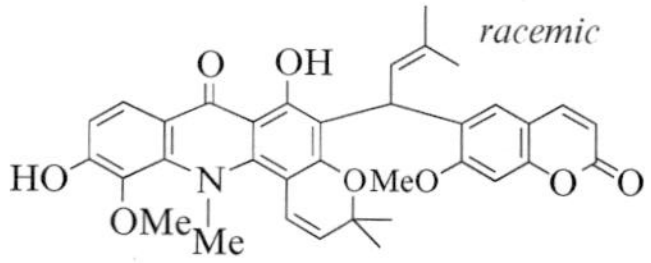 *racemic*	$C_{35}H_{33}NO_8$	oil			*Citrus paradisi* × *C. tangerina*	(242)

#	Name	Year	Structure	Formula	mp (°C)	[α]	Solvent	Source	Ref
44	Acrimarine-D	1990	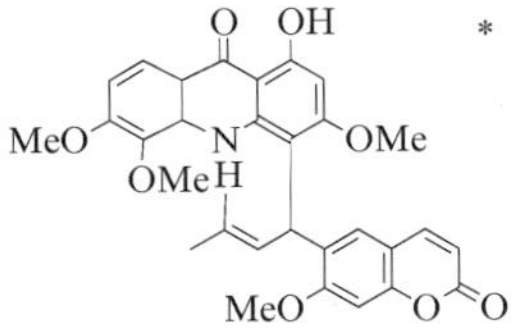 *	$C_{31}H_{29}NO_8$	oil	−3.0	$CHCl_3$	Citrus funadoko	(72)
45		1991	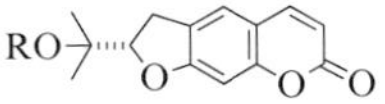 R = α-L-rhamnosyl	$C_{20}H_{24}O_8$	140–142			Aegle marmelos	(175)
46		1990	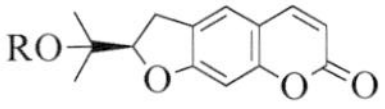 R = 6′-feruloyl-β-D-glucosyl	$C_{30}H_{32}O_{12}$	138–140	49.8	aq. EtOH	Notopterygium forbesii	(75)
47		1993	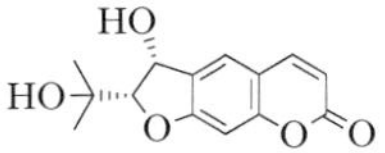	$C_{14}H_{14}O_5$	173–175	34^{25}_{578}	MeOH	Brosimum gaudichaudi	(259)
48	Decumbensol	1993	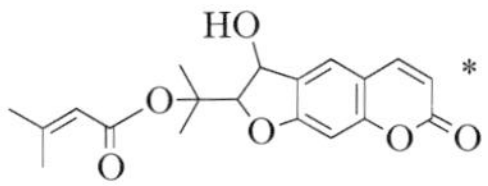 *	$C_{19}H_{20}O_6$	182–182.5	202^{20}	$CHCl_3$	Peucedanum decumbens	(202)
49		1995	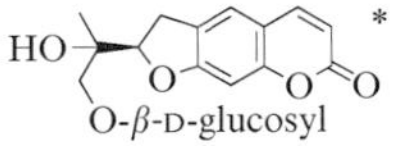 O-β-D-glucosyl *	$C_{20}H_{24}O_{10}$	152–154.5	$-22^{20.5}$	MeOH	Diplolophium buchananii	(142)
50		1992	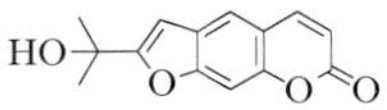	$C_{14}H_{12}O_4$	140–143			Phebalium anceps	(196)

Table 1.1 (continued)

	Trivial name(s)	Year isolated	Structure	Formula	M.p.	$[\alpha]_\lambda^t$	Solvent	Plant sources	Leading references
51	Ll-2	1991	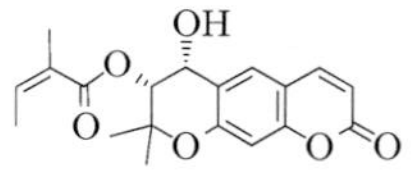	$C_{19}H_{20}O_6$		-55.5^{20}	EtOH	Libanotis laticalycina	(14)
52	Ll-1	1991	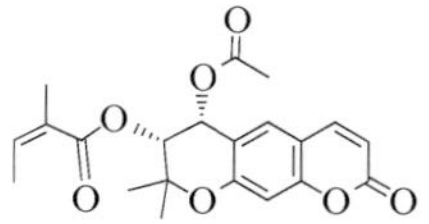	$C_{21}H_{22}O_7$		-63.3^{20}	EtOH	Libanotis laticalycina	(14)
53		1995	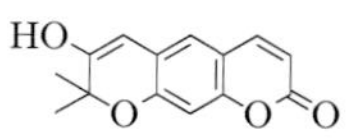	$C_{14}H_{12}O_4$	amorph.			Boronia algida	(222)
54	Qianhucoumarin F	1993	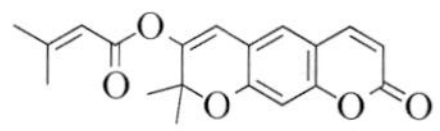	$C_{19}H_{18}O_5$	126–128			Peucedanum praeruptorum	(125, 128)

References pp. 105–119

Table 1.2. 8-Substituted-7-Oxygenated Coumarins

Trivial name(s)	Year isolated	Structure	Formula	M.p.	$[\alpha]_\lambda^t$	Solvent	Plant sources	Leading references
55	1990	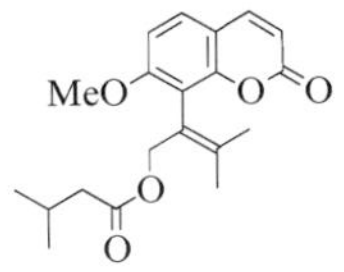 (structure with MeO)	$C_{20}H_{24}O_5$	syrup			*Murraya paniculata*	(104)
56	1990	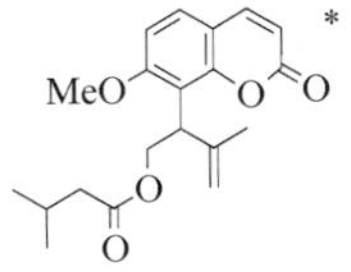 * (structure with MeO)	$C_{20}H_{24}O_5$	syrup	14.5	CHCl$_3$	*Murraya paniculata*	(104)
57	1992	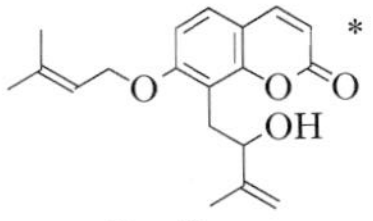 * (structure with OH)	$C_{19}H_{22}O_4$	gum	15	CHCl$_3$	*Eriostemon spicatus*	(209)
58 Isoarnottinin	1993	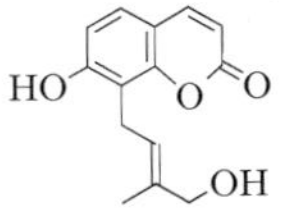 (structure with HO)	$C_{14}H_{14}O_4$	155–157			*Ammi majus*	(64)
59	1993	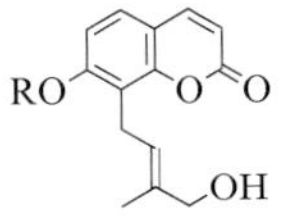 (structure with RO, OH) R = β-D-glucosyl	$C_{20}H_{14}O_9$	202–204			*Ammi majus*	(64)

Table 1.2 (continued)

	Trivial name(s)	Year isolated	Structure	Formula	M.p.	$[\alpha]_\lambda^t$	Solvent	Plant sources	Leading references
60		1993	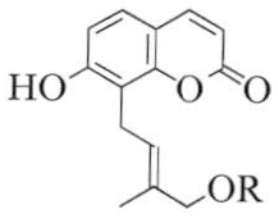	$C_{20}H_{14}O_9$	215–217			Ammi majus	(64)
61		1992	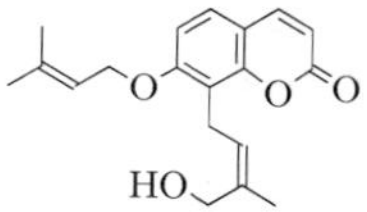	$C_{19}H_{22}O_4$	gum			Eriostemon spicatus	(209)
62		1994	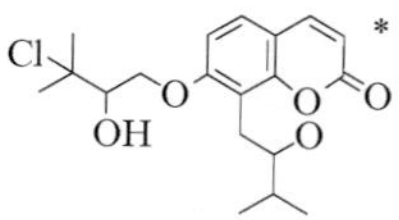	$C_{19}H_{23}ClO_5$	amorph.			Triphasia trifolia	(1)
63	Yuehgesin-B	1994	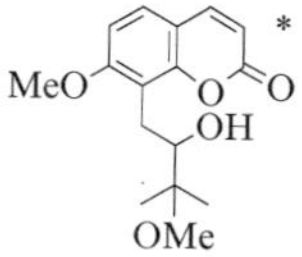	$C_{16}H_{20}O_5$	syrup	-2^{25}	$CHCl_3$	Murraya paniculata	(143)
64	Yuehgesin-C	1994	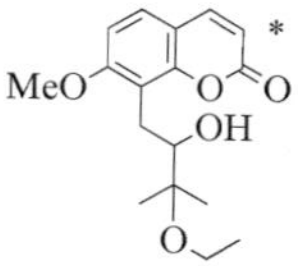	$C_{17}H_{22}O_5$	syrup			Murraya paniculata	(143)

#	Name	Year	Structure	Formula	mp (°C)	[α]	Solvent	Source	Ref
65	Yuehgesin-A	1994	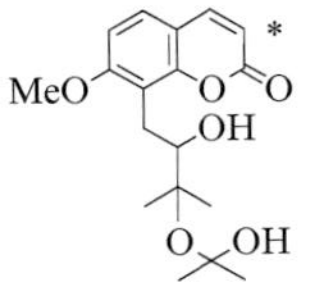	$C_{18}H_{24}O_6$	167–169	2^{25}	$CHCl_3$	*Murraya paniculata*	(143)
66		1991	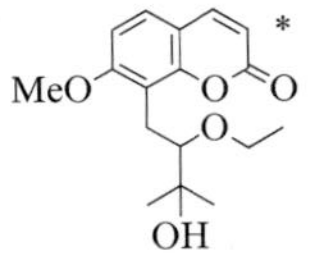	$C_{17}H_{22}O_5$				*Murraya exotica*	(46)
67		1990	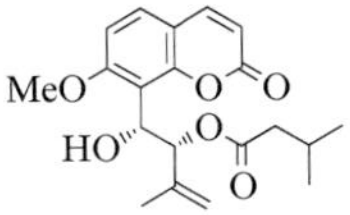	$C_{20}H_{24}O_6$	100–103	−2.97	$CHCl_3$	*Murraya paniculata*	(104)
68		1995	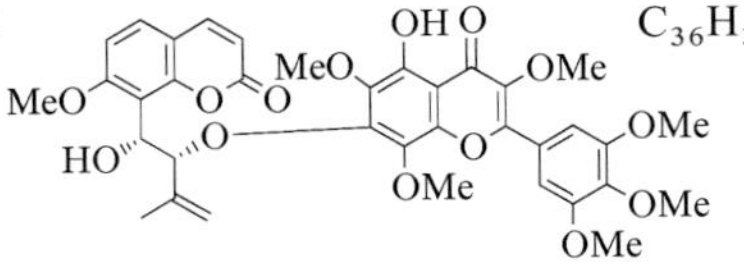	$C_{36}H_{36}O_{14}$	168–170			*Murraya exotica*	(56)
69		1990	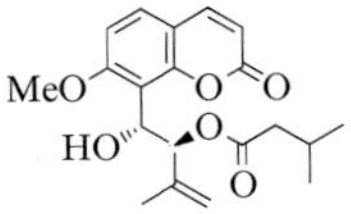	$C_{20}H_{24}O_6$	oil	40.9	$CHCl_3$	*Murraya paniculata*	(104)
70	Chloculol	1990	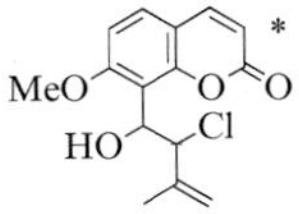	$C_{15}H_{15}ClO_4$	149–151	−23	$CHCl_3$	*Murraya paniculata*	(104)

Table 1.2 (continued)

	Trivial name(s)	Year isolated	Structure	Formula	M.p.	$[\alpha]_\lambda^t$	Solvent	Plant sources	Leading references
71	Albiflorin-2	1993	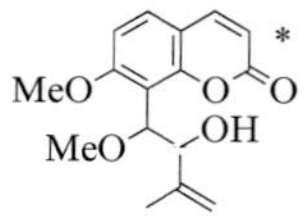 *	$C_{16}H_{18}O_5$	94–95	49.04	EtOH	Boenninghausenia albiflora	(116)
72	Albiflorin-3	1993	Diastereoisomer of (71)*	$C_{16}H_{18}O_5$	144–145			Boenninghausenia albiflora	(116)
73		1992	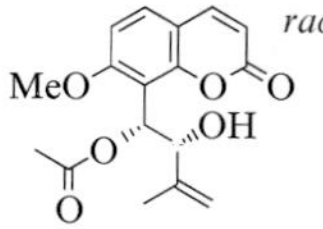 racemic	$C_{17}H_{18}O_6$				Phebalium phylicifolium	(196)
74	cis-Casegravol	1991	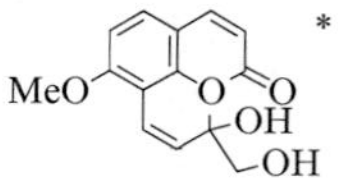 *	$C_{15}H_{16}O_5$	oil	−12.1	$CHCl_3$	Citrus sulcata	(103)
75	Dioxinoacrimarine-A	1994	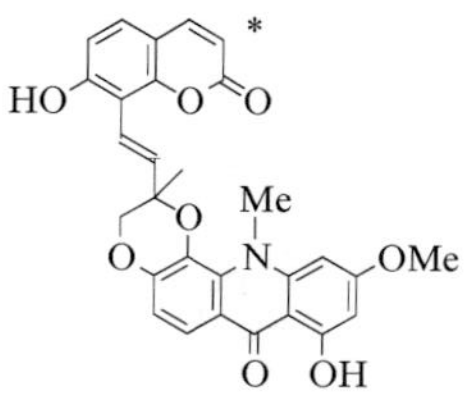	$C_{29}H_{23}NO_8$	179–181	18	$CHCl_3$	Citrus paradisi × C. tangerina	(246)

76	1992	[structure with MeO, OH, isobutyrate ester]	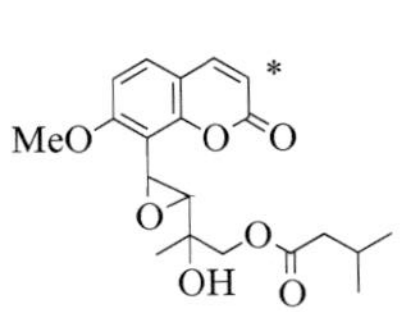 $C_{20}H_{24}O_7$	oil	25	$CHCl_3$	*Phebalium coxii*	(196)
77	1992	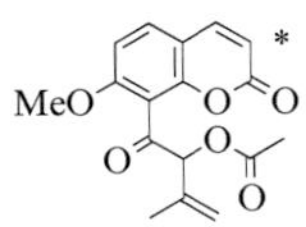 [structure with MeO, acetate]	$C_{17}H_{16}O_6$	oil			*Phebalium ralstonii*	(196)
78	1992	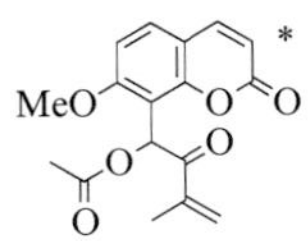 [structure with MeO, OAc]	$C_{17}H_{16}O_6$	oil	120	$CHCl_3$	*Phebalium bilobum*	(196)
79	1992	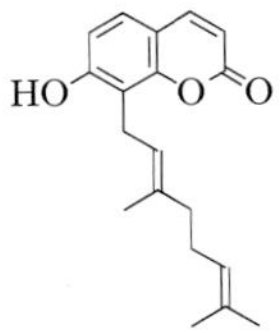 [structure with HO, geranyl]	$C_{19}H_{22}O_3$	gum			*Eriostemon tomentellus*	(208)
80 Pummeloquinone	1993	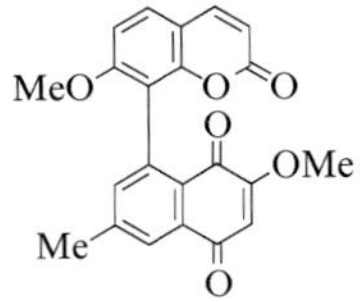 [structure with MeO, OMe, Me, quinone]	$C_{22}H_{16}O_6$	>25			*Citrus grandis* × *C. paradisi*	(109)

Table 1.2 (continued)

	Trivial name(s)	Year isolated	Structure	Formula	M.p.	$[\alpha]_\lambda^t$	Solvent	Plant sources	Leading references
81	Angelidiol	1993	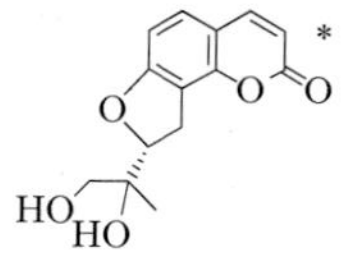	$C_{14}H_{14}O_5$				*Angelica pubescens*	(145)
82	Daucoidin B	1991	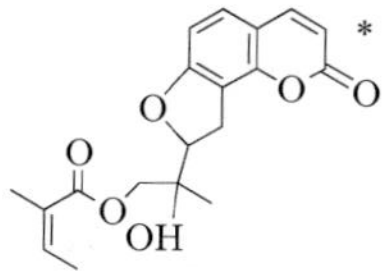	$C_{19}H_{20}O_6$	140–141	-18.2^{20}	$CHCl_3$	*Ligusticum daucoides*	(200, 202)
83	Microminutinin	1993	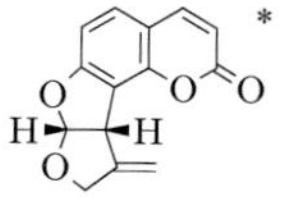	$C_{14}H_{10}O_4$	115–116	81	$CHCl_3$	*Micromelum minutum*	(197, 198)
84	Edulisin IV	1994	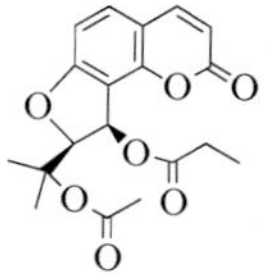	$C_{19}H_{20}O_7$	119–120d	71.2^{25}	$CHCl_3$	*Angelica edulis*	(167)

85	Edulisin III	1994	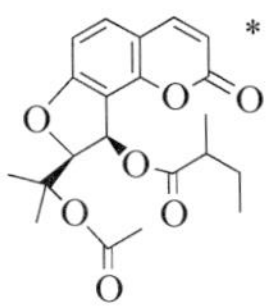	$C_{21}H_{24}O_7$	143–144d	166^{25}	CHCl$_3$	Angelica edulis	(167)
86	Daucoidin A	1991	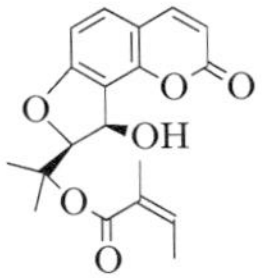	$C_{19}H_{20}O_6$		46^{20}	CHCl$_3$	Ligusticum daucoides	(200, 202)
87	Edulisin V	1994	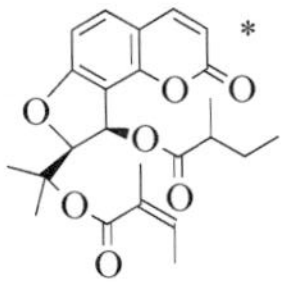	$C_{24}H_{28}O_7$	113.5–115	63.4^{25}	CHCl$_3$	Angelica edulis	(167)
88		1992	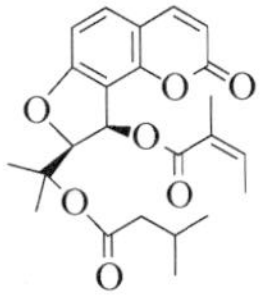	$C_{24}H_{28}O_7$	67–70			Angelica archangelica ssp. archangelica	(82)
89	Edulisin VI	1994	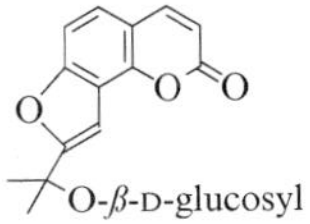	$C_{20}H_{22}O_9$				Angelica edulis	(166)

Table 1.2 (continued)

	Trivial name(s)	Year isolated	Structure	Formula	M.p.	$[\alpha]_\lambda^t$	Solvent	Plant sources	Leading references
90	(−)-cis-Khellactone	1992	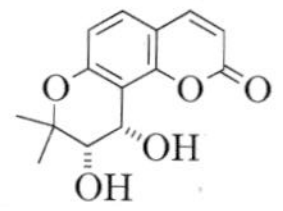	$C_{14}H_{14}O_5$	174–175	-82.4^{18}	$CHCl_3$	Peucedanum japonicum	(41,63,215)
91	(+)-trans-Khellactone	1991	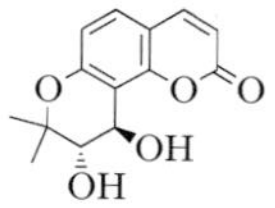	$C_{14}H_{14}O_5$	184–185	18.1^{18}	$CHCl_3$	Peucedanum japonicum	(62)
92	Qianhucoumarin C	1993	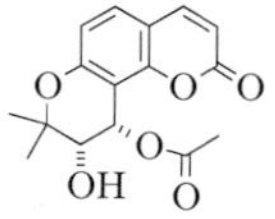	$C_{16}H_{16}O_6$	cryst.	7.6^{20}	$CHCl_3$	Peucedanum praeruptorum	(127)
93	d-Laserpitin	1990	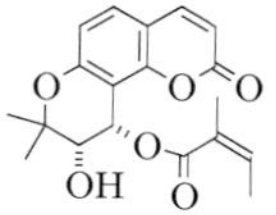	$C_{19}H_{20}O_6$		92.3^{21}	$CHCl_3$	Peucedanum zhongdia-nensis	(201, 202)
94	Qianhucoumarin A	1993	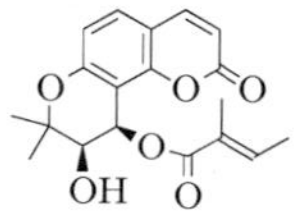	$C_{19}H_{20}O_6$	123.5–125.5	209.6^{20}	$CHCl_3$	Peucedanum praeruptorum	(128)

References pp. 105–119

95	1996	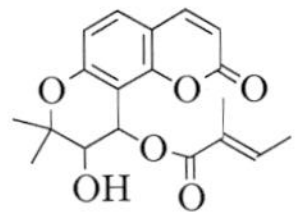 3′,4′-cis racemic	$C_{19}H_{20}O_6$	188–192			*Peucedanum japonicum* (41)
96	1996	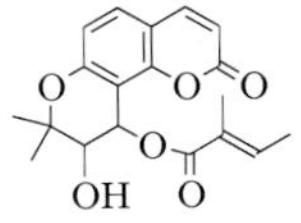 3′,4′-trans*	$C_{19}H_{20}O_6$	125–127	254.1^{25}	$CHCl_3$	*Peucedanum japonicum* (41)
97	Peujaponisinol B 1993	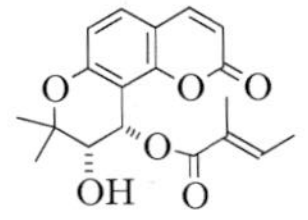	$C_{19}H_{20}O_6$	oil	-86.4^{22}	$CHCl_3$	*Peucedanum japonicum* (89)
98	Neoacrimarine-C 1993	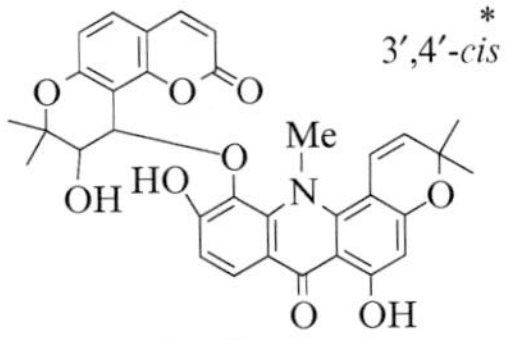 3′,4′-cis*	$C_{33}H_{29}NO_9$	111–113	42.3	$CHCl_3$	*Citrus hassaku* (247)
99	Qianhucoumarin B 1993	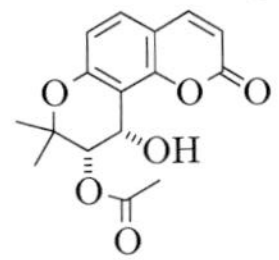	$C_{16}H_{16}O_6$	cryst.	3.9^{20}	$CHCl_3$	*Peucedanum praeruptorum* (127)
100	Qianhucoumarin D 1994	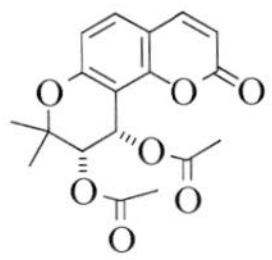	$C_{18}H_{18}O_7$	160.5–162.5	4.0^{20}	$CHCl_3$	*Peucedanum praeruptorum* (126)

Table 1.2 (continued)

Trivial name(s)	Year isolated	Structure	Formula	M.p.	$[\alpha]_\lambda^t$	Solvent	Plant sources	Leading references
101 Peucedanocoumarin III	1990	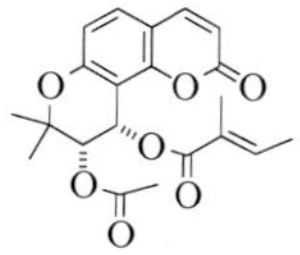	$C_{21}H_{22}O_7$	123–124	28.6^{24}	$CHCl_3$	Peucedanum praeruptorum	(241)
	1991			112–114	14.2^{18}	$CHCl_3$	Peucedanum japonicum	(62)
	1996			136–138	23.5^{25}	$CHCl_3$	Peucedanum japonicum	(41)
102	1996	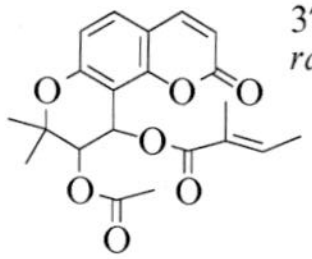 3',4'-cis racemic	$C_{21}H_{22}O_7$	124–126			Peucedanum japonicum	(41)
103 Peucedanocoumarin II	1990	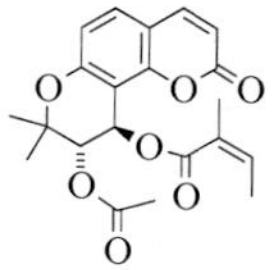	$C_{21}H_{22}O_7$	134.5–136d	7.0^{24}	$CHCl_3$	Peucedanum praeruptorum	(241)
104	1996	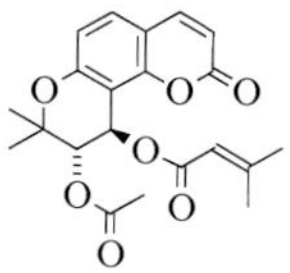	$C_{21}H_{22}O_7$	oil	-7.0^{25}	$CHCl_3$	Peucedanum japonicum	(41)
105 (−)-Visnadin	1992	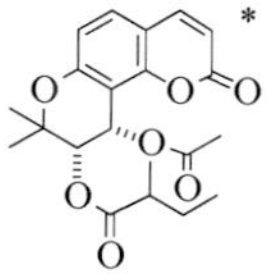 *	$C_{21}H_{24}O_7$	oil	-12.9^{22}	$CHCl_3$	Peucedanum japonicum	(88)

106	Peucedanocoumarin I	1990	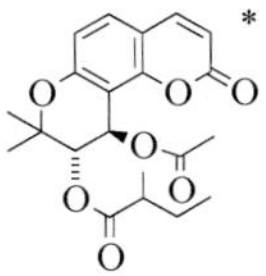 $C_{21}H_{24}O_7$	155–156.5d	24.2^{24} CHCl$_3$	*Peucedanum praeruptorum*	(241)
107	Badycoumarin A	1991	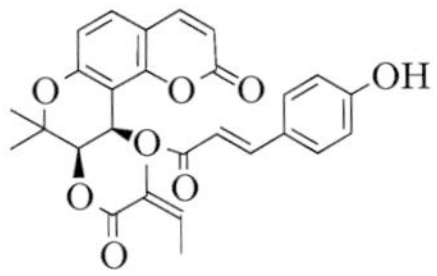 $C_{28}H_{26}O_8$	90–92	77.7^{27}	*Angelica flaccida*	(228)
108		1992	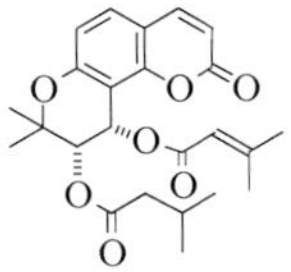 $C_{24}H_{28}O_7$	103–104	-13.5^{23} CHCl$_3$	*Peucedanum japonicum*	(114)
109	Peujaponisinol A	1993	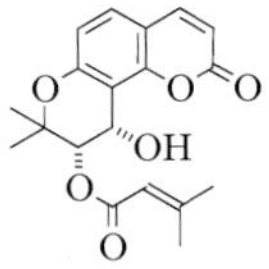 $C_{19}H_{20}O_6$	oil	-16.4^{22} CHCl$_3$	*Peucedanum japonicum*	(89)
110	Peujaponisin	1992	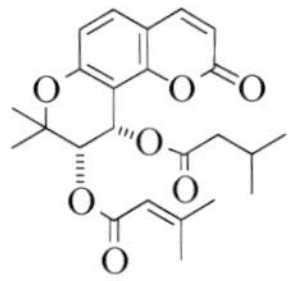 $C_{24}H_{28}O_7$	oil	-7.0^{25} CHCl$_3$	*Peucedanum japonicum*	(88)

Table 1.2 (continued)

	Trivial name(s)	Year isolated	Structure	Formula	M.p.	$[\alpha]_\lambda^t$	Solvent	Plant sources	Leading references
111	Badycoumarin B	1991	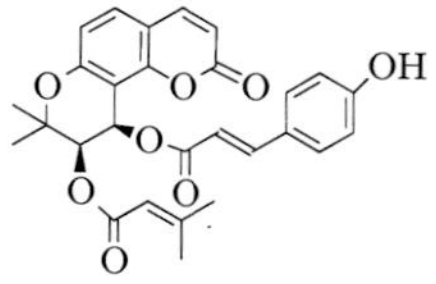	$C_{28}H_{26}O_8$	93–95	68.0^{27}	MeOH	*Angelica flaccida*	(228)
112	Qianhucoumarin E	1993	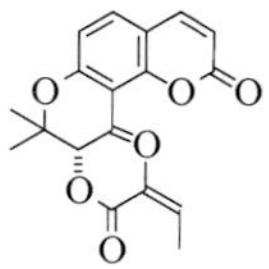	$C_{19}H_{18}O_6$	103.5–105.5	19.6^{20}	$CHCl_3$	*Peucedanum praeruptorum*	(126, 128)
113	Seselinol	1991	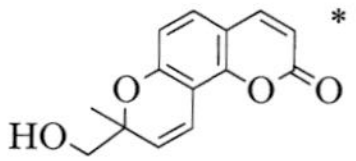	$C_{14}H_{12}O_4$	oil	−111.6	$CHCl_3$	*Citrus hassaku*	(117)
114	Seselinol isovalerate	1991	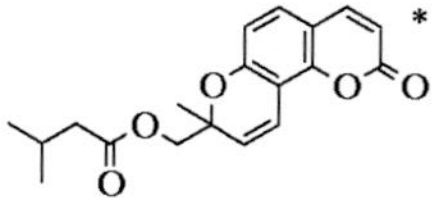	$C_{19}H_{20}O_5$	oil	50.84	$CHCl_3$	*Citrus hassaku*	(117)
115	5-Deoxyproto-bruceol-I regioisomer	1994	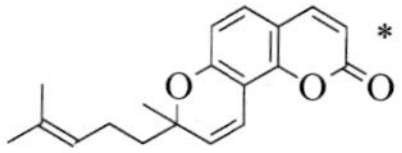	$C_{19}H_{20}O_3$	gum	75	$CHCl_3$	*Boronia lanceolata*	(6)

116	5-Deoxyproto-bruceol-III hydroperoxide regioisomer	1994	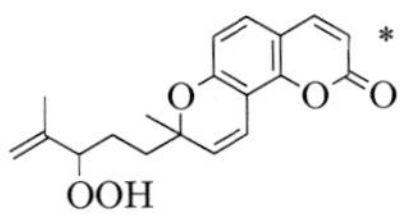 [structure]	$C_{19}H_{20}O_5$	gum		*Boronia lanceolata*	(6)
117	5-Deoxyproto-bruceol-II hydroperoxide regioisomer	1994	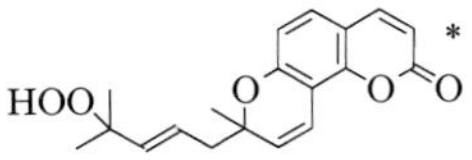 [structure]	$C_{19}H_{20}O_5$	gum	84 CHCl$_3$	*Boronia lanceolata*	(6)

Table 1.3. *5,6-Disubstituted-7-Oxygenated Coumarins*

Trivial name(s)	Year isolated	Structure	Formula	M.p.	$[\alpha]_\lambda^t$	Solvent	Plant sources	Leading references
		No new entries in this class [cf. ref. (172)]						

Table 1.4. *6,8-Disubstituted-7-Oxygenated Coumarins*

Trivial name(s)	Year isolated	Structure	Formula	M.p.	$[\alpha]_\lambda^t$	Solvent	Plant sources	Leading references
118	1989	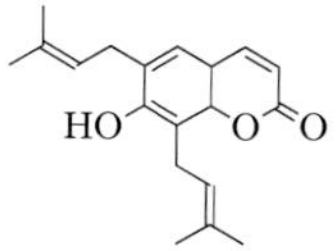	$C_{19}H_{22}O_3$	132–134			*Clausena indica*	(193)

Table 2. *5,7-Dioxygenated Coumarins*

	Trivial name(s)	Year isolated	Structure	Formula	M.p.	$[\alpha]_\lambda^t$	Solvent	Plant sources	Leading references
119		1992	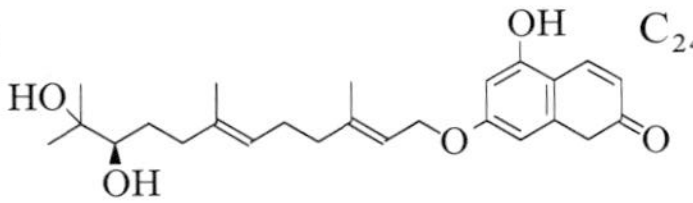	$C_{24}H_{32}O_6$	oil	-25.1^{25}	$CHCl_3$	*Heptaptera anisoptera*	(11)
120		1992	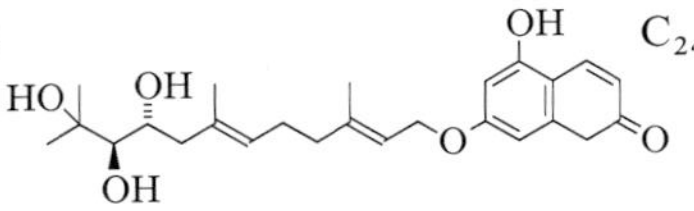	$C_{24}H_{32}O_7$	93–95	-44.6^{25}	$CHCl_3$	*Heptaptera anisoptera*	(11)
121		1992	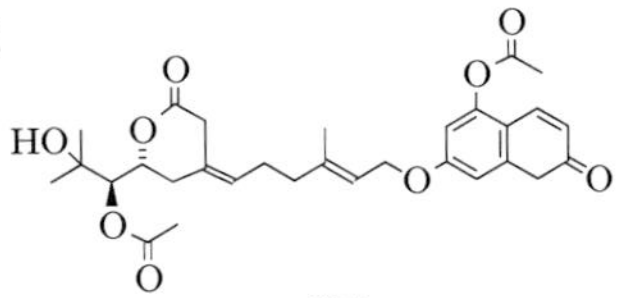	$C_{30}H_{38}O_{10}$	oil	23.4^{25}	$CHCl_3$	*Heptaptera anisoptera*	(11)
122	Toddalenol	1991	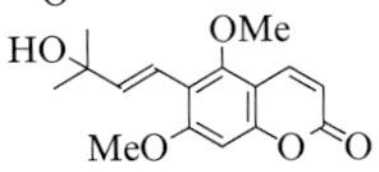	$C_{16}H_{18}O_5$	117–118			*Toddalia asiatica*	(103)
123		1991	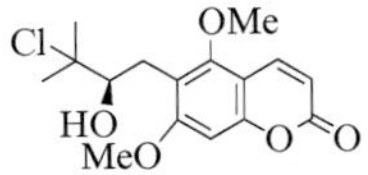	$C_{16}H_{19}ClO_5$	150–152	-73.5^{15}	$CHCl_3$	*Toddalia asiatica*	(97, 103)
124	(+)-Elisin	1990	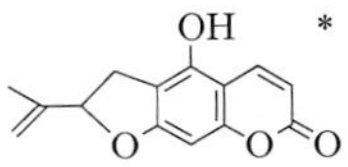	$C_{14}H_{12}O_4$	238–240	12.8	MeOH	*Pilocarpus goudotianus*	(8)

Table 2 (continued)

	Trivial name(s)	Year isolated	Structure	Formula	M.p.	$[\alpha]_\lambda^t$	Solvent	Plant sources	Leading references
125		1992	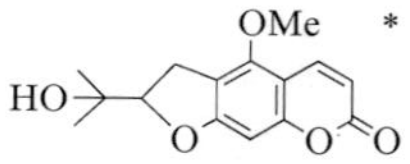	$C_{15}H_{16}O_5$	oil	−8	$CHCl_3$	Phebalium anceps	(196)
126	Ethylnotopterol	1994	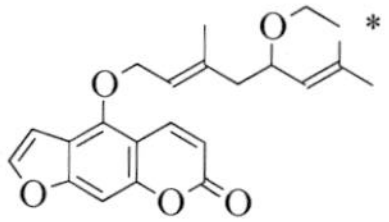	$C_{23}H_{26}O_5$	oil			Notopterygium incisum	(263, 264)
127	Notoptolide	1994	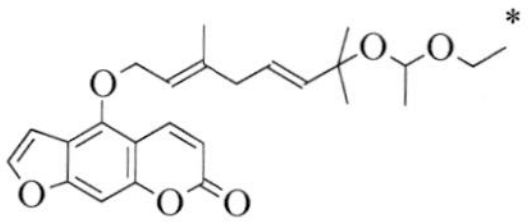	$C_{25}H_{30}O_6$	oil			Notopterygium incisum	(263, 264)
128	Anhydronotoptoloxide	1994	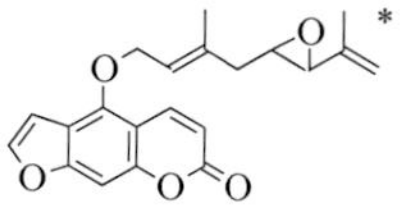	$C_{21}H_{20}O_5$	oil			Notopterygium incisum	(263, 264)
129		1992	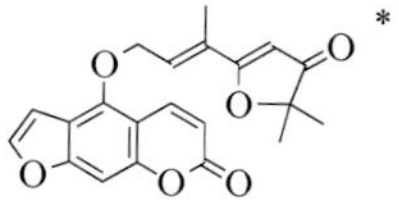	$C_{21}H_{20}O_6$	132–135	0		Dorstenia cayapiaa	(148)
		1994			140	1.4	$CHCl_3$	Dorstenia brasiliensis	(134)
130		1991	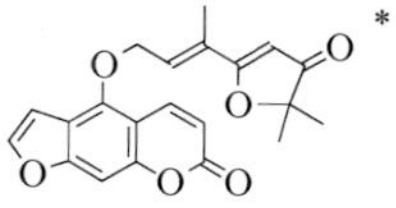	$C_{21}H_{18}O_6$	136–138			Dorstenia contrajerva	(240, 253)
		1994						Dorstenia brasiliensis	(134)

References pp. 105–119

131	1990	O-β-D-glucosyl (structure)	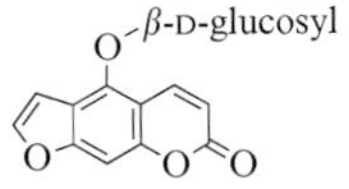 $C_{17}H_{16}O_9$	254–256			*Notopterygium forbesii* (75)
132	1993	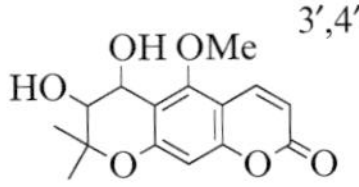 3',4'-*trans* * (structure)	$C_{15}H_{16}O_6$	oil	−30.0	$CHCl_3$	*Citrus funadoko* (249)
133	1993	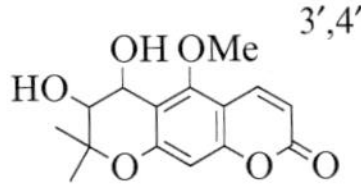 3',4'-*trans* * (structure)	$C_{15}H_{16}O_6$	162–165	−2.2	MeOH	*Citrus funadoko* (249)
				162–165	−49.9	$CHCl_3$	*Citrus hassaku*
				198–201	−4.7	$CHCl_3$	*Citrus hassaku*
134	4'β-Hydroxy-eriobrucinol	1992	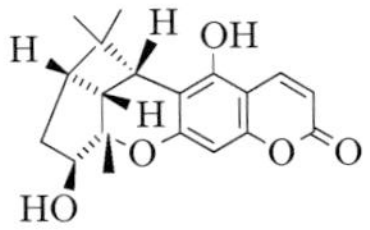 (structure)	$C_{19}H_{20}O_5$	269–271	−66	MeOH	*Eriostemon brucei* (207)
135	Pseudobruceol-II	1992	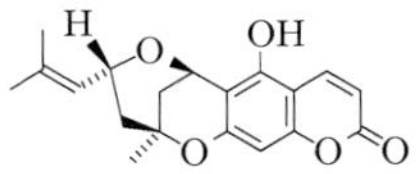 (structure)	$C_{19}H_{20}O_5$	gum	49	$CHCl_3$	*Eriostemon brucei* (207)
136	Pseudobruceol-I	1992	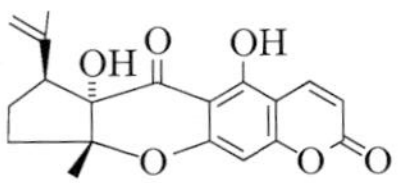 (structure)	$C_{19}H_{18}O_6$	gum	−35	$CHCl_3$	*Eriostemon brucei* (207)

Table 2 (continued)

	Trivial name(s)	Year isolated	Structure	Formula	M.p.	$[\alpha]_\lambda^t$	Solvent	Plant sources	Leading references
137	Protobruceol-IV	1992	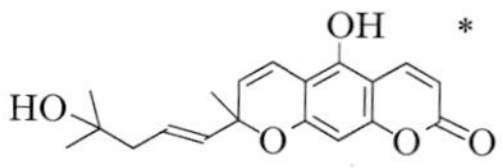	$C_{19}H_{20}O_5$	gum	7	$CHCl_3$	*Eriostemon brucei* var. *cinereus*	(204)
138	Phyllocoumarin	1989	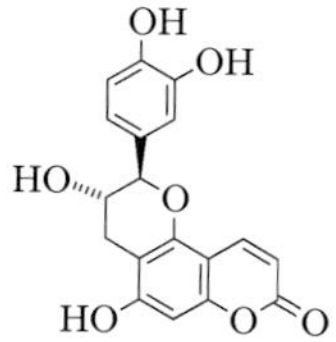	$C_{18}H_{14}O_7$	solid	−400	MeOH	*Phyllocladus trichomanoides*	(68)
139	Epiphyllocoumarin	1989	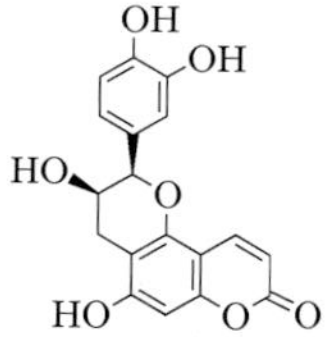	$C_{18}H_{14}O_7$	solid	−100	MeOH	*Phyllocladus trichomanoides*	(68)
140	Eriobrucinol regioisomer-A	1992	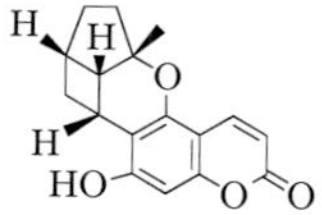	$C_{19}H_{20}O_4$	gum	47	$CHCl_3$	*Eriostemon brucei*	(207)

141	Alloxanthoxyletol	1990	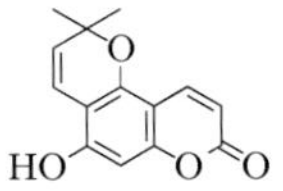	$C_{14}H_{12}O_4$	217–219d		*Pilocarpus goudotianus*	(8)
142	Protobruceol-I	1992	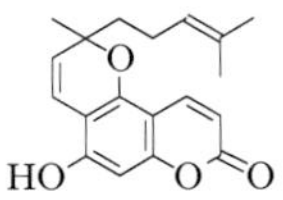	$C_{19}H_{20}O_4$	amorph.	27	$CHCl_3$	*Eriostemon brucei* var. *cinereus*	(204)
143	Protobruceol-III	1992	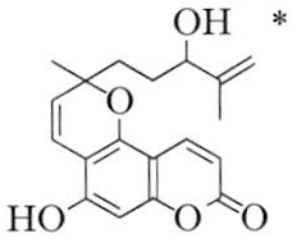	$C_{19}H_{20}O_5$	gum	5	$CHCl_3$	*Eriostemon brucei* var. *cinereus*	(204)
144	Protobruceol-II	1992	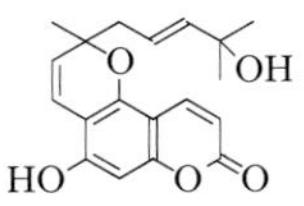	$C_{19}H_{20}O_5$	gum	30	$CHCl_3$	*Eriostemon brucei* var. *cinereus*	(204)
145	Protobruceol-III hydroperoxide	1992	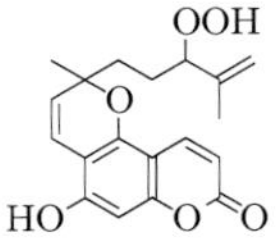	$C_{19}H_{20}O_6$	gum	19	$CHCl_3$	*Eriostemon brucei* var. *cinereus*	(204)
146	Protobruceol-II hydroperoxide	1992	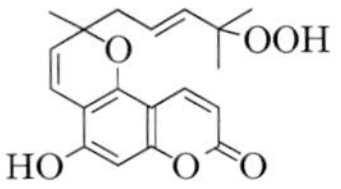	$C_{19}H_{20}O_6$	gum	15	$CHCl_3$	*Eriostemon brucei* var. *cinereus*	(204)

Table 2 (continued)

Trivial name(s)	Year isolated	Structure	Formula	M.p.	$[\alpha]_\lambda^t$	Solvent	Plant sources	Leading references
147	1994	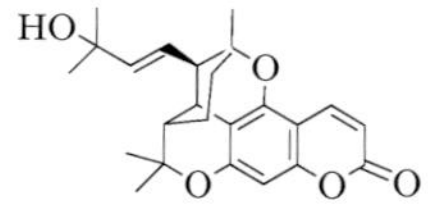	$C_{24}H_{28}O_5$	powder	−99.5	$CHCl_3$	*Eriostemon myoporoides*	(221)
148	1994	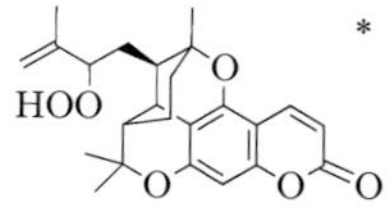	$C_{24}H_{28}O_6$	amorph.	−101.0	$CHCl_3$	*Eriostemon myoporoides*	(221)
149	1994	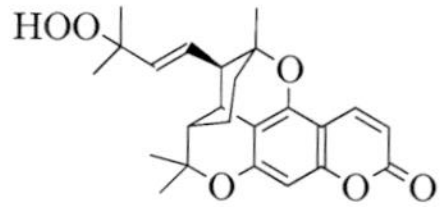	$C_{24}H_{28}O_6$	amorph.	−123.9	$CHCl_3$	*Eriostemon myoporoides*	(221)
150	1994	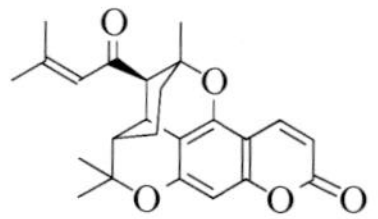	$C_{24}H_{26}O_5$	amorph.	−70.0	$CHCl_3$	*Eriostemon myoporoides*	(221)
151	1994	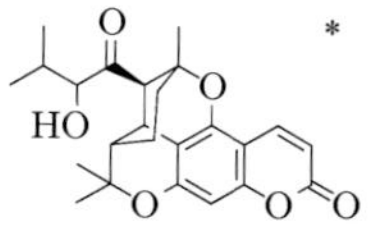	$C_{24}H_{28}O_6$	powder	−135.0	$CHCl_3$	*Eriostemon myoporoides*	(221)
152	1994	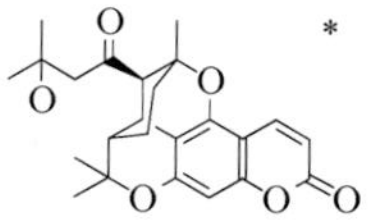	$C_{24}H_{26}O_6$	powder	−146.0	$CHCl_3$	*Eriostemon myoporoides*	(221)

References pp. 105–119

153	1994	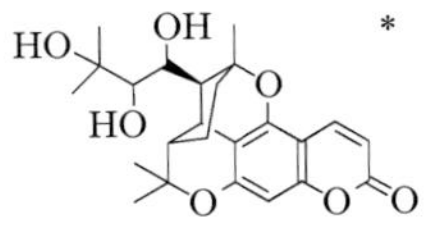	$C_{24}H_{30}O_7$	amorph.	−85.8	$CHCl_3$	*Eriostemon myoporoides* (221)
154	1991	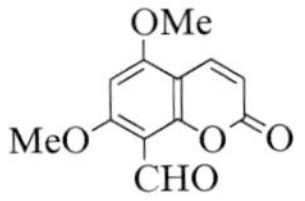	$C_{12}H_{10}O_5$	274–275			*Toddalia asiatica* (95)
155 Toddalenone	1991	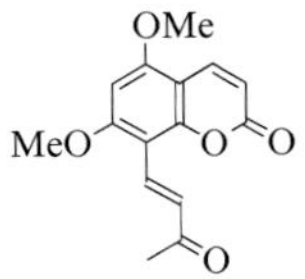	$C_{15}H_{14}O_5$	243–245			*Toddalia asiatica* (95)
156 Rubricauloside	1993	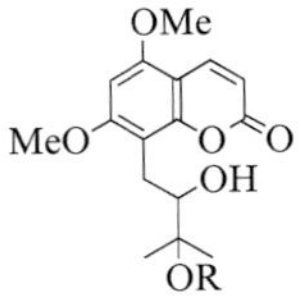 R=6'-β-D-apiosyl-β-D-glucosyl	$C_{27}H_{38}O_{15}$	123–127	−88.5^{16}	DMSO	*Peucedanum rubricaule* (202)
157	1994	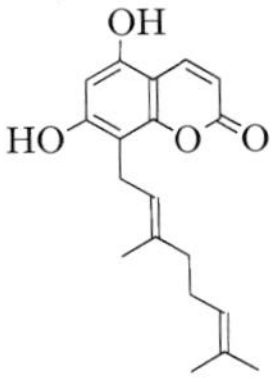	$C_{19}H_{22}O_4$	amorph.			*Eriostemon myoporoides* (221)

Table 2 (continued)

Trivial name(s)	Year isolated	Structure	Formula	M.p.	$[\alpha]_\lambda^t$	Solvent	Plant sources	Leading references
158 Toddacoumaquinone	1992	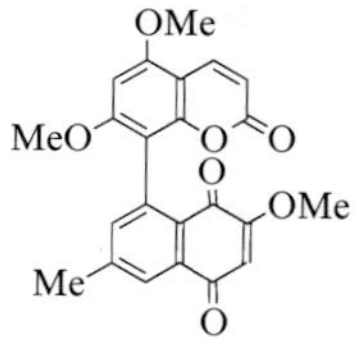	$C_{23}H_{18}O_7$	278–281			*Toddalia asiatica*	(98, 102)
159 Toddacoumalone	1991	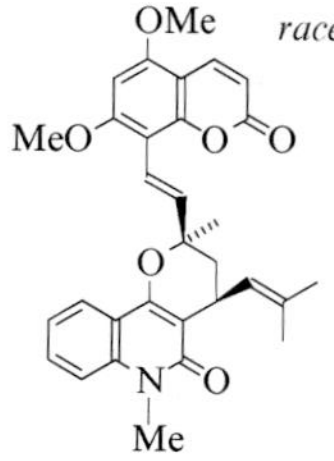 racemic	$C_{31}H_{31}NO_6$	202–204			*Toddalia asiatica*	(96)
160 Ptilin	1993	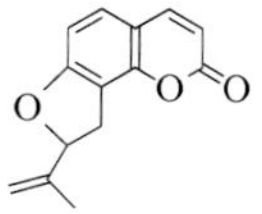 *	$C_{15}H_{14}O_4$	amorph.			*Haplophyllum ptilostylum*	(256)
161 Ptilostol	1993	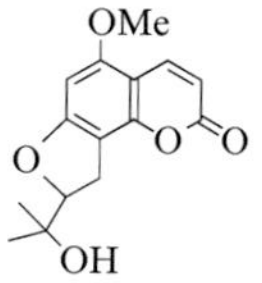 *	$C_{15}H_{16}O_5$		29.8	MeOH	*Haplophyllum ptilostylum*	(255)

162	Ptilostin	1993	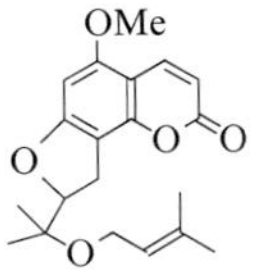	$C_{20}H_{24}O_5$		30.4	MeOH	*Haplophyllum ptilostylum*	(255)
163	*trans*-Grandmarin	1990	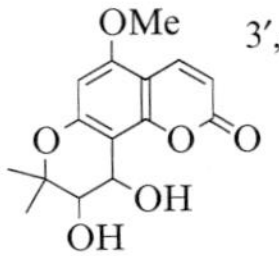	$C_{15}H_{16}O_6$	oil	7.69	$CHCl_3$	*Citrus tamurana × C. kinokuni*	(110)
164	Eriobrucinol regioisomer-B	1992	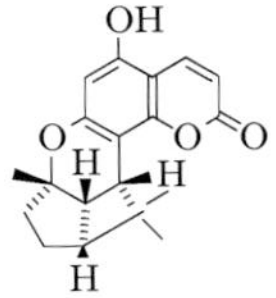	$C_{19}H_{20}O_4$	gum	−74	$CHCl_3$	*Eriostemon brucei*	(207)
165	Neoacrimarine-D	1993	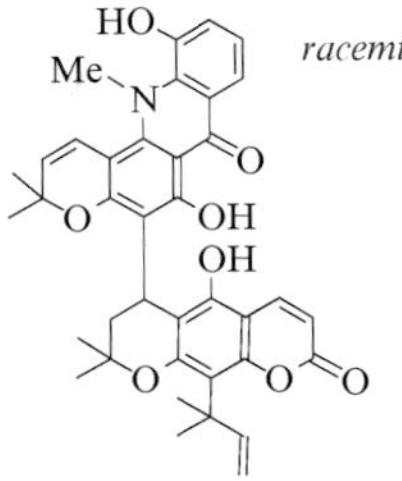	$C_{38}H_{37}NO_8$	oil			*Citrus hassaku*	(247)

Table 2 (continued)

	Trivial name(s)	Year isolated	Structure	Formula	M.p.	$[\alpha]_\lambda^t$	Solvent	Plant sources	Leading references
166	Neoacrimarine-E	1994	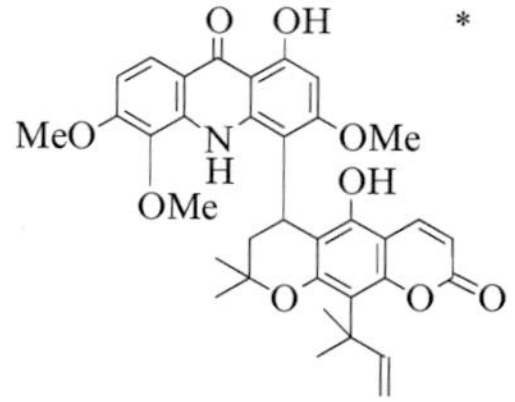 *	$C_{35}H_{35}NO_9$	212–215	−21.6	$CHCl_3$	Citrus paradisi × C. tangerina	(246)
167	Oxanordentatin	1991	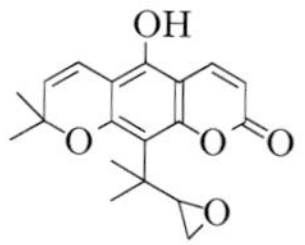 *	$C_{19}H_{20}O_5$	oil	−59.0	EtOH	Citrus hassaku	(117)
168		1995	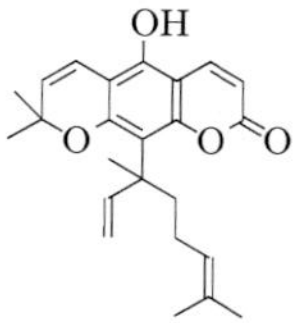 *	$C_{24}H_{28}O_4$	oil	2.9^{22}	$CHCl_3$	Paramignya monophylla	(133)
169		1995	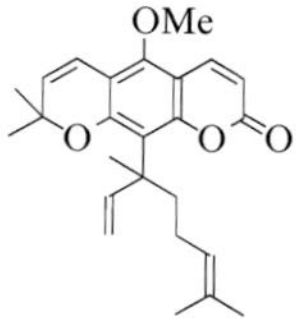 *	$C_{25}H_{30}O_4$	oil	12.7^{22}	$CHCl_3$	Paramignya monophylla	(133)

No.	Structure	Year	Formula	State	Value	Solvent	Source	Ref.
170	*	1991	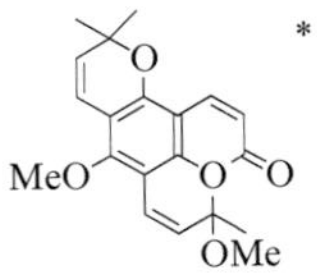 $C_{21}H_{24}O_5$				*Philotheca citrina*	(205)
171		1991	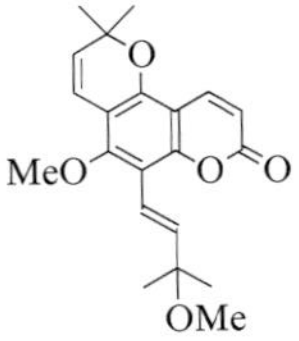 $C_{21}H_{24}O_5$				*Philotheca citrina*	(205)
172		1991	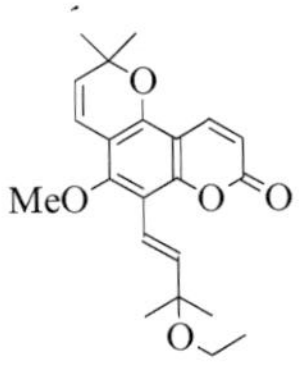 $C_{22}H_{26}O_5$	gum			*Philotheca citrina*	(205)
173 Citrusarin-B	*	1991	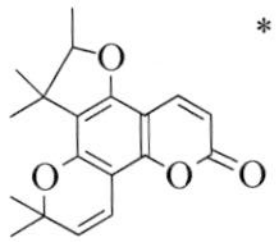 $C_{19}H_{20}O_4$	oil	3.0	CHCl$_3$	*Citrus hassaku*	(103)
174 Citrusarin-A	*	1991	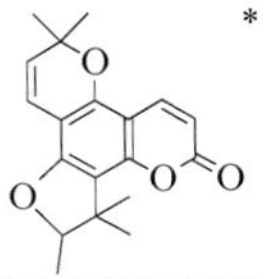 $C_{19}H_{20}O_4$	oil	5.2	CHCl$_3$	*Citrus hassaku*	(103)

Table 3. 6,7-Dioxygenated Coumarins

	Trivial name(s)	Year isolated	Structure	Formula	M.p.	$[\alpha]_\lambda^t$	Solvent	Plant sources	Leading references
175	Prionanthoside	1994	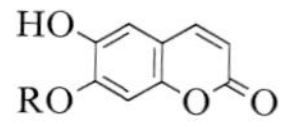 R=6'-acetyl-β-D-glucosyl	$C_{17}H_{18}O_{10}$				*Viola prionantha*	(194, 195)
176	Isobaisseoside	1995	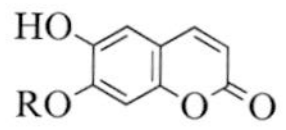 R=6'-α-L-rhamnosyl-β-D-glucosyl	$C_{21}H_{26}O_{13}$	gum			*Eriostemon cymbiformis*	(225)
177		1995	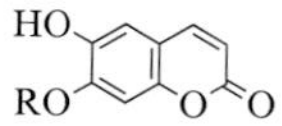 R=4''-p-coumaroyl-6'-α-L-rhamnosyl-β-D-glucosyl	$C_{30}H_{32}O_{15}$	gum			*Eriostemon cymbiformis*	(225)
178	Frachinoside	1992	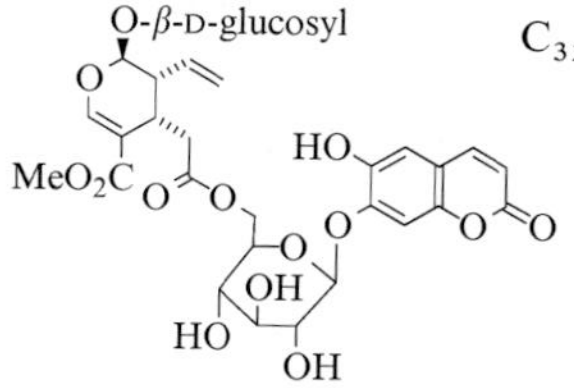 O-β-D-glucosyl	$C_{32}H_{38}O_{19}$	powder	-114.1^{16}	MeOH	*Fraxinus chinensis*	(135)
179		1992	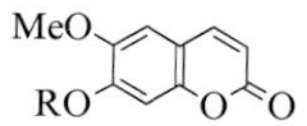 R=3''-α-L-arabinosyl-6'-β-D-galactosyl-β-D-galactosyl	$C_{26}H_{36}O_{18}$	180			*Pterocarpus santalinus*	(233)

180	1992	(structure with MeO, O, O)	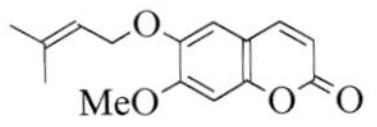 $C_{15}H_{16}O_4$	115–117		*Carduus tenuiflorus*	(36)
181	1992	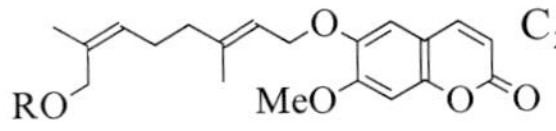 (structure) R=β-D-glucosyl	$C_{26}H_{34}O_{10}$	amorph.		*Coptis trifolia*	(168, 169)
182	1992	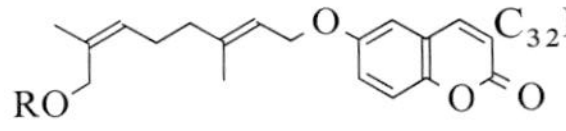 (structure) R=rutinosyl or 6′-fucosyl-β-D-glucosyl	$C_{32}H_{44}O_{14}$	amorph.		*Coptis trifolia*	(169)
183	Hemidesminine 1991	(structure) *racemic* 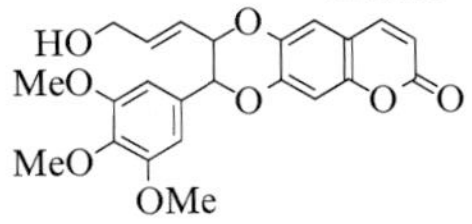	$C_{23}H_{22}O_8$	amorph.		*Hemidesmus indicus*	(155)
184	1995	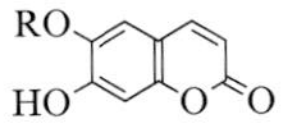 (structure) R=6′-β-D-apiosyl-β-D-glucosyl	$C_{20}H_{24}O_{13}$	amorph.	−68.2 MeOH	*Lonicera gracilipes* var. *glandulosa*	(161)
185	Palustroside 1990	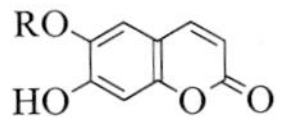 (structure) R=6′-(3-hydroxy-3-methylglutaryl)-β-D-glucosyl	$C_{21}H_{24}O_{13}$	205d	15.1^{20} aqMeOH	*Ledum palustre*	(61)

Table 3 (continued)

Trivial name(s)	Year isolated	Structure	Formula	M.p.	$[\alpha]_\lambda^t$	Solvent	Plant sources	Leading references
186	1990	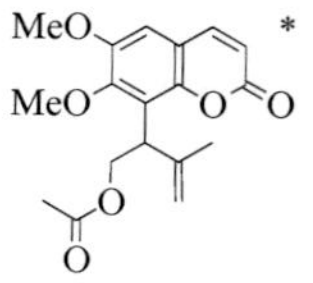	$C_{18}H_{20}O_6$	oil	8	$CHCl_3$	Phebalium elatius ssp. beckleri	(66)
187	1990	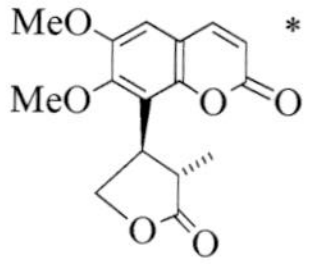	$C_{16}H_{16}O_6$	110	−10	$CHCl_3$	Phebalium elatius ssp. beckleri	(66)
188	1990		$C_{16}H_{18}O_5$	114				Phebalium elatius ssp. beckleri	(66)
189	1990	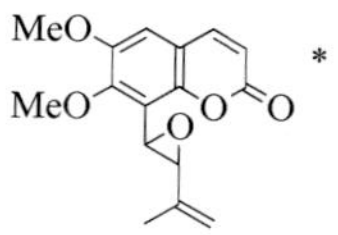	$C_{16}H_{16}O_5$	118	22	$CHCl_3$	Phebalium elatius ssp. beckleri	(66)

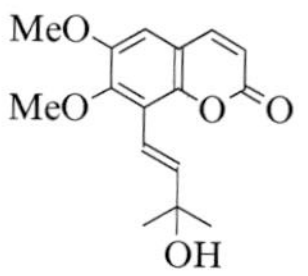

190	1990	* structure	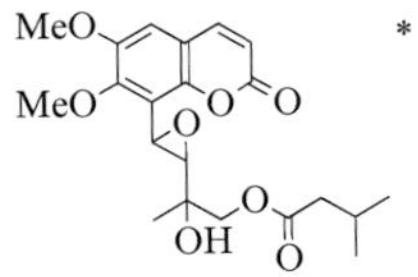 $C_{21}H_{26}O_8$	oil	−14 CHCl$_3$	Phebalium elatius ssp. beckleri	(66)
191	1990	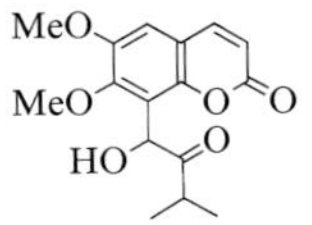 * structure	$C_{16}H_{18}O_6$	oil	65 CHCl$_3$	Phebalium elatius ssp. beckleri	(66)
192	1990	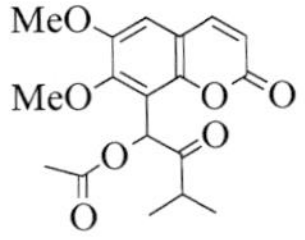 * structure	$C_{18}H_{20}O_7$			Phebalium elatius ssp. beckleri	(67)
193	1990	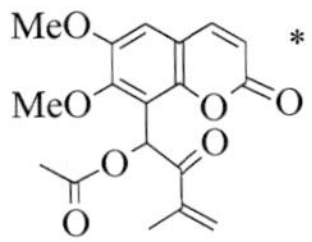 * structure	$C_{18}H_{18}O_7$	oil	166 CHCl$_3$	Phebalium elatius ssp. beckleri	(67)
194	1990	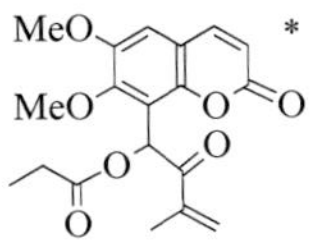 * structure	$C_{19}H_{20}O_7$	oil		Phebalium elatius ssp. beckleri	(67)

Table 3 (continued)

	Trivial name(s)	Year isolated	Structure	Formula	M.p.	$[\alpha]_\lambda^t$	Solvent	Plant sources	Leading references
195		1990	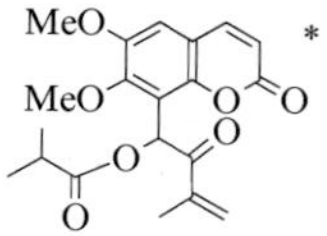	$C_{20}H_{22}O_7$	oil	−126	$CHCl_3$	Phebalium elatius ssp. beckleri	(67)
196		1990	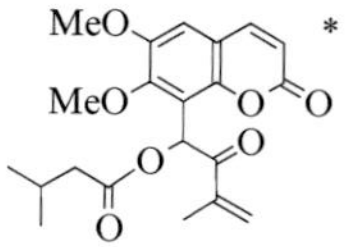	$C_{21}H_{24}O_7$	oil			Phebalium elatius ssp. beckleri	(67)
197		1994	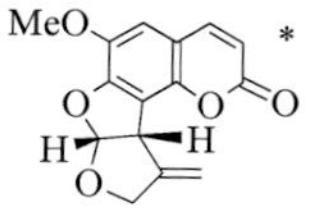	$C_{15}H_{12}O_5$	239–241	113	$CHCl_3$	Micromelum minutum	(198)
198	Pyracanthin B	1992	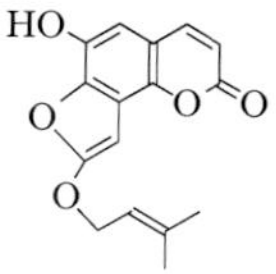	$C_{16}H_{14}O_5$				Pyracantha coccinea	(25)
199	Pyracanthin A	1992	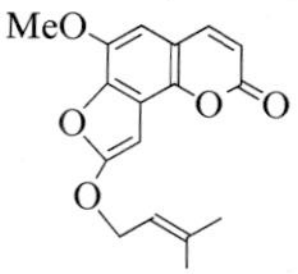	$C_{17}H_{16}O_5$				Pyracantha coccinea	(25)

References pp. 105–119

Table 4. 7,8-Dioxygenated Coumarins

	Trivial name(s)	Year isolated	Structure	Formula	M.p.	$[\alpha]_\lambda^t$	Solvent	Plant sources	Leading references
200	Epoxycollinin	1995	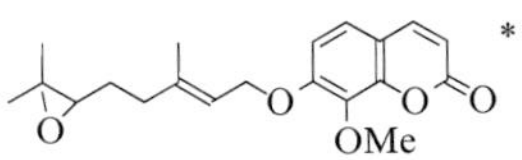	*$C_{20}H_{24}O_5$	oil			Zanthoxylum schinifolium	(42)
201		1992	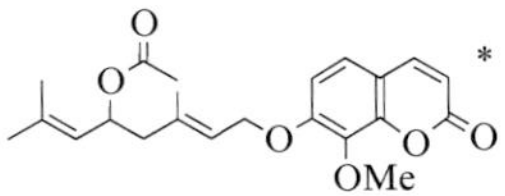	*$C_{22}H_{22}O_6$		-32.8^{23}	$CHCl_3$	Zanthoxylum schinifolium	(113)
		1995						Zanthoxylum schinifolium	(42)
202	Schininallylol	1995	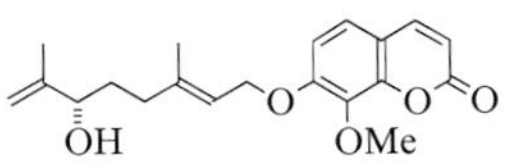	$C_{20}H_{24}O_5$	78–80	-16.4^{22}	$CHCl_3$	Zanthoxylum schinifolium	(42)
203	Schinilenol	1995	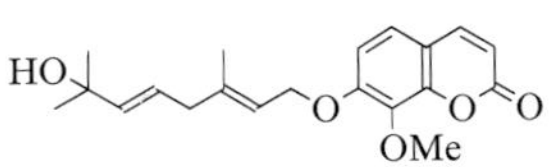	$C_{20}H_{24}O_5$				Zanthoxylum schinifolium	(42)
204	Schinindiol	1995	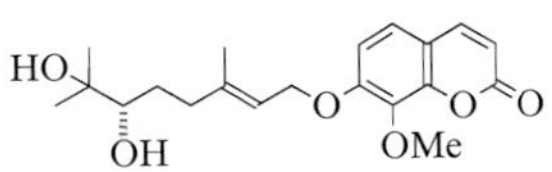	$C_{20}H_{26}O_6$	63–65	-22.0^{22}	$CHCl_3$	Zanthoxylum schinifolium	(42)
205	Anisocoumarin I	1991	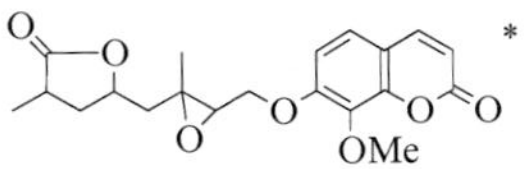	*$C_{20}H_{22}O_7$	127–128	33^{20}	$CHCl_3$	Clausena anisata	(177)
206	Anisocoumarin J	1991	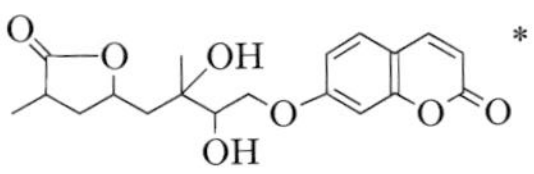	*$C_{20}H_{24}O_8$	154	-125^{20}	$CHCl_3$	Clausena anisata	(177)

Table 4 (*continued*)

	Trivial name(s)	Year isolated	Structure	Formula	M.p.	$[\alpha]_\lambda^t$	Solvent	Plant sources	Leading references
207	Daphneside	1991	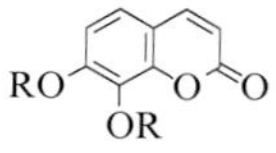 R=D-β-D-glucosyl	$C_{21}H_{26}O_{14}$	237d	17.1	H_2O	*Daphne arisanensis*	(178)
208	Qianhucoumarin G	1993	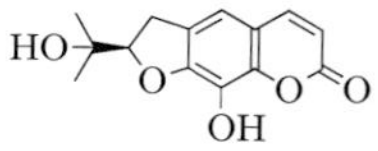	$C_{14}H_{14}O_5$	cubes	22^{20}	$CHCl_3$	*Peucedanum praeruptorum*	(128)
209		1994	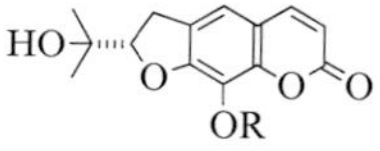 R=6′-α-L-rhamnosyl-β-D-glucosyl	$C_{26}H_{34}O_{14}$	150–152	45^{25}	EtOH	*Ruta graveolens*	(238)
210		1992	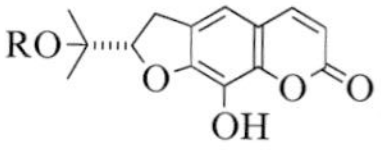 R=6′-feruloyl-β-D-glucosyl	$C_{30}H_{32}O_{13}$	150–153	37.9^{20}	MeOH	*Skimmia japonica*	(212)
211		1992	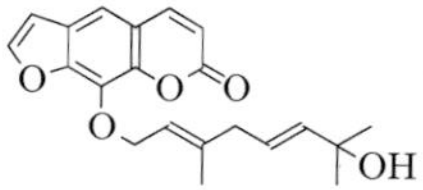	$C_{21}H_{22}O_5$	gum			*Phebalium filifolium*	(210)
212		1992	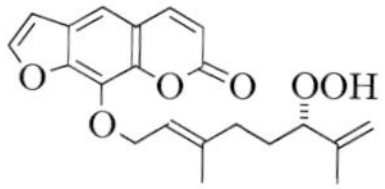	$C_{21}H_{22}O_6$	gum	−13	$CHCl_3$	*Phebalium filifolium*	(210)

213	1994	[structure with α-L-arabinosyl]	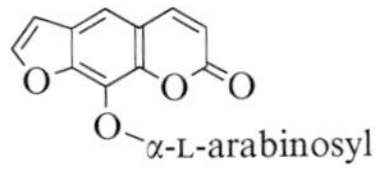 $C_{16}H_{14}O_8$	180–182			*Ruta graveolens* (238)
214	1992	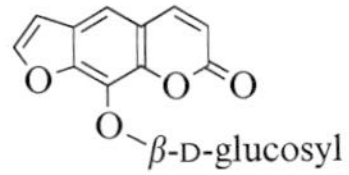 [structure with β-D-glucosyl]	$C_{17}H_{16}O_9$	244–245			*Angelica dahurica* (136)
215 Donatin	1990	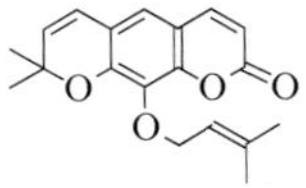 [structure]	$C_{19}H_{20}O_4$	oil	12^{25}	MeOH	*Pilocarpus goudotianus* (8)
216	1994	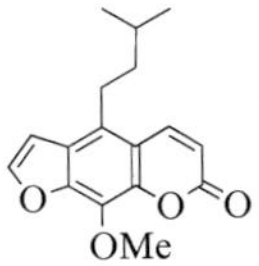 [structure, OMe]	$C_{17}H_{18}O_4$	215–217			*Elsholtzia densa* (230)
217	1994	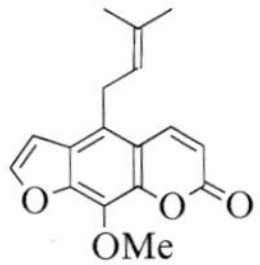 [structure, OMe]	$C_{17}H_{16}O_4$	149–151			*Elsholtzia densa* (230)
218	1994	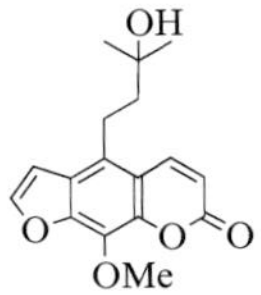 [structure, OH, OMe]	$C_{17}H_{18}O_5$	224–226			*Elsholtzia densa* (230)

Table 5. *5,6,7-Trioxygenated Coumarins*

Trivial name(s)		Year isolated	Structure	Formula	M.p.	$[\alpha]_\lambda^t$	Solvent	Plant sources	Leading references
219	*	1995	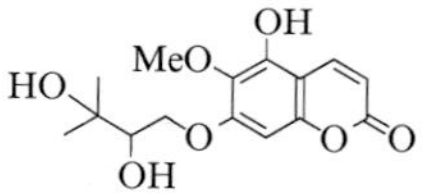	$C_{15}H_{18}O_7$	semisolid			*Pterocaulon alopecuroides*	(258)
220		1994	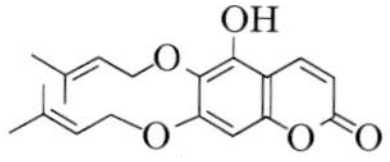	$C_{19}H_{22}O_5$	amorph.			*Drummondita hassellii*	(206)
221		1994	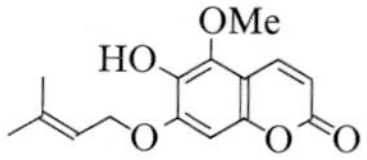	$C_{15}H_{16}O_5$	amorph.			*Drummondita hassellii*	(206)
222	*	1990	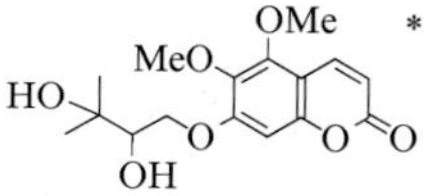	$C_{16}H_{20}O_7$	110	-60.6^{20}	aq. EtOH	*Artemisia laciniata*	(39)
223		1995	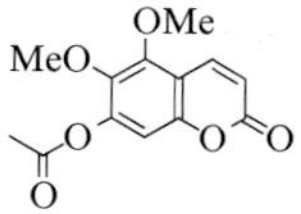	$C_{13}H_{12}O_6$	123–125			*Pelargonium sidoides*	(123)
224		1994	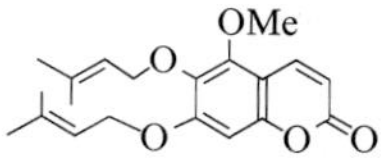	$C_{20}H_{24}O_5$	gum			*Drummondita hassellii*	(206)

Naturally Occurring Plant Coumarins

	Year	Structure	Formula	mp	$[\alpha]$	Solvent	Source	Ref.
225	1993	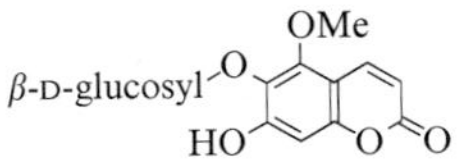 β-D-glucosyl-O, OMe, HO	$C_{16}H_{18}O_{10}$	amorph.	-32.9^{20}	MeOH	*Prunus prostrata*	(24)
226	1993	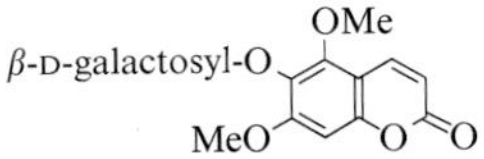 β-D-galactosyl-O, OMe, MeO	$C_{17}H_{20}O_{10}$	amorph.	-24.5^{20}	MeOH	*Prunus prostrata*	(24)
227	1992	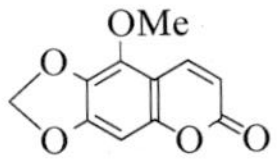 OMe	$C_{11}H_8O_5$	200–202			*Simsia cronquistii*	(152)
228	1981	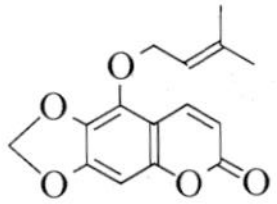	$C_{15}H_{14}O_5$	129–130			*Pterocaulon lanatum*	(51, 149)

Table 6. 5,7,8-Trioxygenated Coumarins

	Trivial name(s)	Year isolated	Structure	Formula	M.p.	$[\alpha]_\lambda^t$	Solvent	Plant sources	Leading references
229		1990	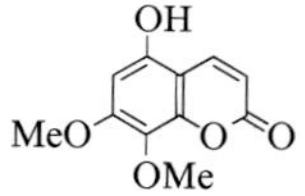 OH, MeO, OMe	$C_{11}H_{10}O_5$	242			*Artemisia laciniata*	(39)
230		1995	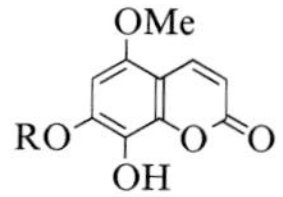 OMe, RO, OH; R=β-D-sophorosyl	$C_{22}H_{28}O_{15}$				*Tetraphis pellucida*	(120)
231		1994	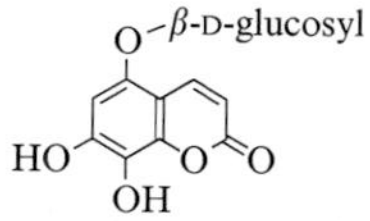 O-β-D-glucosyl, HO, OH	$C_{15}H_{16}O_{10}$				*Atrichum undulatum*	(121)
232		1994	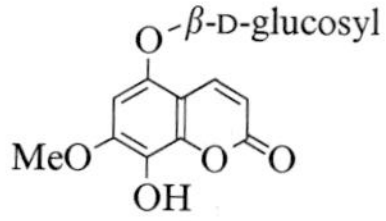 O-β-D-glucosyl, MeO, OH	$C_{16}H_{18}O_{10}$				*Atrichum undulatum*	(121)
233		1994	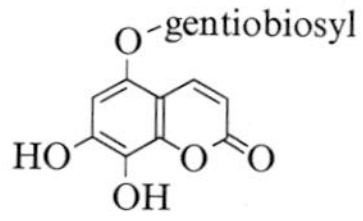 O-gentiobiosyl, HO, OH	$C_{21}H_{26}O_{15}$				*Polytrichum formosum*	(121)

References pp. 105–119

234	1994	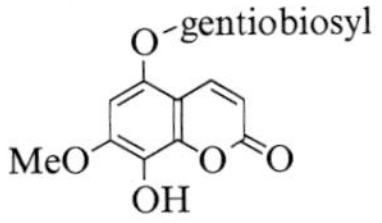 (structure with O-gentiobiosyl, MeO, OH substituents on coumarin)	$C_{22}H_{28}O_{15}$	*Polytrichum formosum*	(121)
235	1995	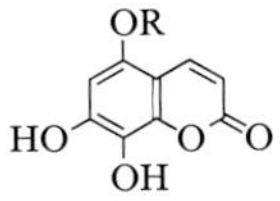 (structure with OR, HO, OH substituents on coumarin; R=6′-malonyl-β-D-glucosyl)	$C_{18}H_{18}O_{13}$	*Tetraphis pellucida*	(120)

Table 7. 6,7,8-Trioxygenated Coumarins

	Trivial name(s)	Year isolated	Structure	Formula	M.p.	$[\alpha]_\lambda^t$	Solvent	Plant sources	Leading references
236		1995	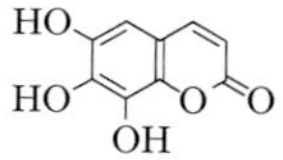	$C_9H_6O_5$	237–239			Pelargonium sidoides	(123)
237		1995	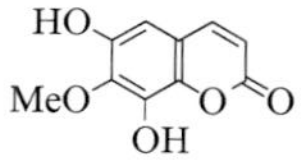	$C_{10}H_8O_5$	198–200			Polytrichum formosum	(123)
238		1989 1994	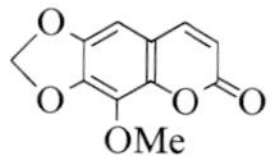	$C_{11}H_8O_5$	216–218 amorph.			Artemisia sacrorum Asterolasia trymalioides	(269) (224)
239	Purpurasol	1992	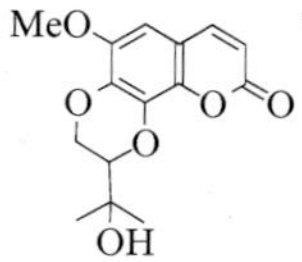 *	$C_{15}H_{16}O_6$	148–149	80.0^{20}	EtOH	Pterocaulon purpurascens	(50)
240	Hemidesmin-1	1992	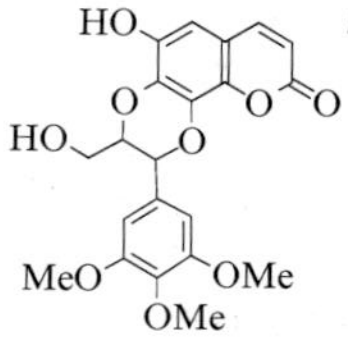	racemic* $C_{21}H_{20}O_9$	232–234			Hemidesmus indicus	(48)

References pp. 105–119

| 241 | Hemidesmin-2 | 1992 | 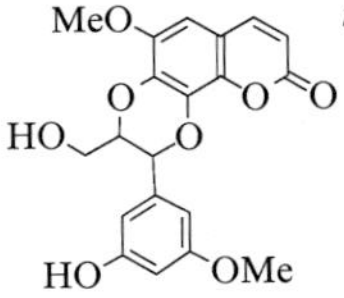 racemic* $C_{20}H_{18}O_8$ | 245–250 | *Hemidesmus indicus* | (48) |

Table 8. 5,6,7,8-Tetraoxygenated Coumarins

	Trivial name(s)	Year isolated	Structure	Formula	M.p.	$[\alpha]_\lambda^t$	Solvent	Plant sources	Leading references
242		1995	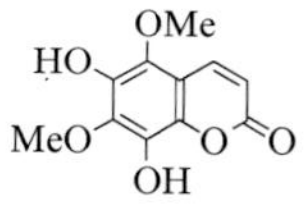	$C_{11}H_{10}O_6$	144–147			*Pelargonium sidoides*	(123)
243	Purpurenol	1991	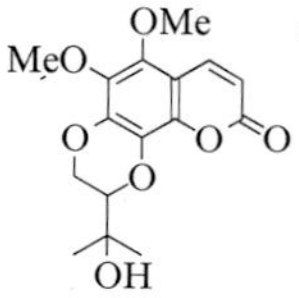	$C_{16}H_{18}O_7$	110–112			*Pterocaulon purpurascens*	(49)
244		1994	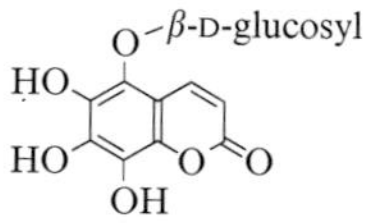	$C_{15}H_{16}O_{11}$				*Atrichum undulatum*	(121)
245		1995	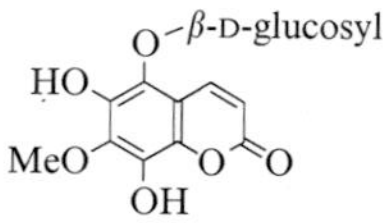	$C_{16}H_{18}O_{11}$				*Tetraphis pellucida*	(120)
246		1994	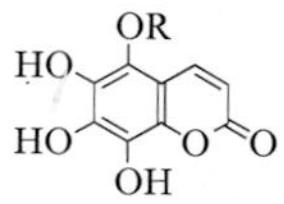 R=6′-acetyl-β-D-glucosyl	$C_{17}H_{18}O_{12}$				*Atrichum undulatum*	(121)

References pp. 105–119

247	1994	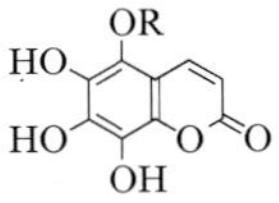 ![structure with OR, HO, HO, OH on coumarin] R=6′-malonyl-β-D-glucosyl	$C_{18}H_{18}O_{14}$	*Atrichum undulatum*	(121)
248	1994	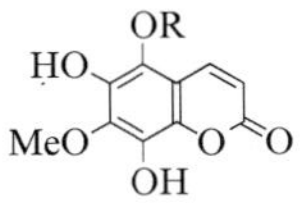 ![structure with OR, HO, MeO, OH on coumarin] R=6′-malonyl-β-D-glucosyl	$C_{19}H_{20}O_{14}$	*Atrichum undulatum*	(121)

Table 9. 3-Substituted Coumarins

	Trivial name(s)	Year isolated	Structure	Formula	M.p.	$[\alpha]_\lambda^t$	Solvent	Plant sources	Leading references
249		1994	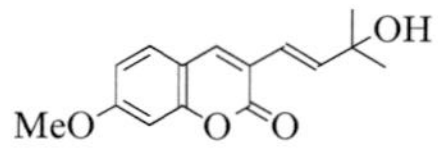	$C_{15}H_{16}O_4$	amorph.			*Asterolasia squamiligera*	(223)
250		1993	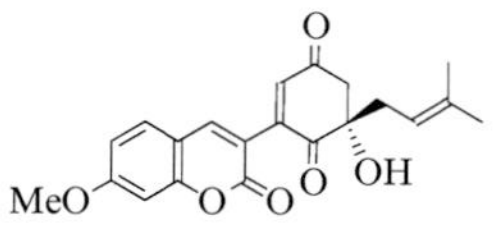	$C_{21}H_{20}O_6$				*Pueraria mirifica*	(45)
251		1992	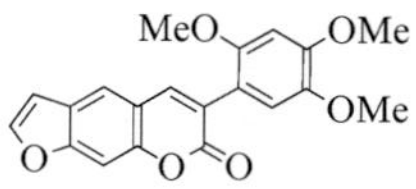	$C_{20}H_{16}O_6$	190–192			*Pachyrhizus tuberosus*	(150)
252		1996	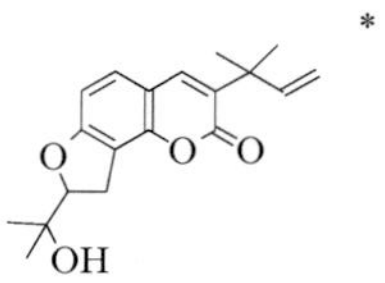 *	$C_{19}H_{22}O_4$	oil	52.9^{21}	$CHCl_3$	*Esenbeckia grandiflora*	(186)
253	Neoacrimarine-B	1993	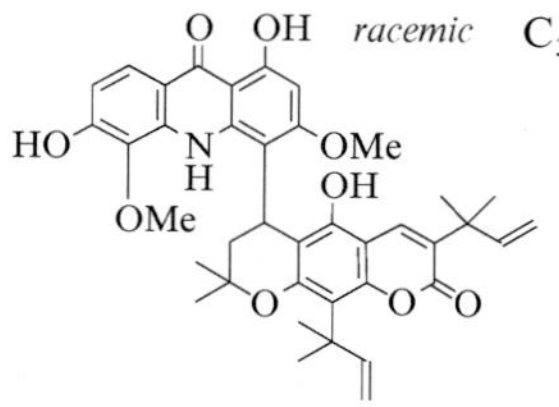 racemic	$C_{39}H_{41}NO_9$	240–243			*Citrus paradisi × C. tangerina*	(248)

254	Neoacrimarine-A	1993	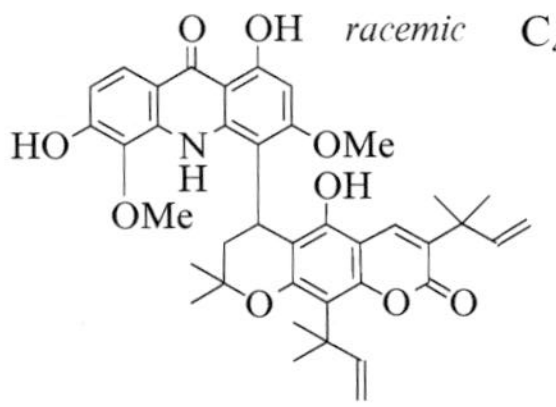 [structure] racemic	$C_{40}H_{43}NO_9$	225–230			Citrus paradisi × C. tangerina	(248)
255	Oxaclausarin	1991	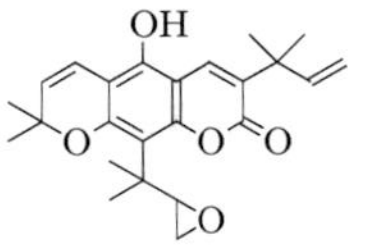 [structure] *	$C_{24}H_{28}O_5$	oil	−32.3	EtOH	Citrus hassaku	(117)

Table 9.1. *3-Aryl-Substituted Coumarins*

	Trivial name(s)	Year isolated	Structure	Formula	M.p.	$[\alpha]_\lambda^t$	Solvent	Plant sources	Leading references
256	Gancaoin W	1993	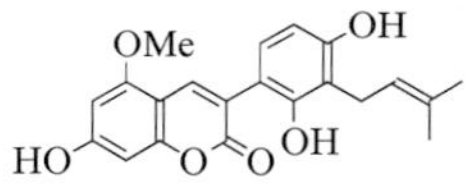	$C_{21}H_{20}O_6$	205–211			*Glycyrrhiza* sp.	(69)
257	Glyasperin L	1994	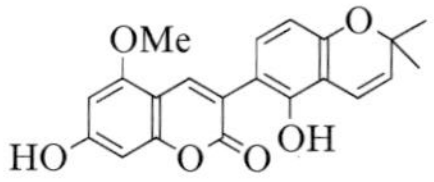	$C_{21}H_{18}O_6$	182–184			*Glycyrrhiza aspera*	(70)
258	Isoglycycoumarin	1994	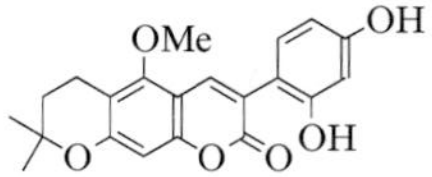	$C_{21}H_{20}O_6$	236–237			*Glycyrrhiza aspera*	(267, 268)
259	Licoarylcoumarin	1989	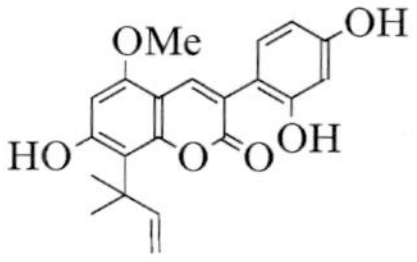	$C_{21}H_{20}O_6$	160			*Glycyrrhiza* sp.	(85)

Table 10. *4-Substituted Coumarins*

	Trivial name(s)	Year isolated	Structure	Formula	M.p.	$[\alpha]_\lambda^t$	Solvent	Plant sources	Leading references
260	Pseudocalanolide C	1992	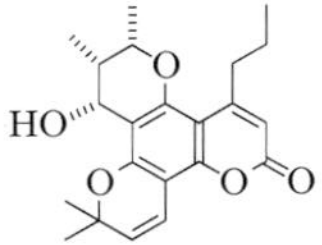	$C_{22}H_{26}O_5$		68	$CHCl_3$	*Calophyllum lanigerum*	*(122,163,191)*
261	Pseudocalanolide D	1992	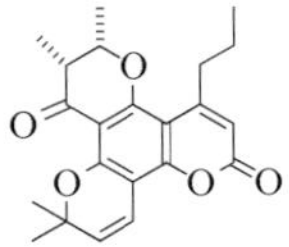	$C_{22}H_{24}O_5$		60	$CHCl_3$	*Calophyllum lanigerum*	*(122,163,191)*
262	Calanolide B	1992	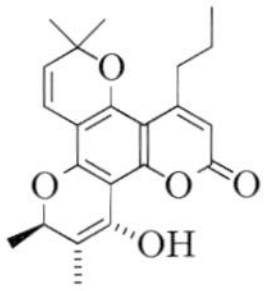	$C_{22}H_{26}O_5$		10	$CHCl_3$	*Calophyllum lanigerum*	*(122)*
263		1992	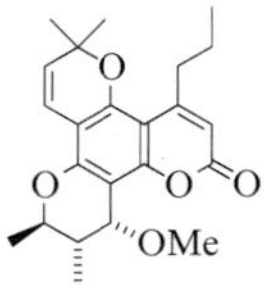	$C_{23}H_{28}O_5$		34	$CHCl_3$	*Calophyllum lanigerum*	*(122)*

Table 10 (continued)

	Trivial name(s)	Year isolated	Structure	Formula	M.p.	$[\alpha]_D$	Solvent	Plant sources	Leading references
264	Calanolide A	1992	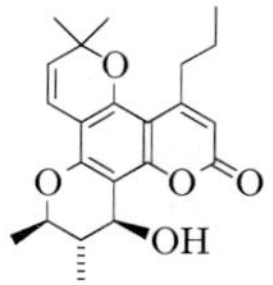	$C_{22}H_{26}O_5$	oil	60	$CHCl_3$	*Calophyllum lanigerum*	(35, 44, 71, 122)
265		1992	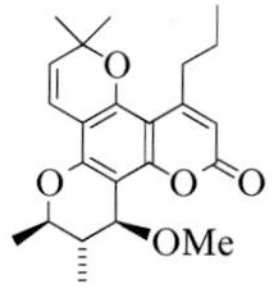	$C_{23}H_{28}O_5$		32	$CHCl_3$	*Calophyllum lanigerum*	(122)
266		1992	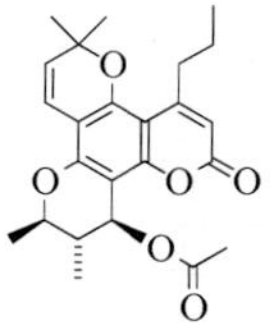	$C_{24}H_{28}O_6$		20	$CHCl_3$	*Calophyllum lanigerum*	(122)

Table 10.1. *4-Aryl-Substituted Coumarins*

	Trivial name(s)	Year isolated	Structure	Formula	M.p.	$[\alpha]_\lambda^t$	Solvent	Plant sources	Leading references
267	Inflacoumarin A	1994	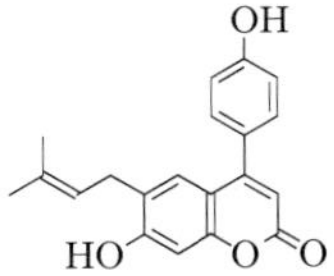	$C_{20}H_{18}O_4$	232–233			*Glycyrrhiza inflata*	(273)
268	Nivetin	1990	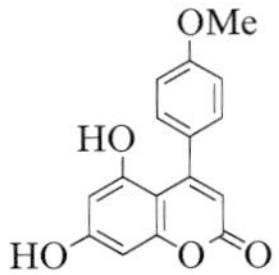	$C_{16}H_{12}O_5$	314–316d			*Echinops niveus*	(232)
269		1990	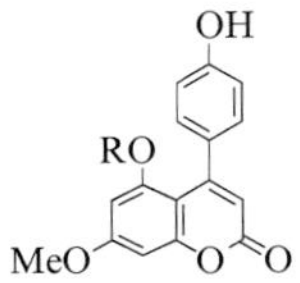 R=β-D-galactosyl	$C_{22}H_{22}O_{10}$	218–221			*Hintonia latiflora*	(159)
270		1992	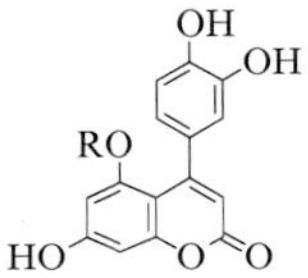 R=6'-acetyl-β-D-glucosyl	$C_{23}H_{22}O_{12}$	205–208	−129	MeOH	*Hintonia latiflora*	(160)

Table 10.1 (*continued*)

	Trivial name(s)	Year isolated	Structure	Formula	M.p.	$[\alpha]_\lambda^t$	Solvent	Plant sources	Leading references
271	Calanone	1994	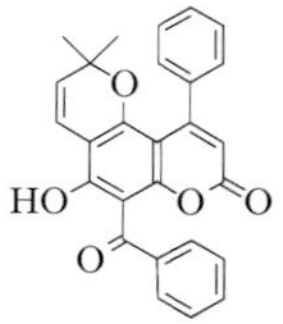	$C_{27}H_{20}O_5$	glass			*Calophyllum teysmannii*	(77)
272		1990	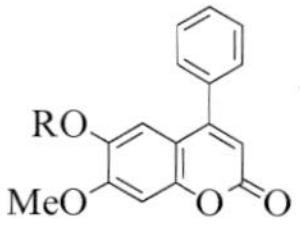 R=β-D-galactosyl	$C_{22}H_{22}O_9$	208–209			*Hesperathusa crenulata*	(132)
273		1990	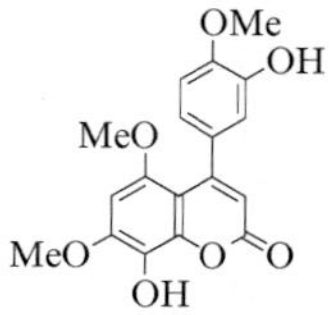	$C_{18}H_{16}O_7$	197–198			*Coutarea hexandra*	(47)
		1990			191–192			*Coutarea hexandra*	(52)
274		1990	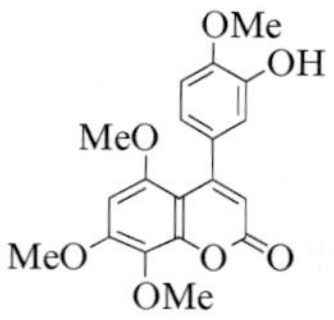	$C_{19}H_{18}O_7$	161–162			*Coutarea hexandra*	(52)

Table 11. *Miscellaneous Coumarins*

	Trivial name(s)	Year isolated	Structure	Formula	M.p.	$[\alpha]_\lambda^t$	Solvent	Plant sources	Leading references
275		1993	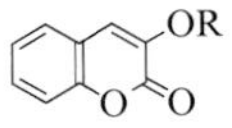 R=5″-ferulyl-6′-β-D-apiosyl-β-D-glucosyl	$C_{30}H_{32}O_{13}$		−111	MeOH	*Alyxia reinwardti* var. *lucida*	(144)
276		1993	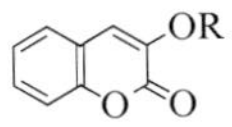 R=5″-sinapyl-6′-β-D-apiosyl-β-D-glucosyl	$C_{31}H_{34}O_{14}$		−101	MeOH	*Alyxia reinwardti* var. *lucida*	(144)
277	Setarin	1991	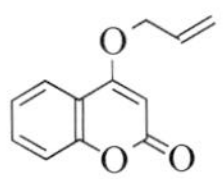	$C_{12}H_{10}O_3$	104–105			*Setaria italica*	(112)
278		1991	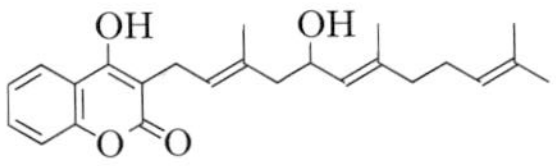	*$C_{24}H_{30}O_4$				*Ferula communis* var. *genuina*	(138)
279	Fercoprolone	1990	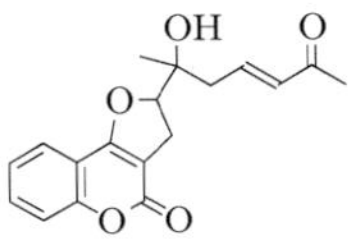 *	$C_{18}H_{18}O_5$	gum			*Ferula communis* ssp. *communis*	(165)

Table 11 (continued)

	Trivial name(s)	Year isolated	Structure	Formula	M.p.	$[\alpha]_\lambda^t$	Solvent	Plant sources	Leading references
280	Fercoprenol	1990	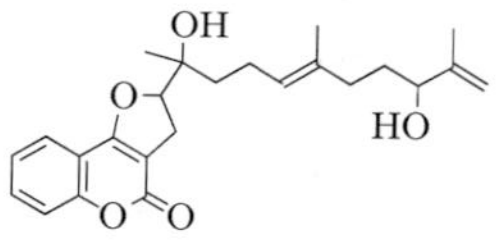 *	$C_{24}H_{30}O_5$	gum			Ferula communis ssp. communis	(165)
281	Isoferprenin	1991	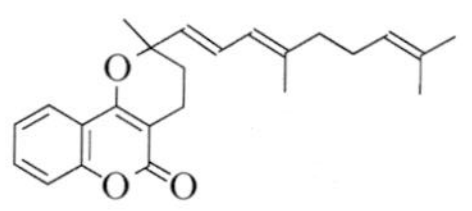 *	$C_{24}H_{28}O_3$	oil	0.4^{22}	$CHCl_3$	Ferula communis var. genuina	(10, 139)
282	Frutinone A	1989	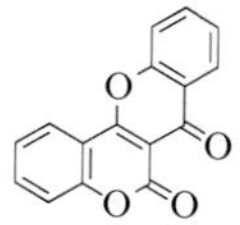	$C_{16}H_8O_4$	235–236			Polygala fruticosa	(58)
283	Frutinone C	1989	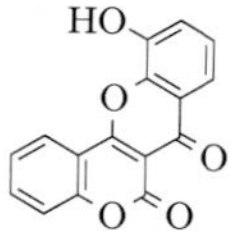	$C_{16}H_8O_5$	240–250d			Polygala fruticosa	(58)
284		1991	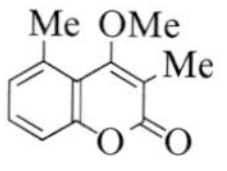	$C_{12}H_{12}O_3$	102–103			Clutia abyssinica	(260)

285	Hoehneliacoumarin	1993	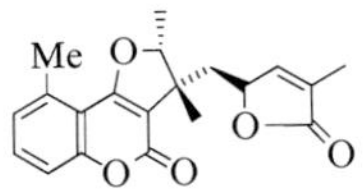	$C_{20}H_{20}O_5$		-28	$CHCl_3$	*Ethulia vernonioides*	(227)
286		1990	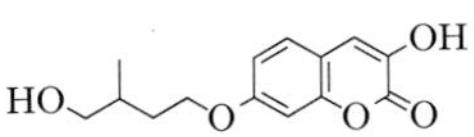	$C_{14}H_{16}O_5$	gum			*Bahia ambrosioides*	(266)
287		1990	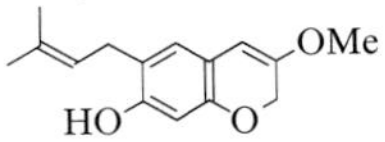	$C_{15}H_{16}O_4$	167			*Bahia ambrosioides*	(266)
288		1990	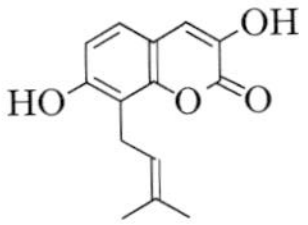	$C_{14}H_{14}O_4$	gum			*Bahia ambrosioides*	(266)
289		1990	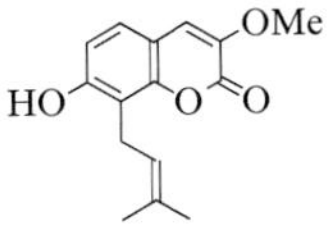	$C_{15}H_{16}O_4$	gum			*Bahia ambrosioides*	(266)
290		1991	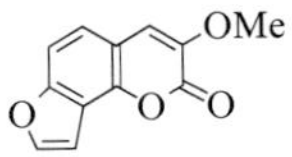	$C_{12}H_8O_4$	138–140			*Pilocarpus riedelianus*	(170)
291		1991	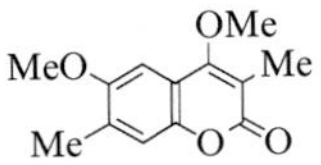	$C_{13}H_{14}O_4$	amorph.			*Colchicum decaisnei*	(7)
292	Thesiolen	1993	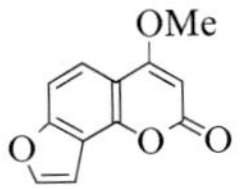	$C_{12}H_8O_4$	amorph.			*Haplophyllum thesioides*	(254)

Table 11 (continued)

	Trivial name(s)	Year isolated	Structure	Formula	M.p.	$[\alpha]_\lambda^t$	Solvent	Plant sources	Leading references
293	Frutinone B	1989	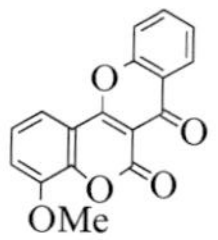	$C_{17}H_{10}O_5$	279–280			*Polygala fruticosa*	(58)
294		1991	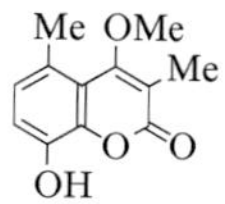	$C_{12}H_{12}O_4$	200–201			*Clutia abyssinica*	(260)
295		1991	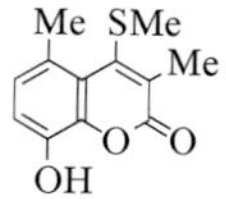	$C_{12}H_{12}O_3S$	174–175			*Clutia abyssinica*	(260)
296		1989	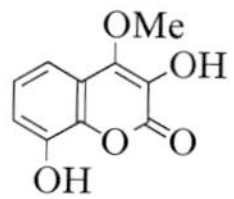	$C_{10}H_8O_5$	205–207			*Gerbera anandria*	(74)
297		1994	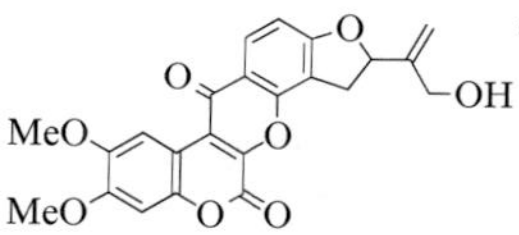	$C_{23}H_{18}O_8$	298d			*Dalbergia sissoides*	(237)

298	1993	structure with MeO, OMe, furan ring	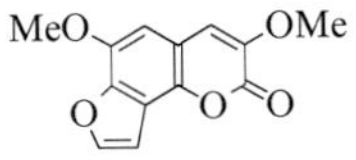 $C_{13}H_{10}O_5$	amorph.	*Pilocarpus riedelianus*	(170)
299 Schinicoumarin	1995	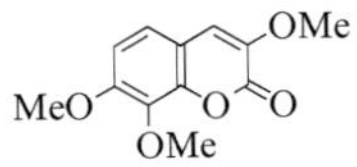 structure with MeO, OMe, OMe	$C_{12}H_{12}O_5$	147–151	*Zanthoxylum schinifolium*	(42)
300	1991	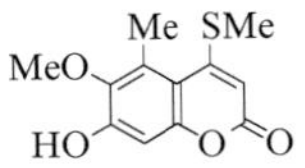 structure with Me, SMe, MeO, HO	$C_{12}H_{12}O_4S$	224–225	*Clutia abyssinica*	(260)
301	1991	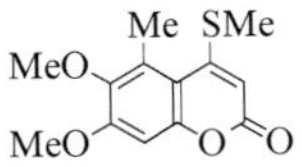 structure with Me, SMe, MeO, MeO	$C_{13}H_{14}O_4S$	146–147	*Clutia abyssinica*	(260)
302	1991	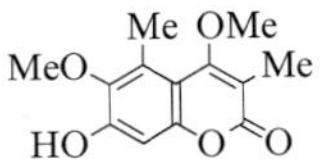 structure with Me, OMe, Me, MeO, HO	$C_{13}H_{14}O_5$	199–200	*Clutia abyssinica*	(260)
303	1991	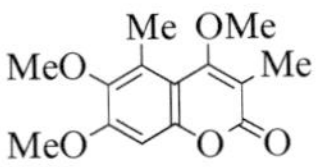 structure with Me, OMe, Me, MeO, MeO	$C_{14}H_{16}O_5$	123–124	*Clutia abyssinica*	(260)

Table 11.1. *3-Aryl-Oxygenated Coumarins*

Trivial name(s)	Year isolated	Structure	Formula	M.p.	$[\alpha]_\lambda^t$	Solvent	Plant sources	Leading references
304	1994	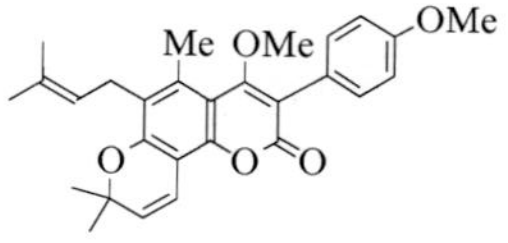	$C_{28}H_{30}O_6$	124			*Derris scandens*	(203)

References pp. 105–119

Table 11.2. *Coumestans*

	Trivial name(s)	Year isolated	Structure	Formula	M.p.	$[\alpha]_\lambda^t$	Solvent	Plant sources	Leading references
305	Bavacoumestan B	1990	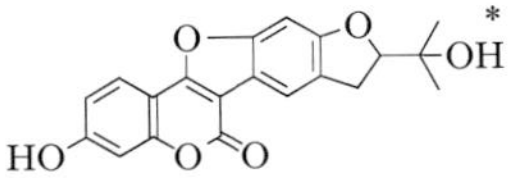	$C_{20}H_{16}O_6$				*Psoralea corylifolia*	(76)
306	Bavacoumestan A	1990	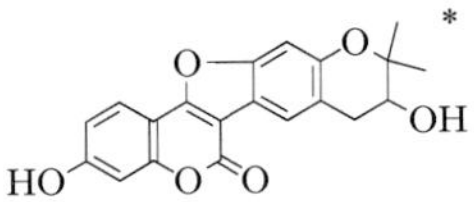	$C_{20}H_{16}O_6$				*Psoralea corylifolia*	(76)
307	Plicadin	1991	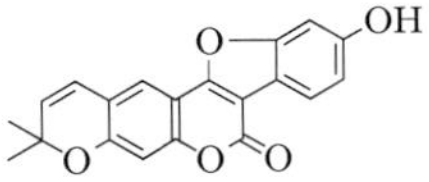	$C_{20}H_{14}O_5$	127			*Psoralea plicata*	(211)
308		1990	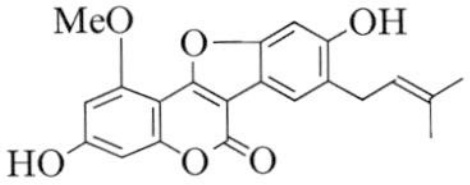	$C_{21}H_{18}O_6$	285–287			*Lotus creticus*	(151)
309	Neoglycyrol	1991	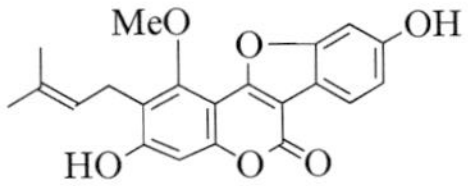	$C_{21}H_{18}O_6$	263.5–265			*Glycyrrhiza uralensis*	(261)
310		1994	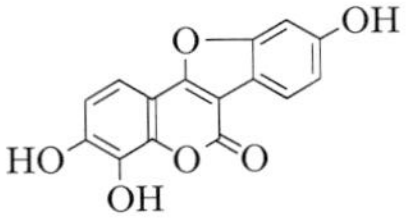	$C_{15}H_8O_6$	287			*Erythrina sigmoidea*	(179)

Table 12. *Biscoumarins*

	Trivial name(s)	Year isolated	Structure	Formula	M.p.	$[\alpha]_\lambda^t$	Solvent	Plant sources	Leading references
311	Edgeworoside C	1990	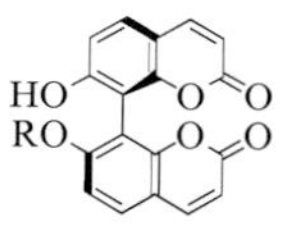 (structure; R=α-L-rhamnosyl)	$C_{24}H_{20}O_{10}$	194–196			*Edgeworthia chrysantha*	(15)
312		1989	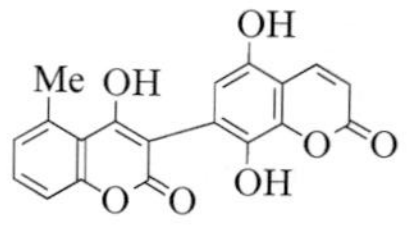 (structure)	$C_{19}H_{12}O_7$	>300			*Gerbera anandria*	(74)
313	Isokotanin C	1994	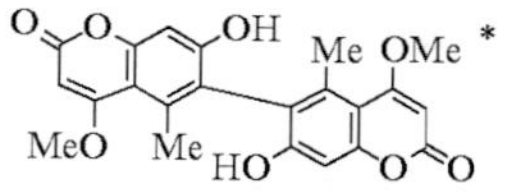 (structure)	$C_{22}H_{18}O_8$	219–222d	−29.1	MeOH	*Aspergillus alliaceus*	(137)
314	Isokotanin B	1994	(structure)	$C_{23}H_{20}O_8$	213–216	40.8	$CHCl_3$	*Aspergillus alliaceus*	(137)
		1994			>290			*Petromyces alliaceus*	(180)
315	Isokotanin A	1994	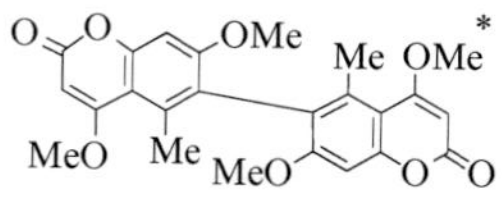 (structure)	$C_{24}H_{22}O_8$	223–226d	21.4	$CHCl_3$	*Aspergillus alliaceus*	(137)

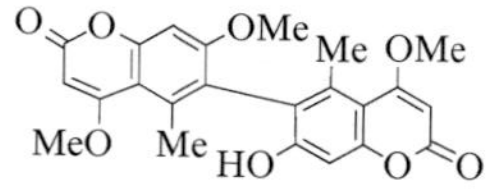

#	Name	Year	Structure	Formula	mp (°C)	[α]	Solvent	Source	Ref
316		1994	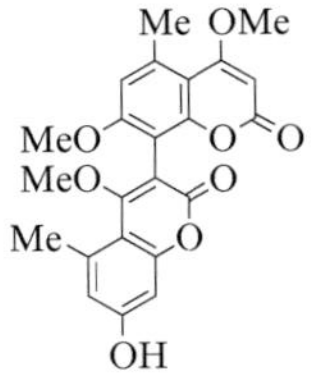	$C_{23}H_{20}O_8$	>290			*Petromyces alliaceus*	(180)
317	Aflavarin	1992	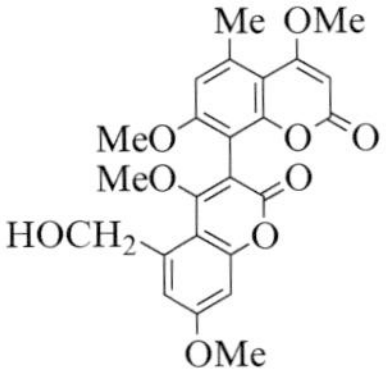	$C_{24}H_{22}O_9$	178–180d	−48.5	MeOH	*Aspergillus flavus*	(252)
318	Chamaejasmoside	1994	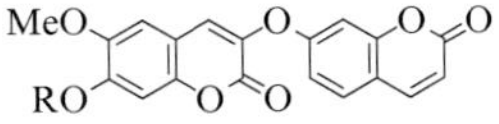 R=β-D-glucosyl	$C_{25}H_{22}O_{12}$				*Stellera chamaejasme*	(174)
319		1996	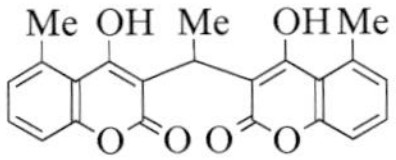	$C_{22}H_{18}O_6$	210–217			*Diospyros kaki*	(190)
320	Gigasol	1991	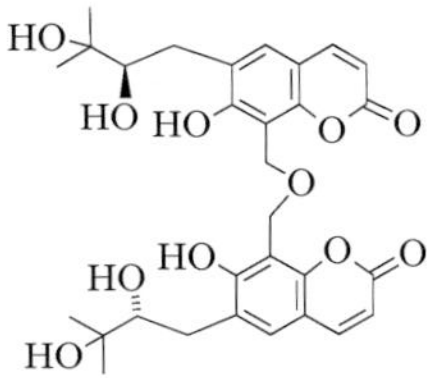	$C_{30}H_{34}O_{11}$	159–160	27.9^{24}	MeOH	*Angelica gigas*	(119)

Table 12 (continued)

	Trivial name(s)	Year isolated	Structure	Formula	M.p.	$[\alpha]_\lambda^t$	Solvent	Plant sources	Leading references
321	Bisosthenon	1990	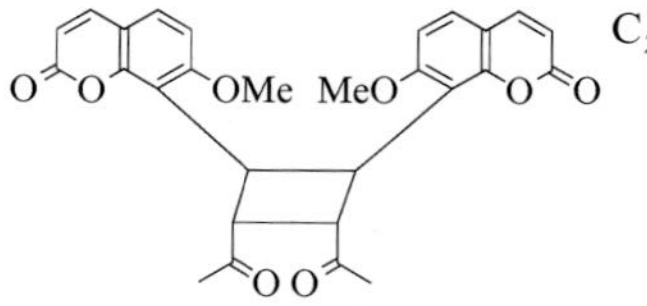	$C_{28}H_{24}O_8$	234–237			*Citrus funadoko*	(106)
322	Khelmarin-B	1990	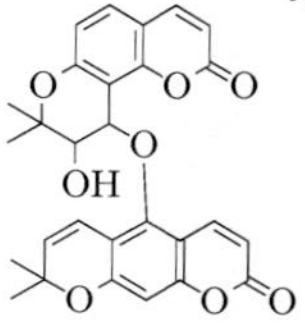 9,10-*cis* *	$C_{28}H_{24}O_8$	oil	11.2	$CHCl_3$	*Citrus canariculata*	(105)
323	Khelmarin-B	1990	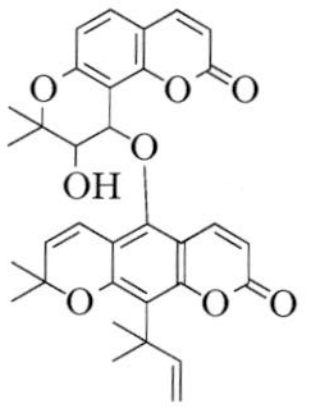 9,10-*cis* *	$C_{33}H_{32}O_8$	oil	−10.6	$CHCl_3$	*Poncirus trifoliata*	(105)

324 Khelmarin-C	1990	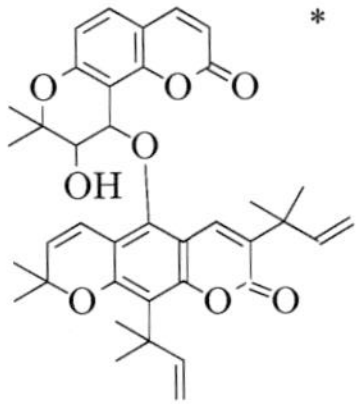	$C_{38}H_{40}O_8$	oil	−32.5 CHCl$_3$	*Citrus hassaku*	(250)
325 Nordenletin	1991	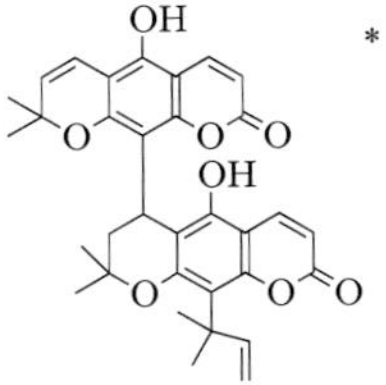	$C_{33}H_{32}O_8$	265–270	−61.8 Pyr	*Citrus hassaku*	(117)
326 Bisnorponcitrin	1990	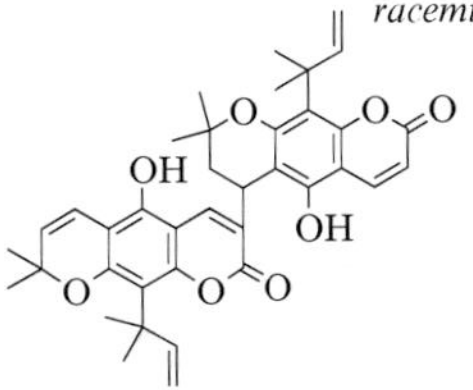	$C_{38}H_{40}O_8$	220–225		*Citrus hassaku*	(250)
327 Bisclausarin	1991	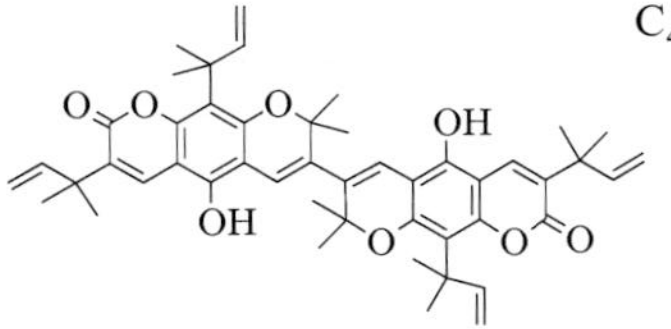	$C_{48}H_{54}O_8$	oil		*Citrus hassaku*	(118)

Table 12 (continued)

	Trivial name(s)	Year isolated	Structure	Formula	M.p.	$[\alpha]_\lambda^t$	Solvent	Plant sources	Leading references
328	Bishassanidin	1994	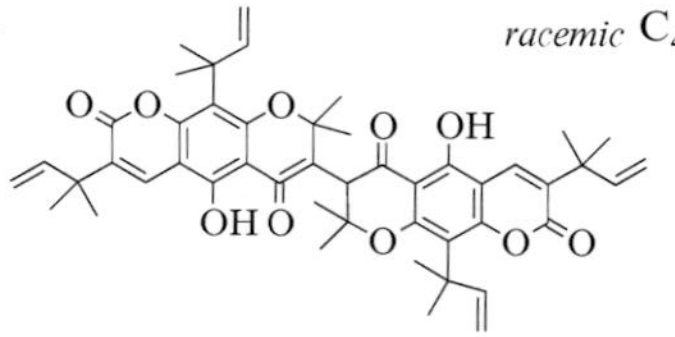	racemic $C_{48}H_{54}O_{10}$	oil			*Citrus hassaku*	*(250)*
329	Ferulenoloxy-ferulenol	1991	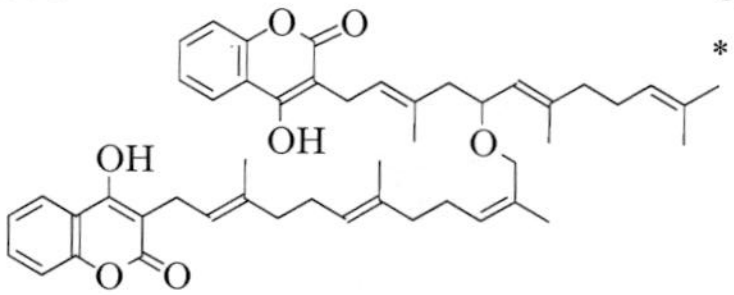	$C_{48}H_{60}O_7$ *	oil	-1.1^{20}	$CHCl_3$	*Ferula communis* var. *genuina*	*(138)*
330	Bisparasin	1993	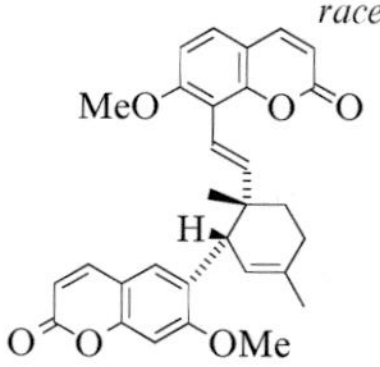	racemic $C_{30}H_{28}O_6$	oil			*Citrus paradisi* × *C. sinensis*	*(107)*

References pp. 105–119

331	Microcybin	1993	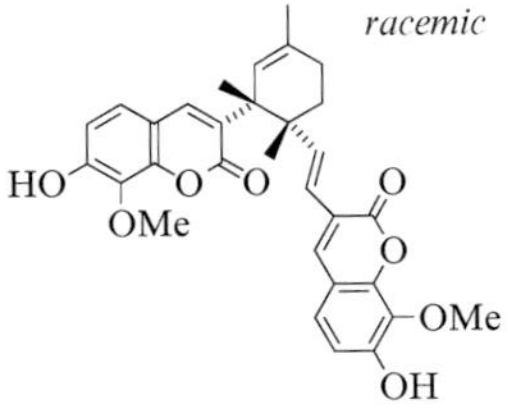 racemic structure	$C_{30}H_{28}O_8$	amorph.		*Microcybe multiflorus*	(84)
332	Toddalosin	1991	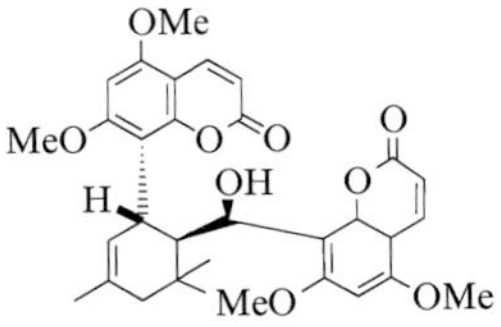 structure	$C_{32}H_{34}O_9$	265–270		*Toddalia asiatica*	(95, 99)
333	Hassmarin	1993	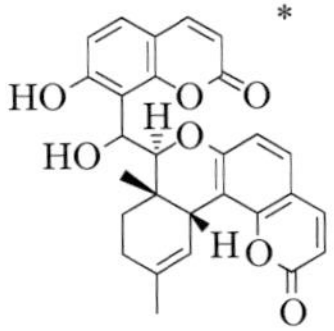 * structure	$C_{28}H_{24}O_7$	235–238	−3.9 CHCl$_3$	*Citrus hassaku*	(108)
334	Citrumarin-B	1993	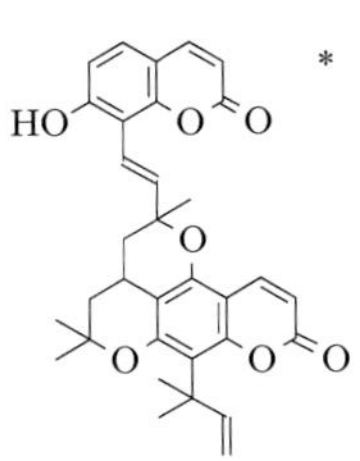 * structure	$C_{33}H_{32}O_7$	243–245	2.9 CHCl$_3$	*Poncirus trifoliata* × *Citrus paradisi*	(245)

Table 12 (continued)

Trivial name(s)	Year isolated	Structure	Formula	M.p.	$[\alpha]_\lambda^t$	Solvent	Plant sources	Leading references
335 Citrumarin-A	1993	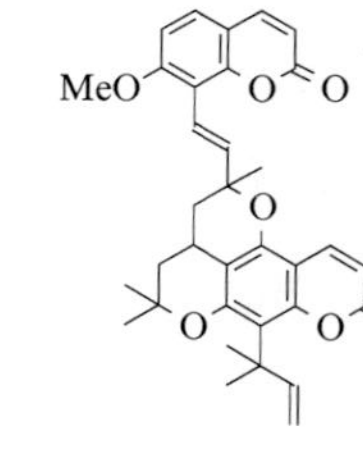 *racemic*	$C_{34}H_{34}O_7$	amorph.			*Poncirus trifoliata* × *Citrus paradisi*	(245)
336 Citrumarin-C	1993	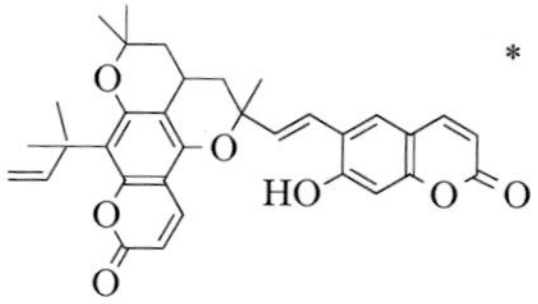 *	$C_{33}H_{32}O_7$	oil	7.8	$CHCl_3$	*Poncirus trifoliata* × *Citrus paradisi*	(245)
337 Citrumarin-D	1993	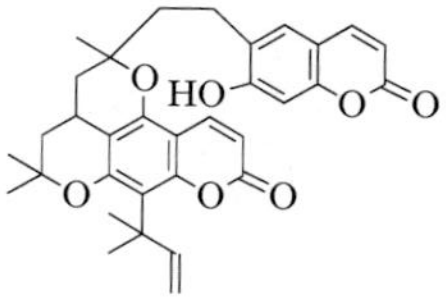 *	$C_{33}H_{32}O_7$	oil	−15.4	$CHCl_3$	*Poncirus trifoliata* × *Citrus paradisi*	(245)

References pp. 105–119

338	Furobinordentatin 1994	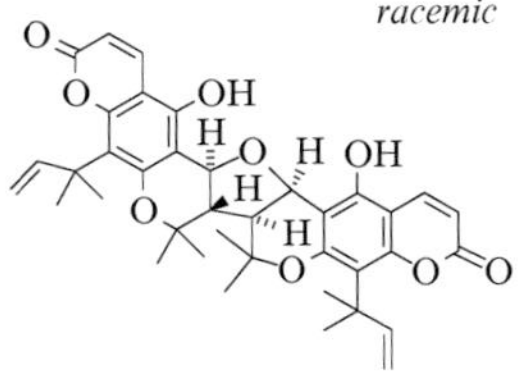 racemic	$C_{38}H_{40}O_9$	225–227	*Citrus yuko* (243, 244)
339	Furobiclausarin 1994	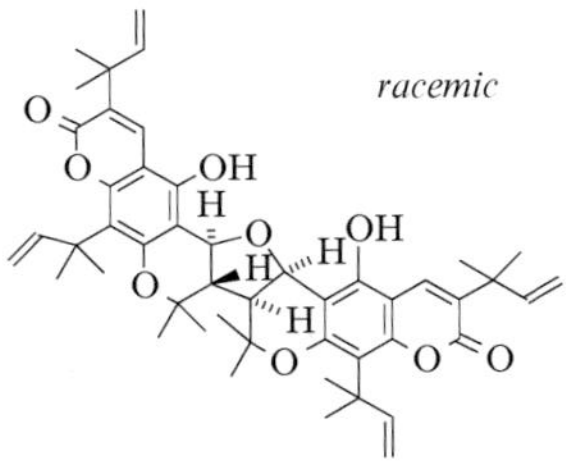 racemic	$C_{48}H_{56}O_9$	oil	*Citrus hassaku* (244)
340	Claudimerin-A 1993	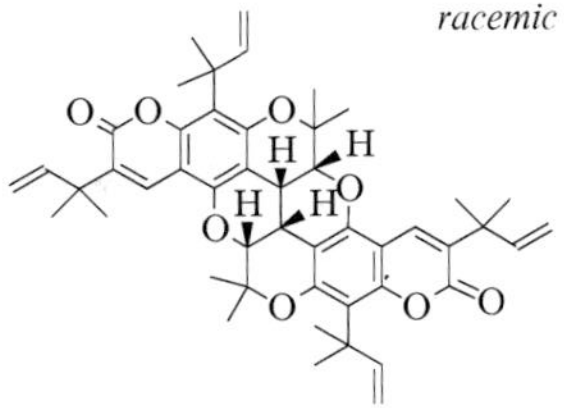 racemic	$C_{48}H_{54}O_8$	318–320	*Citrus hassaku* (251)

Table 13. *Triscomarins*

	Trivial name(s)	Year isolated	Structure	Formula	M.p.	$[\alpha]_\lambda^t$	Solvent	Plant sources	Leading references
341	Triumbelletin	1990	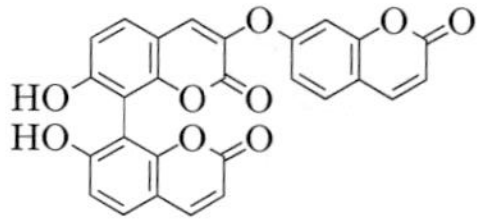	$C_{27}H_{14}O_9$				*Daphne mezereum*	*(131)*
342	Edgeworoside B	1990	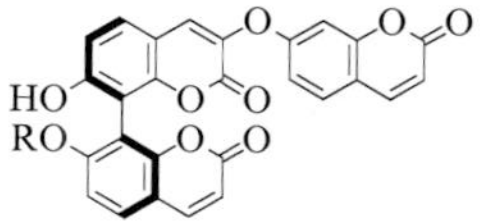 R=β-D-apiosyl	$C_{32}H_{22}O_{13}$	212.5–213			*Edgeworthia chrysantha*	*(15)*
343	Triumbellin	1990	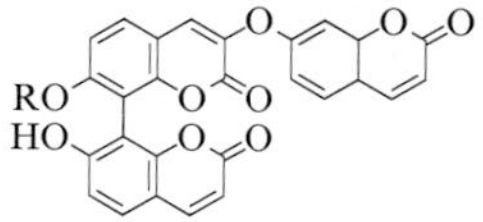 R=α-L-rhamnosyl	$C_{33}H_{24}O_{13}$	185–188d			*Daphne mezereum*	*(131)*

Amendments/Additions to Entries in Table 1 in Reference (171) and/or Reference [172]

	Trivial name(s)	Structure	Formula	Amendment/addition	Leading references
344 (1)	Coumarin	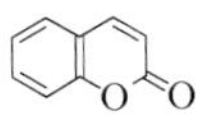	$C_9H_6O_2$	Synthesis	(271, 272)
345 (2)	Umbelliferone	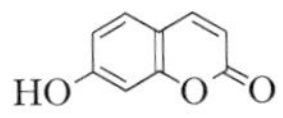	$C_9H_6O_3$	Synthesis Phytoalexin	(80, 87) (65)
346 (3)	Herniarin	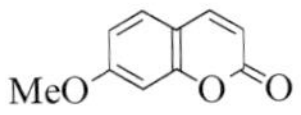	$C_{10}H_8O_3$	Synthesis	(22, 93, 100)
347 (8) [786]		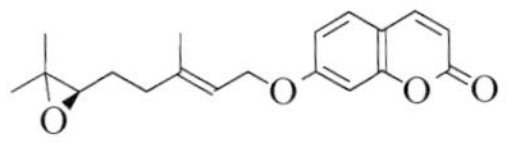	$C_{19}H_{22}O_4$	Synthesis	(270)
348 (9) [787]		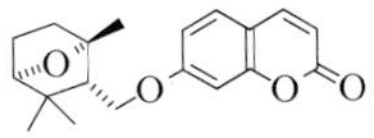	$C_{19}H_{22}O_4$	Synthesis	(57)
349 (18) [791]	Marmin	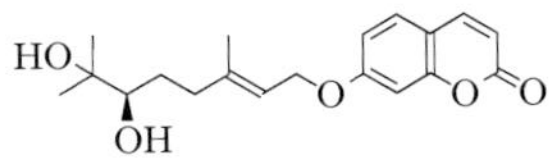	$C_{19}H_{24}O_5$	Synthesis	(270)
350 [27]	Asacoumarin B	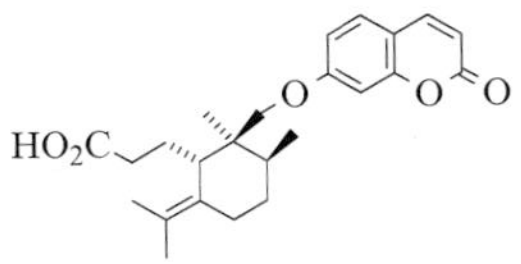	$C_{24}H_{30}O_5$	Shown to be identical with galbanic acid [804]	(13)

Amendments/Additions to Entries in Table 1.1 in Reference (171) and/or Reference [172]

Trivial name(s)	Structure	Formula	Amendment/addition	Leading references
351 [76] Tenuidin	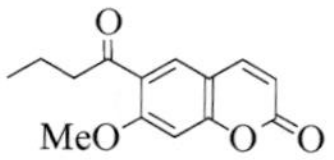	$C_{14}H_{14}O_4$	Synthesis	(192)
352 (86) [836] Demethylsuberosin	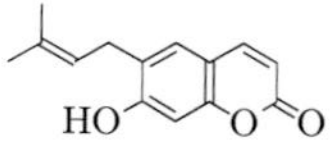	$C_{14}H_{14}O_3$	Synthesis	(33)
353 (87) [837] Suberosin	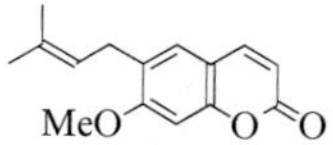	$C_{15}H_{16}O_3$	Synthesis	(33, 154)
354 (89) Suberenol	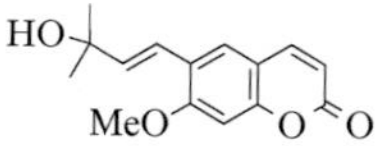	$C_{15}H_{16}O_4$	Synthesis	(213)
355 (90) [838] Geijerin	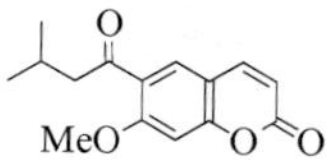	$C_{15}H_{16}O_4$	Synthesis	(32, 192)
356 (92) [839] Dehydrogeijerin	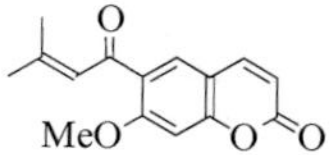	$C_{15}H_{14}O_4$	Synthesis	(32)
357 (100) [844] Ostruthin	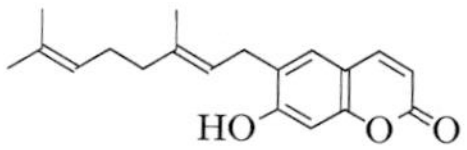	$C_{19}H_{22}O_3$	Synthesis	(33)

References pp. 105–119

Naturally Occurring Plant Coumarins

358 [112]	Acrimarine-A	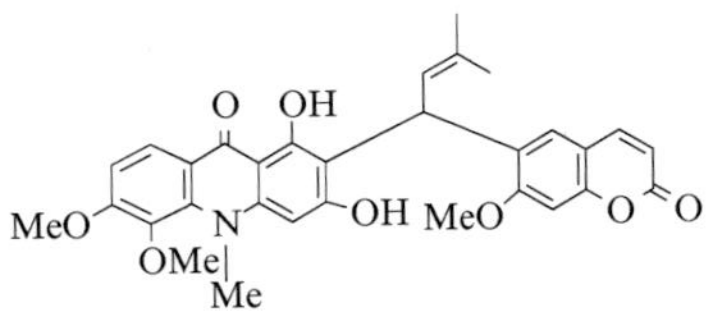	$C_{31}H_{29}NO_8$	Structure	(72)
359 [113]	Acrimarine-B	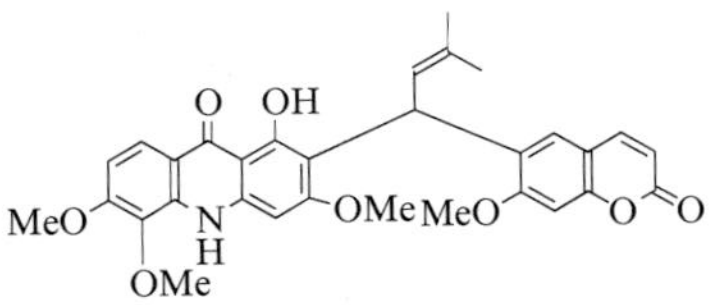	$C_{31}H_{29}NO_8$	Structure	(72)
360 [114]	Acrimarine-C (Acrimarine-L)	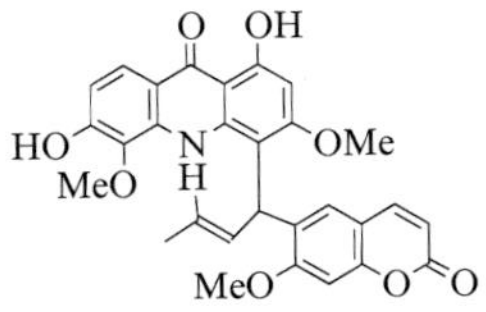	$C_{30}H_{27}NO_8$	Structure	(72, 242, 246)
361 (122) [849]	Psoralen	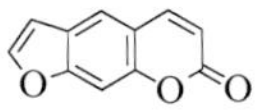	$C_{11}H_6O_3$	Synthesis	(274)

Amendments/Additions to Entries in Table 1.2 in Reference (171) and/or Reference [172]

Trivial name(s)	Structure	Formula	Amendment/addition	Leading references
362 [164]	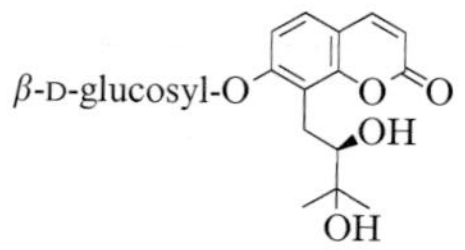	$C_{20}H_{26}O_{10}$	Absolute configuration	(142)
363 [166] Tortuoside	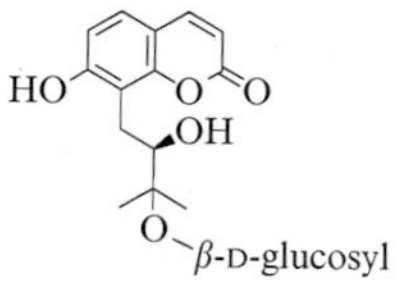	$C_{20}H_{26}O_{10}$	Structure	(38)
364 [177] Minumicrolin	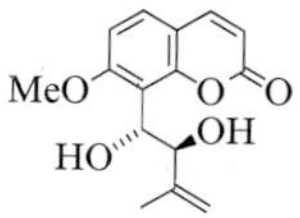	$C_{15}H_{16}O_5$	Stereochemistry Synthesis	(104) (17)
365 (164) [859] Murrangatin	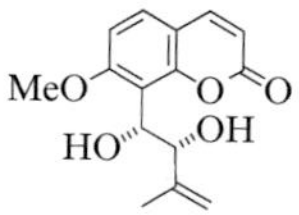	$C_{15}H_{16}O_5$	Stereochemistry Synthesis	(104) (17)
366 (166) Columbianetin	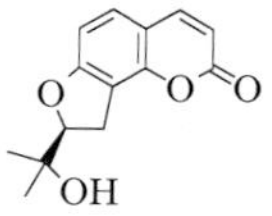	$C_{14}H_{14}O_4$	Phytoalexin	(5)

367 (199) [863]	Angelicin	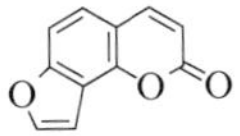	$C_{11}H_6O_3$	Bakuchicin shown to have the same structure	(124, 164)
368 (200) [864]	Oroselol	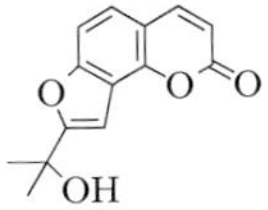	$C_{14}H_{12}O_4$	Synthesis	(274)
369 (202) [865]	Oroselone	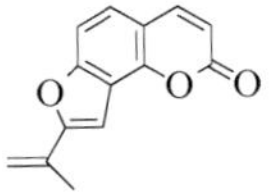	$C_{14}H_{10}O_3$	Synthesis	(274)
370 (203) [866]	Lomatin	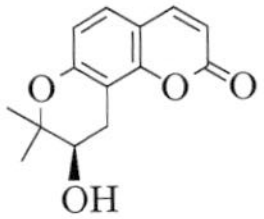	$C_{14}H_{14}O_4$	Resolution of racemate	(111)
371 (214)	(−)-*trans*-Khellactone	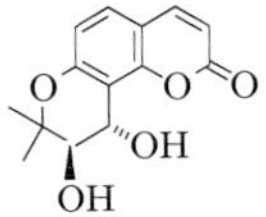	$C_{14}H_{14}O_5$	Synthesis	(215)
372 [225]	(±)-Praeruptorin A Pd-1a	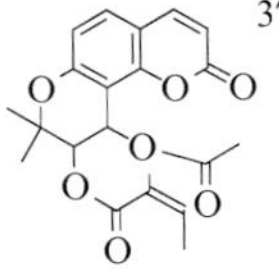 3′4′-*cis*	$C_{21}H_{22}O_7$	Racemic, not dextrorotatory First isolated in 1979 Another trivial name Synthesis X-ray structure and n.m.r.	(185) (43) (185) (16) (129)

Amendments/Additions (continued)

Trivial name(s)	Structure	Formula	Amendment/addition	Leading references
373 (230) [868] (+)-Anomalin (+)-Praeruptorin B Pd-II	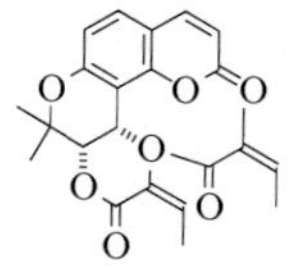	$C_{24}H_{26}O_7$	Another trivial name	(185)
374 (241) Seselin	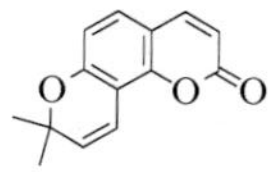	$C_{14}H_{12}O_3$	Synthesis	(153)

Amendments/Additions to Entries in Table 2 in Reference (171) and/or References [172]

	Trivial name(s)	Structure	Formula	Amendment/addition	Leading references
375 (243)	Limettin	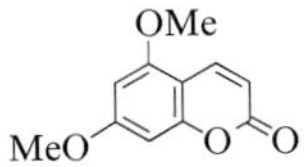	$C_{11}H_{10}O_4$	Synthesis	(79, 189)
376 [248]		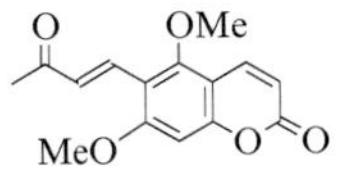	$C_{15}H_{14}O_5$	Structure revision required	(92)
377 [249]	(−)-Aculeatin	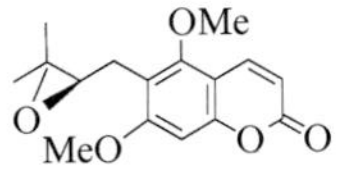	$C_{16}H_{18}O_5$	Absolute configuration	(97)
378 [249]	Toddanol	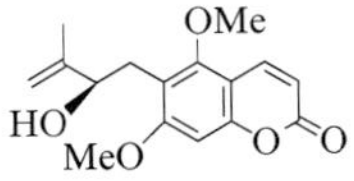	$C_{16}H_{18}O_5$	Absolute configuration	(97)
379 (250)	Toddalolactone	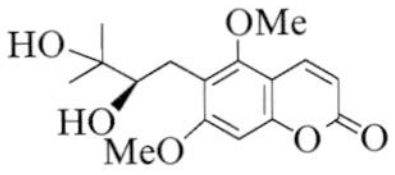	$C_{16}H_{20}O_6$	Absolute configuration Genuine natural coumarin	(97) (101)
380 (251)		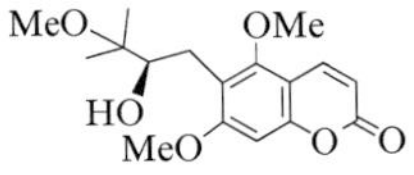	$C_{17}H_{22}O_6$	Absolute configuration	(97)
381 (255)	Oxypeucedanin	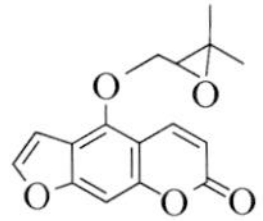	$C_{16}H_{14}O_5$	Antifungal	(157)

Amendments/Additions to Entries in Table 2 in Reference (171) and/or References [172] (continued)

Trivial name(s)	Structure	Formula	Amendment/addition	Leading references
382 (262) Oxypeucedanin hydrate	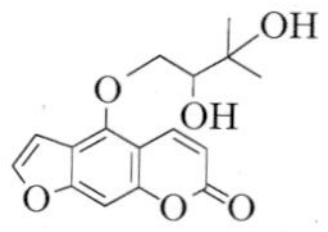	$C_{16}H_{16}O_6$	Antifungal	(157)
383 (277) [874] Deoxybruceol	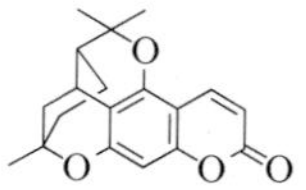	$C_{19}H_{20}O_4$	Full n.m.r.	(73)
384 (278) [875] Bruceol	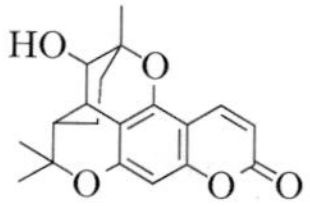	$C_{19}H_{20}O_5$	Full n.m.r.	(73)
385 [271] Gleinene	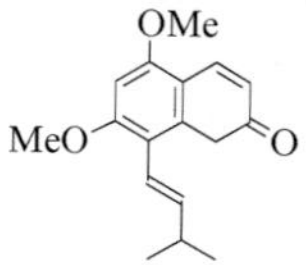	$C_{16}H_{18}O_4$	Synthesis	(217)
386 [278] Seselinal	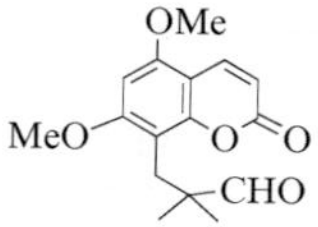	$C_{16}H_{18}O_5$	X-ray structure	(199)

387 [280]	Gleinadiene	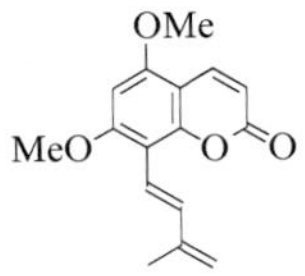	$C_{16}H_{16}O_4$	Synthesis (217)
388 (288)	Sesibiricin	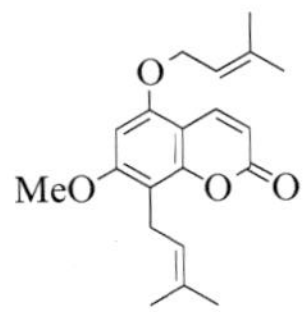	$C_{20}H_{24}O_4$	X-ray structure (188)

Amendments/Additions to Entries in Table 3 in Reference (171) and/or References [172]

	Trivial name(s)	Structure	Formula	Amendment/addition	Leading references
389 (303)	Aesculetin	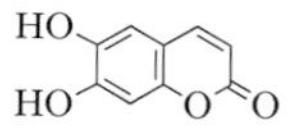	$C_9H_6O_4$	Biomimetic synthesis Synthesis	(28) (111)
390 (305)	Isoscopoletin	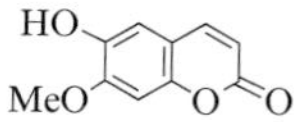	$C_{10}H_8O_4$	Synthesis	(189)
391 (310)	Scopoletin	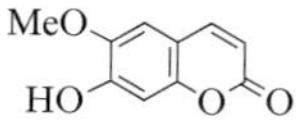	$C_{10}H_8O_4$	Tissue culture product Platelet inhibition Phytoalexin	(18) (184) (65)
392 (313)	Scoparone	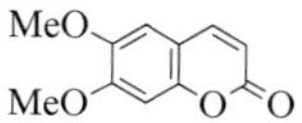	$C_{11}H_{10}O_4$	Synthesis Platelet inhibition Phytoalexin	(79, 189) (184) (239)
393 (318)	Ayapin	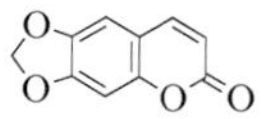	$C_{10}H_6O_4$	Synthesis	(189)
394 (329)	Heratomol	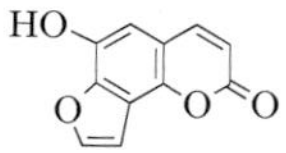	$C_{11}H_6O_4$	Synthesis	(162)
395 (330) [889]	Sphondin	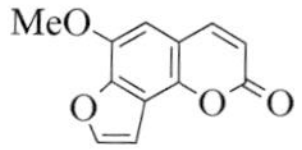	$C_{12}H_8O_4$	Synthesis	(162)

Amendments/Additions to Entries in Table 4 in Reference (171) and/or Reference [172]

Trivial name(s)	Structure	Formula	Amendment/addition	Leading references
396 (338)	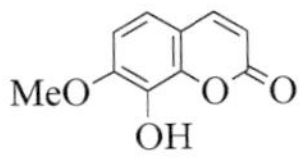	$C_{10}H_8O_4$	Synthesis	(189)
397 [340]	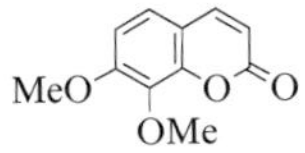	$C_{11}H_{10}O_4$	Not 5,8-dimethoxycoumarin by synthesis Synthesis	(59, 94) (219)
398 [342]	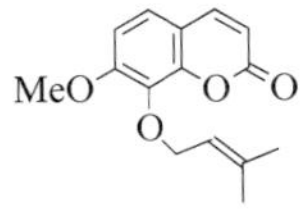	$C_{15}H_{16}O_4$	Not 5-methoxy-8-prenyloxy-coumarin by synthesis	(60, 80)
399 (341) Collinin Schinifolin	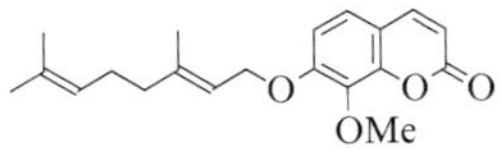	$C_{20}H_{24}O_4$	New trivial name	(113)
400 (346) [891] Rutaretin	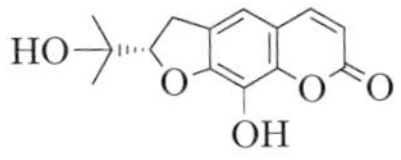	$C_{14}H_{14}O_5$	Synthesis	(215)
401 (351) Xanthotoxin	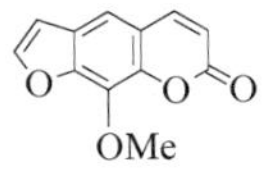	$C_{12}H_8O_4$	Fungal metabolism Epoxidation Phytoalexin	(236) (4) (55)

Amendments/Additions to Entries in Table 4 in Reference (171) and/or Reference [172] (continued)

Trivial name(s)	Structure	Formula	Amendment/addition	Leading references
402 (352) [895] Imperatorin	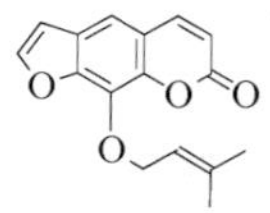	$C_{16}H_{14}O_4$	Photo-oxygenation Epoxidation	(2) (3)

Amendments/Additions to Entries in Table 5 in Reference (171) and/or References [172]

Trivial name(s)	Structure	Formula	Amendment/addition	Leading references
403 (379) Fraxinol	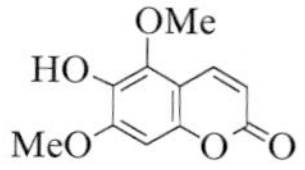	$C_{11}H_{10}O_5$	Synthesis	(189)
404 (381)	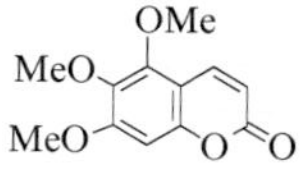	$C_{12}H_{12}O_5$	Synthesis	(189)

Amendments/Additions to Entries in Table 6 in Reference (171) and/or Reference [172]

	Trivial name(s)	Structure	Formula	Amendment/addition	Leading references
405 [382]	Leptodactylone	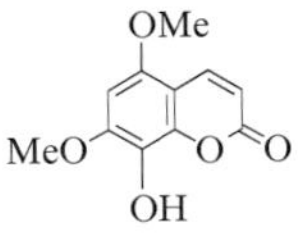 (structure: coumarin with OMe, MeO, OH)	$C_{11}H_{10}O_5$	Synthesis	(189)

Amendments/Additions to Entries in Table 7 in Reference (171) and/or Reference [172]

	Trivial name(s)	Structure	Formula	Amendment/addition	Leading references
406 (413)	Isofraxidin	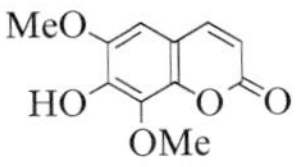 (structure: coumarin with MeO, HO, OMe)	$C_{11}H_{10}O_5$	Synthesis Tissue culture product Methyl transferase on fraxetin	(218) (18) (187)
407 (420) [906]	Obtusifol	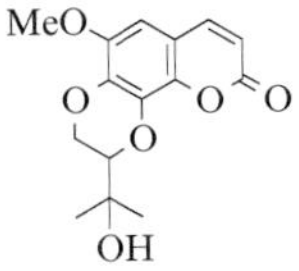 (structure: coumarin with MeO, O, OH)	$C_{15}H_{16}O_6$	Structure revised again	(27)

Amendments/Additions to Entries in Table 9 in Reference (171) and/or Reference [172]

	Trivial name(s)	Structure	Formula	Amendment/addition	Leading references
408 [452]	Angustifolin	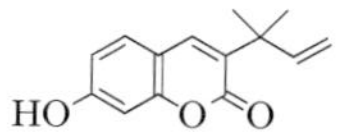	$C_{14}H_{14}O_3$	Synthesis	(86)
409 (430) [909]	Gravelliferone	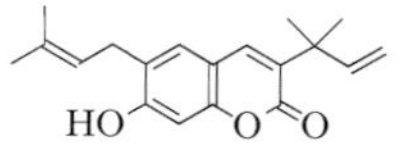	$C_{19}H_{22}O_3$	Synthesis	(33)
410 [456]	Balsamiferone	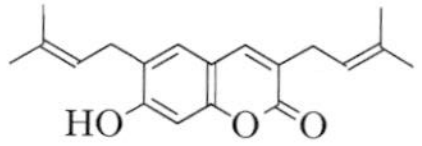	$C_{19}H_{22}O_3$	Synthesis	(33)
411 (434) [911]	Rutamarin	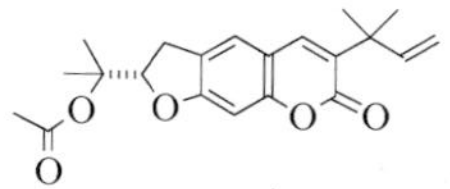	$C_{21}H_{24}O_5$	Absolute configuration	(20)
412 [469]	Ramosinin	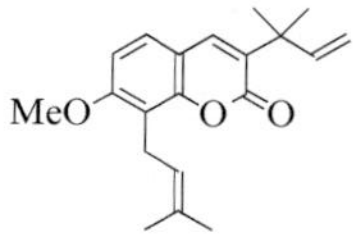	$C_{20}H_{24}O_3$	Synthesis	(220)

References pp. 105–119

Amendments/Additions to Entries in Table 9.1 in Reference [172]

	Trivial name(s)	Structure	Formula	Amendment/addition	Leading references
413 [477]	Glycycoumarin	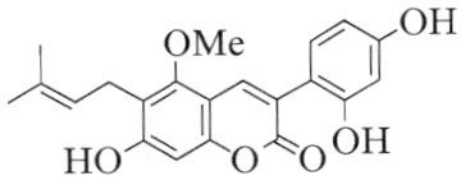	$C_{21}H_{20}O_6$	Separation	(267)
414 [479]	Licopyranocoumarin	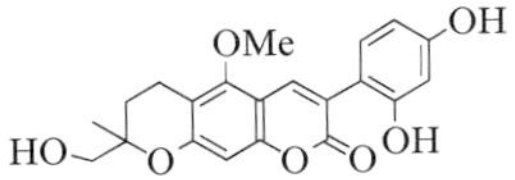	$C_{21}H_{20}O_7$	Structure	(85)

Amendments/Additions to Entries in Table 10 in Reference (171) and/or Reference [172]

Trivial name(s)	Structure	Formula	Amendment/addition	Leading references
415 [481]	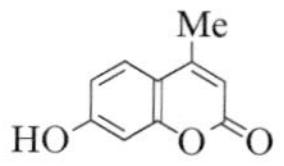 (4-methyl-7-hydroxycoumarin)	$C_{10}H_8O_3$	Synthesis O-Alkylation	(26) (53)
416 [485] Oblongulide	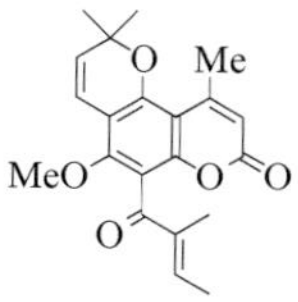 (structure)	$C_{21}H_{22}O_5$	Revised structure	(191)
417 (476) Costatolide	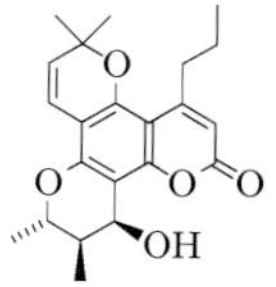 (structure)	$C_{22}H_{26}O_5$	Absolute configuration HIV-1 inhibitor	(122) (35, 71)

Amendments/Additions to Entries in Table 10.1 in Reference [172]

Trivial name(s)	Structure	Formula	Amendment/addition	Leading references
418 [500]	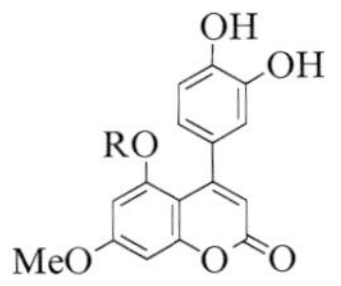 (structure; R=β-D-galactosyl)	$C_{22}H_{22}O_{11}$	X-ray structure	(235)
419 Seshadrin [505]	(structure)	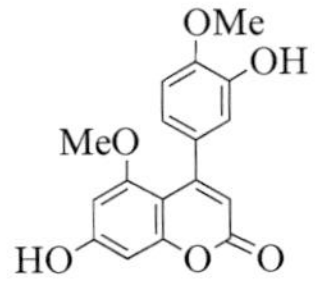 $C_{17}H_{14}O_6$	Structure revision required	(30)
420 Calophyllolide [526]	(structure)	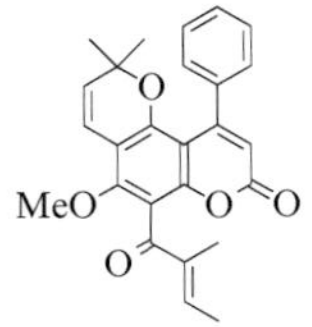 $C_{26}H_{24}O_5$	Synthesis	(191)
421 [548]	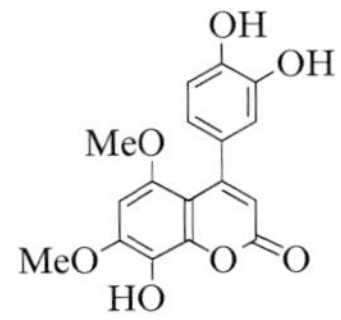 (structure)	$C_{17}H_{14}O_7$	m.p. 253–255°	(52)

Amendments/Additions to Entries in Table 11 in Reference (171) and/or Reference [172]

Trivial name(s)	Structure	Formula	Amendment/addition	Leading references
422 (481) Ferulenol	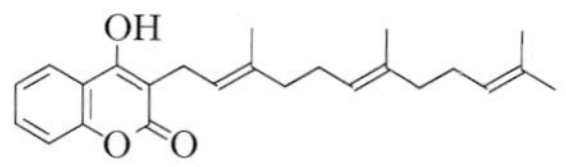	$C_{24}H_{30}O_3$	Synthesis	(9)
423 [567]	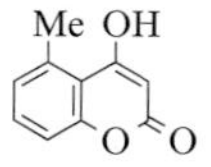	$C_{10}H_8O_3$	Biosynthesis	(91)
424 [624] Ethuliacoumarin A	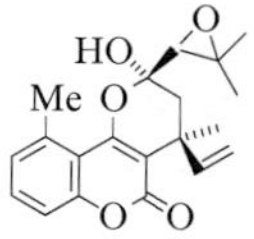	$C_{20}H_{22}O_5$	Absolute configuration	(140)
425 [625] Isoethuliacoumarin A	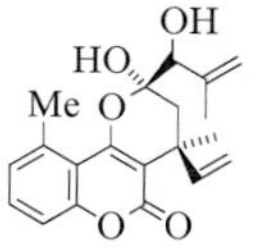	$C_{20}H_{22}O_5$	Absolute configuration	(140)
426 [689]	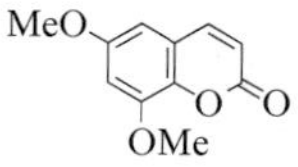	$C_{11}H_{10}O_4$	Not re-isolable from *Gomortega keule*	(257)

Amendments/Additions to Entries in Table 11.1 in Reference [172]

	Trivial name(s)	Structure	Formula	Amendment/addition	Leading references
427 [701]	Derrusnin	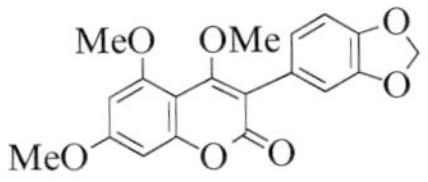	$C_{19}H_{16}O_7$	Synthesis	(19)

Amendments/Additions to Entries in Table 11.2 in Reference [172]

	Trivial name(s)	Structure	Formula	Amendment/addition	Leading references
428 [714]	Coumestrol	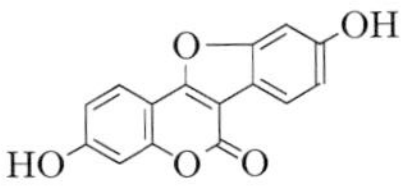	$C_{15}H_{18}O_5$	Synthesis	(141)
429 [741]	Glycyrol	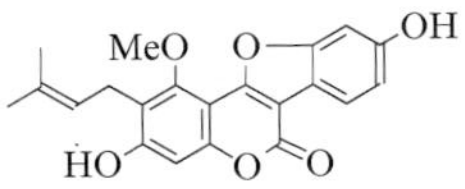	$C_{21}H_{18}O_6$	Separation	(267)
430 [743]	Isoglycyrol	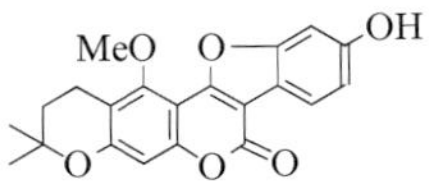	$C_{21}H_{18}O_6$	Separation	(267)

Amendments/Additions to Entries in Table 12 in Reference [172]

	Trivial name(s)	Structure	Formula	Amendment/addition	Leading references
431 [764]	Edgeworin	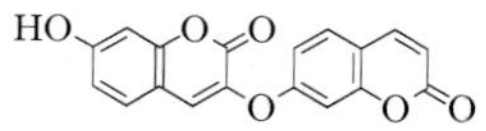	$C_{18}H_{10}O_6$	Synthesis	(54)
432 [766]	Fatagarin	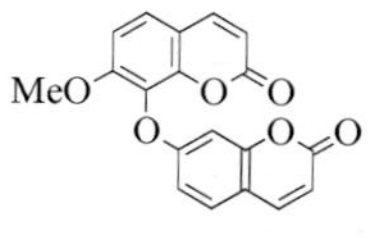	$C_{19}H_{12}O_6$	Structure revision required	(216)
433 [773]	Oreojasmin	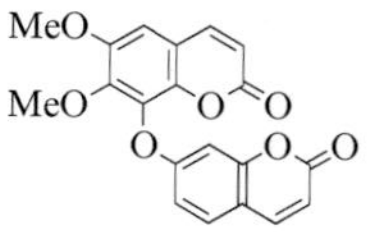	$C_{20}H_{14}O_7$	Structure revision required	(216)
434 [779]	Cyclobisuberodiene	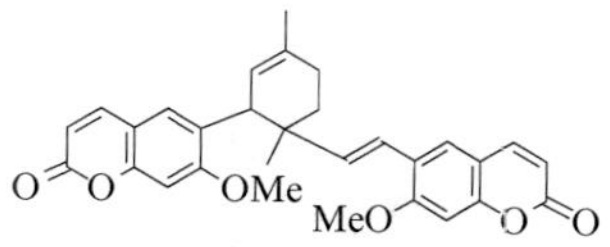	$C_{30}H_{28}O_6$	Synthesis	(213)

References pp. 105–119

Formula Index

Formula	Compound number
$C_9H_6O_2$	(344)
$C_9H_6O_3$	(345)
$C_9H_6O_4$	(389)
$C_9H_6O_5$	(236)
$C_{10}H_6O_4$	(393)
$C_{10}H_8O_3$	(346), (415), (423)
$C_{10}H_8O_4$	(390), (391), (396)
$C_{10}H_8O_5$	(237), (296)
$C_{11}H_6O_3$	(361), (367)
$C_{11}H_6O_4$	(394)
$C_{11}H_8O_5$	(227), (238)
$C_{11}H_{10}O_4$	(375), (392), (397), (426)
$C_{11}H_{10}O_5$	(229), (403), (405), (406)
$C_{11}H_{10}O_6$	(242)
$C_{12}H_8O_4$	(290), (292), (395), (401)
$C_{12}H_{10}O_3$	(277)
$C_{12}H_{10}O_5$	(154)
$C_{12}H_{12}O_3$	(284)
$C_{12}H_{12}O_3S$	(295)
$C_{12}H_{12}O_4$	(294)
$C_{12}H_{12}O_4S$	(300)
$C_{12}H_{12}O_5$	(299), (404)
$C_{13}H_{10}O_5$	(298)
$C_{13}H_{12}O_6$	(223)
$C_{13}H_{14}O_4$	(291)
$C_{13}H_{14}O_4S$	(301)
$C_{13}H_{14}O_5$	(302)
$C_{14}H_{10}O_3$	(369)
$C_{14}H_{10}O_4$	(83)
$C_{14}H_{12}O_3$	(374)
$C_{14}H_{12}O_4$	(50), (53), (113), (124), (141), (368)
$C_{14}H_{14}O_3$	(352), (408)
$C_{14}H_{14}O_4$	(22), (58), (288), (351), (366), (370)
$C_{14}H_{14}O_5$	(21), (47), (81), (90), (91), (208), (371), (400)
$C_{14}H_{16}O_5$	(286), (303)
$C_{15}H_8O_6$	(310)
$C_{15}H_{12}O_5$	(197)
$C_{15}H_{14}O_4$	(160), (356)
$C_{15}H_{14}O_5$	(155), (228), (376)
$C_{15}H_{15}ClO_4$	(70)
$C_{15}H_{16}O_3$	(353)
$C_{15}H_{16}O_4$	(180), (249), (287), (289), (354), (355), (398)
$C_{15}H_{16}O_5$	(23), (74), (125), (161), (221), (364), (365)
$C_{15}H_{16}O_6$	(132), (133), (163), (239), (407)
$C_{15}H_{16}O_{10}$	(231)
$C_{15}H_{16}O_{11}$	(244)
$C_{15}H_{18}O_5$	(428)
$C_{15}H_{18}O_6$	(26), (27)
$C_{15}H_{18}O_7$	(219)
$C_{16}H_8O_4$	(282)
$C_{16}H_{12}O_5$	(268)
$C_{16}H_{14}O_4$	(402)
$C_{16}H_{14}O_5$	(198), (381)
$C_{16}H_{14}O_8$	(213)
$C_{16}H_{16}O_4$	(387)
$C_{16}H_{16}O_5$	(189)
$C_{16}H_{16}O_6$	(92), (99), (187), (382)
$C_{16}H_{18}O_4$	(385)
$C_{16}H_{18}O_5$	(71), (72), (122), (188), (283), (377), (378), (386)
$C_{16}H_{18}O_6$	(191)
$C_{16}H_{18}O_7$	(243)
$C_{16}H_{18}O_{10}$	(225), (232)
$C_{16}H_{18}O_{11}$	(245)
$C_{16}H_{19}ClO_5$	(123)
$C_{16}H_{20}O_5$	(63)
$C_{16}H_{20}O_6$	(379)
$C_{16}H_{20}O_7$	(222)
$C_{17}H_{10}O_5$	(293)
$C_{17}H_{14}O_6$	(419)
$C_{17}H_{14}O_7$	(421)
$C_{17}H_{16}O_4$	(217)
$C_{17}H_{16}O_5$	(199)
$C_{17}H_{16}O_6$	(77), (78)
$C_{17}H_{16}O_9$	(131), (214)
$C_{17}H_{18}O_4$	(1), (216)
$C_{17}H_{18}O_5$	(218)
$C_{17}H_{18}O_6$	(25), (73)
$C_{17}H_{18}O_{10}$	(175)

$C_{17}H_{18}O_{12}$	(246)	$C_{20}H_{22}O_5$	(424), (425)
$C_{17}H_{20}O_{10}$	(226)	$C_{20}H_{22}O_7$	(195), (205)
$C_{17}H_{22}O_5$	(64), (66)	$C_{20}H_{22}O_9$	(89)
$C_{17}H_{22}O_6$	(28), (380)	$C_{20}H_{24}O_3$	(412)
		$C_{20}H_{24}O_4$	(388), (399)
$C_{18}H_{10}O_6$	(431)	$C_{20}H_{24}O_5$	(55), (56), (162), (200), (202), (203), (224)
$C_{18}H_{14}O_7$	(138), (139)		
$C_{18}H_{16}O_7$	(273)	$C_{20}H_{24}O_6$	(67), (69)
$C_{18}H_{18}O_5$	(279)	$C_{20}H_{24}O_7$	(29), (76)
$C_{18}H_{18}O_7$	(100), (193)	$C_{20}H_{24}O_8$	(45), (206)
$C_{18}H_{18}O_{13}$	(235)	$C_{20}H_{24}O_{10}$	(49)
$C_{18}H_{18}O_{14}$	(247)	$C_{20}H_{24}O_{13}$	(184)
$C_{18}H_{20}O_6$	(186)	$C_{20}H_{26}O_6$	(204)
$C_{18}H_{20}O_7$	(192)	$C_{20}H_{26}O_7$	(30)
$C_{18}H_{24}O_6$	(65)	$C_{20}H_{26}O_{10}$	(362), (363)
$C_{19}H_{12}O_6$	(432)	$C_{21}H_{18}O_6$	(130), (257), (308), (309), (429), (430)
$C_{19}H_{12}O_7$	(312)		
$C_{19}H_{16}O_7$	(427)	$C_{21}H_{20}O_5$	(128)
$C_{19}H_{18}O_5$	(54)	$C_{21}H_{20}O_6$	(129), (250), (256), (258), (259), (413)
$C_{19}H_{18}O_6$	(112), (136)		
$C_{19}H_{18}O_7$	(274)	$C_{21}H_{20}O_7$	(414)
$C_{19}H_{20}O_3$	(115)	$C_{21}H_{20}O_9$	(240)
$C_{19}H_{20}O_4$	(140), (142), (164), (173), (174), (215), (383)	$C_{21}H_{22}O_5$	(211), (416)
		$C_{21}H_{22}O_6$	(212)
$C_{19}H_{20}O_5$	(114), (116), (117), (134), (135), (137), (143), (144), (146), (167), (384)	$C_{21}H_{22}O_7$	(52), (101), (102), (103), (104), (372)
		$C_{21}H_{24}O_5$	(3), (170), (171), (411)
$C_{19}H_{20}O_6$	(48), (51), (82), (86), (93), (94), (95), (96), (97), (109), (145)	$C_{21}H_{24}O_7$	(85), (105), (106), (196)
		$C_{21}H_{24}O_{13}$	(185)
		$C_{21}H_{26}O_8$	(190)
$C_{19}H_{20}O_7$	(84), (194)	$C_{21}H_{26}O_{13}$	(176)
$C_{19}H_{20}O_{14}$	(248)	$C_{21}H_{26}O_{14}$	(207)
$C_{19}H_{22}O_3$	(31), (79), (118), (357), (409), (410)	$C_{21}H_{26}O_{15}$	(233)
$C_{19}H_{22}O_4$	(2), (4), (6), (32), (33), (57), (61), (157), (252), (347), (348)	$C_{22}H_{16}O_6$	(80)
		$C_{22}H_{18}O_6$	(319)
		$C_{22}H_{18}O_8$	(313)
		$C_{22}H_{22}O_9$	(272)
$C_{19}H_{22}O_5$	(5), (34), (220)	$C_{22}H_{22}O_{10}$	(269)
$C_{19}H_{23}ClO_4$	(10)	$C_{22}H_{22}O_{11}$	(418)
$C_{19}H_{23}ClO_5$	(62)	$C_{22}H_{24}O_5$	(260), (261)
$C_{19}H_{24}O_5$	(7), (349)	$C_{22}H_{26}O_5$	(172), (260), (262), (264), (417)
		$C_{22}H_{26}O_6$	(201)
$C_{20}H_{14}O_5$	(307)	$C_{22}H_{28}O_{15}$	(230), (234)
$C_{20}H_{14}O_7$	(433)		
$C_{20}H_{14}O_9$	(59), (60)		
$C_{20}H_{16}O_6$	(251), (305), (306)	$C_{23}H_{18}O_7$	(158)
$C_{20}H_{18}O_4$	(267)	$C_{23}H_{18}O_8$	(297)
$C_{20}H_{18}O_8$	(241)	$C_{23}H_{20}O_8$	(314), (316)
$C_{20}H_{20}O_5$	(285)	$C_{23}H_{22}O_8$	(183)

References, pp. 105–119

$C_{23}H_{22}O_{12}$ (270)
$C_{23}H_{26}O_5$ (126)
$C_{23}H_{28}O_5$ (263), (265)

$C_{24}H_{20}O_{10}$ (311)
$C_{24}H_{22}O_8$ (315)
$C_{24}H_{22}O_9$ (317)
$C_{24}H_{26}O_5$ (150)
$C_{24}H_{26}O_6$ (152)
$C_{24}H_{26}O_7$ (373)
$C_{24}H_{28}O_3$ (281)
$C_{24}H_{28}O_4$ (168)
$C_{24}H_{28}O_5$ (20), (147), (255)
$C_{24}H_{28}O_6$ (148), (149), (151), (266)
$C_{24}H_{28}O_7$ (87), (88), (108), (110)
$C_{24}H_{30}O_3$ (422)
$C_{24}H_{30}O_4$ (12), (13), (278)
$C_{24}H_{30}O_5$ (16), (17), (280), (350)
$C_{24}H_{30}O_7$ (153)
$C_{24}H_{32}O_6$ (15), (119)
$C_{24}H_{32}O_7$ (120)

$C_{25}H_{22}O_{12}$ (318)
$C_{25}H_{30}O_4$ (169)
$C_{25}H_{30}O_6$ (127)
$C_{25}H_{34}O_{10}$ (8)
$C_{25}H_{34}O_{11}$ (11)
$C_{25}H_{34}O_{14}$ (24)

$C_{26}H_{24}O_5$ (420)
$C_{26}H_{32}O_6$ (14), (18), (19)
$C_{26}H_{34}O_{10}$ (181)
$C_{26}H_{34}O_{14}$ (209)
$C_{26}H_{36}O_{18}$ (179)

$C_{27}H_{14}O_9$ (341)
$C_{27}H_{20}O_5$ (271)
$C_{27}H_{38}O_{15}$ (156)

$C_{28}H_{24}O_7$ (333)
$C_{28}H_{24}O_8$ (321), (322)
$C_{28}H_{26}O_8$ (107), (111)
$C_{28}H_{30}O_6$ (304)

$C_{29}H_{23}NO_8$ (75)
$C_{29}H_{25}NO_6$ (35)

$C_{29}H_{25}NO_7$ (36)

$C_{30}H_{27}NO_7$ (37)
$C_{30}H_{27}NO_8$ (38), (39), (360)
$C_{30}H_{28}O_6$ (330), (434)
$C_{30}H_{28}O_8$ (331)
$C_{30}H_{32}O_{12}$ (46)
$C_{30}H_{32}O_{13}$ (210), (275)
$C_{30}H_{32}O_{15}$ (177)
$C_{30}H_{34}O_{11}$ (320)
$C_{30}H_{38}O_{10}$ (121)

$C_{31}H_{29}NO_8$ (40), (44), (358), (359)
$C_{31}H_{31}NO_6$ (159)
$C_{31}H_{34}O_{14}$ (276)

$C_{32}H_{22}O_{13}$ (342)
$C_{32}H_{31}NO_8$ (41)
$C_{32}H_{34}O_9$ (332)
$C_{32}H_{38}O_{19}$ (178)
$C_{32}H_{44}O_{14}$ (182)

$C_{33}H_{24}O_{13}$ (343)
$C_{33}H_{29}NO_9$ (98)
$C_{33}H_{32}O_7$ (334), (336), (337)
$C_{33}H_{32}O_8$ (323), (325)

$C_{34}H_{31}NO_7$ (42)
$C_{34}H_{34}O_7$ (335)

$C_{35}H_{33}NO_8$ (43)
$C_{35}H_{35}NO_9$ (166)

$C_{36}H_{36}O_{14}$ (68)

$C_{38}H_{37}NO_8$ (165)
$C_{38}H_{40}O_8$ (324), (326)
$C_{38}H_{40}O_9$ (338)
$C_{38}H_{46}O_8$ (9)

$C_{39}H_{41}NO_9$ (253)

$C_{40}H_{43}NO_9$ (254)

$C_{48}H_{54}O_8$ (327), (340)
$C_{48}H_{54}O_{10}$ (328)
$C_{48}H_{56}O_9$ (339)
$C_{48}H_{60}O_7$ (329)

Trivial Name Index

Name	Compound number	Name	Compound number
Acetoxyaurapten	(3)	*Bisnorponcitrin*	(326)
Acrimarine-A	(358)	Bisosthenon	(321)
Acrimarine-B	(359)	Bisparasin	(330)
Acrimarine-C	(360)	Bruceol	(384)
Acrimarine-D	(44)	Buntansin B	(21)
Acrimarine-E	(39)	Buntansin C	(27)
Acrimarine-F	(40)		
Acrimarine-G	(36)	Calanolide A	(264)
Acrimarine-H	(37)	Calanolide B	(262)
Acrimarine-I	(42)	Calanone	(271)
Acrimarine-J	(43)	Calophyllolide	(420)
Acrimarine-K	(38)	*cis*-Casegravol	(74)
Acrimarine-L	(360)	Chamaejasmoside	(318)
Acrimarine-M	(35)	Chloculol	(70)
Acrimarine-N	(41)	Chloromarmin	(10)
(−)-Aculeatin	(377)	Citrumarin-A	(335)
Aeglin	(11)	Citrumarin-B	(334)
Aesculetin	(389)	Citrumarin-C	(336)
Aflavarin	(317)	Citrumarin-D	(337)
Albiflorin-1	(25)	Citrusarin-A	(174)
Albiflorin-2	(71)	Citrusarin-B	(173)
Albiflorin-3	(72)	Claudimerin-A	(340)
Alloxanthoxyletol	(141)	Collinin	(399)
Angelicin	(367)	Columbianetin	(366)
Angelidiol	(81)	Costatolide	(417)
Angelitriol	(26)	Coumarin	(344)
Angelol J	(28)	Coumestrol	(428)
Angelol K	(29)	Crellisin-B	(1)
Angelol L	(30)	Cyclobisuberodiene	(434)
Angustifolin	(408)		
Anhydronotoptoloxide	(128)	Daphneside	(207)
Anisocoumarin H	(2)	Daucoidin A	(86)
Anisocoumarin I	(205)	Daucoidin B	(82)
Anisocoumarin J	(206)	Decumbensol	(48)
(+)-Anomalin	(373)	Dehydrogeijerin	(356)
Asacoumarin B	(350)	Demethylsuberosin	(352)
Ayapin	(393)	Deoxybruceol	(383)
		5-Deoxyprotobruceol-I	
Badycoumarin A	(107)	regioisomer	(115)
Badycoumarin B	(111)	5-Deoxyprotobruced-II	
Bakuchicin	(367)	hydroperoxide regioisomer	(117)
Balsamiferone	(410)	5-Deoxyprotobruceol-III	
Bavacoumestan A	(306)	hydroperoxide regioisomer	(116)
Bavacoumestan B	(305)	Derrusnin	(427)
Bisclausarin	(327)	Dioxinoacrimarine-A	(75)
Bishassanidin	(328)	Donatin	(215)

Edgeworin	(431)	Isobaisseoside	(176)
Edgeworoside B	(342)	Isoethuliacoumarin A	(425)
Edgeworoside C	(311)	Isoferprenin	(281)
Edulisin III	(85)	Isofraxidin	(406)
Edulisin IV	(84)	Isoglycycoumarin	(258)
Edulisin V	(87)	Isoglycyrol	(430)
Edulisin VI	(89)	Isokotanin A	(315)
(+)-Elisin	(124)	Isokotanin B	(314)
Epiphyllocoumarin	(139)	Isokotanin C	(313)
Epoxycollinin	(200)	Isoscopoletin	(390)
Eriobrucinol regioisomer-A	(140)		
Eriobrucinol regioisomer-B	(164)	(−)-cis-Khellactone	(90)
Ethuliacoumarin A	(424)	(+)-trans-Khellactone	(91)
Ethylnotopterol	(126)	(−)-trans-Khellactone	(371)
		Khelmarin-A	(323)
Fatagarin	(432)	Khelmarin-B	(322)
Fercoprenol	(280)	Khelmarin-C	(324)
Fercoprolone	(279)		
Ferulenol	(422)	d-Laserpitin	(93)
Ferulenoloxyferulenol	(329)	Leptodactylone	(405)
Frachinoside	(178)	Licoarylcoumarin	(259)
Fraxinol	(403)	Licopyranocoumarin	(414)
Frutinone A	(282)	Ligupersin A	(20)
Frutinone B	(293)	Ligupersin B	(16)
Frutinone C	(283)	Limettin	(375)
Furobiclausarin	(339)	Ll-1	(52)
Furobinordentatin	(338)	Ll-2	(51)
		Lomatin	(370)
Gancaoin W	(256)		
Geijerin	(355)	Marmin	(349)
Gigasol	(320)	(−)-(S)-trans-Marmin	(7)
Gleinadiene	(387)	Microcybin	(331)
Gleinene	(385)	Minumicrolin	(364)
Glyasperin L	(257)	Microminutinin	(83)
Glycycoumarin	(413)	Murrangatin	(365)
Glycyrol	(429)		
trans-Grandmarin	(163)	Neoacrimarine-A	(254)
Gravelliferone	(409)	Neoacrimarine-B	(253)
		Neoacrimarine-C	(98)
Hassmarin	(333)	Neoacrimarine-D	(165)
Hemidesmin-1	(240)	Neoacrimarine-E	(166)
Hemidesmin-2	(241)	Neoglycyrol	(309)
Hemidesminine	(183)	Nivetin	(268)
Heratomol	(394)	Nordenletin	(325)
Herniarin	(346)	Notoptolide	(127)
Hoehneliacoumarin	(285)		
4′ β-Hydroxyeriobrucinol	(134)	Oblongulide	(416)
Imperatorin	(402)	Obtusifol	(407)
Inflacoumarin A	(267)	Oreojasmin	(433)
Isoarnottinin	(58)	Oroselol	(368)

Oroselone	(369)	Qianhucoumarin C	(92)
Ostruthin	(357)	Qianhucoumarin D	(100)
Oxaclausarin	(255)	Qianhucoumarin E	(112)
Oxanordentatin	(167)	Qianhucoumarin F	(54)
Oxypeucedanin	(381)	Qianhucoumarin G	(208)
Oxypeucedanin hydrate	(382)		
		Ramosinin	(412)
Palustroside	(185)	Rubricauloside	(156)
Pd-la	(372)	Rutamarin	(411)
Pd-II	(373)	Rutaretin	(400)
Peroxytamarin	(23)	Schinicoumarin	(299)
Peucedanocoumarin I	(106)	Schinifolin	(399)
Peucedanocoumarin II	(103)	Schinilenol	(203)
Peucedanocoumarin III	(101)	Schininallylol	(202)
Peujaponiside	(24)	Schinindiol	(204)
Peujaponisin	(110)	Scoparone	(392)
Peujaponisinol A	(109)	Scopoletin	(391)
Peujaponisinol B	(97)	Seselin	(374)
Phyllocoumarin	(138)	Seselinal	(386)
Pituranthoside	(8)	Seselinol	(113)
Plicadin	(307)	Seshadrin	(419)
(±)-Praeruptorin A	(372)	Sesibiricin	(388)
(±)-Praeruptorin B	(373)	Setarin	(277)
Prionanthoside	(175)	Sphondin	(395)
Protobruceol-I	(142)	Suberenol	(354)
Protobruceol-II	(144)	Suberosin	(353)
Protobruceol-II hydroperoxide	(146)	Tenuidin	(351)
Protobruceol-III	(143)	Thesiolen	(292)
Protobruceol-III hydroperoxide	(145)	Toddacoumalone	(159)
Protobruceol-IV	(137)	Toddacoumaquinone	(158)
Pseudobruceol-I	(136)	Toddalenol	(122)
Pseudobruceol-II	(135)	Toddalenone	(155)
Pseudocalanolide C	(260)	Toddalolactone	(379)
Pseudocalanolide D	(261)	Toddalosin	(332)
Psoralen	(361)	Toddanol	(378)
Ptilin	(160)	Tortuoside	(363)
Ptilostin	(162)	Triumbelletin	(341)
Ptilostol	(161)	Triumbellin	(343)
Pummeloquinone	(80)	Umbelliferone	(345)
Purpurasol	(239)		
Purpurenol	(243)	(−)-Visnadin	(105)
Pyracanthin A	(199)		
Pyracanthin B	(198)	Xanthotoxin	(401)
		Yuehgesin-A	(65)
Qianhucoumarin A	(94)	Yuehgesin-B	(63)
Qianhucoumarin B	(99)	Yuehgesin-C	(64)

Acknowledgements

The reviewer is most grateful to Prof. I.S CHEN (Kaoshiung Medical College, Taiwan), Prof. T. ISHIKAWA (Chiba University, Japan) and Prof. S.K. PAKNIKAR (Goa University, India) for assistance in obtaining copies of a number of the publications.

References

1. ABAUL, J., E. PHILOGENE, P. BOURGEOIS, C. POUPAT, A. AHOND, and P. POTIER: Contribution à la Connaissance des Rutacées Américaines: Étude des Feuilles de *Triphasia trifolia*. J. Nat. Prod. **57**, 846 (1994).
2. ABOU-ELZAHAB, M.M., W. ADAM, and C.R. SAHA-MOELLER: Photooxygenation of some Potentially Skin–Photosensitizing Furocoumarins: Imperatorin, Alloimperatorin and Its Methyl Ether and Acetate Derivatives. Liebigs Ann. Chem. 967 (1991).
3. ABOU-ELZAHAB, M.M., W. ADAM, and C.R. SAHA-MOELLER: Synthesis of Furocoumarin–Type Potential Intercalative Alkylating and Oxidising Agents of DNA through Dimethyldioxirane Epoxidation of Imperatorin and Its Derivatives. Liebigs Ann. Chem. 731 (1992).
4. ADAM, W., and M. SAUTER: Dioxirane Oxidation of Furocoumarins and Naphthofurans. Preparation of the Psoralen Epoxides. Liebigs Ann. Chem. 689 (1994).
5. AFEK, U., S. CARMELI, and N. AHARONI: Columbianetin, a Phytoalexin Associated with Celery Resistance to Pathogens During Storage. Phytochemistry **39**, 1347 (1995).
6. AHSAN, M., A.I. GRAY, G. LEACH, and P.G. WATERMAN: Novel Angular Pyranocoumarins from *Boronia lanceolata*. Phytochemistry **36**, 777 (1994).
7. AL-TEL, T.H., M.H.A. ZARGA, S.S. SABRI, M. FEROZ, N. FATIMA, Z. SHAH, and ATTA-UR-RAHMAN: Phenolics from *Colchicum decaisnei*. Phytochemistry **30**, 3081 (1991).
8. AMARO-LUIS, J.M., G. MASSANET, E. PANDO, F. RODRIGUEZ-LUIS, and E. ZUBIA: New Coumarins from *Pilocarpus goudotianus*. Planta Medica **56**, 304 (1990).
9. APPENDINO, G., G. CRAVOTTO, G.M. NANO, and G. PALMISANO: A Regioselective Synthesis of 3-Isoprenyl-4-hydroxycoumarins. Synth. Commun. **22**, 2205 (1992).
10. APPENDINO, G., G. CRAVOTTO, L. TOMA, R. ANNUZIATA, and G. PALMISANO: Diels-Alder Trapping of 3-Methylene-2,4-chromadione. A New Entry to Substituted Pyrano[3,2-c] coumarins. J. Org. Chem. **59**, 5556 (1994).
11. APPENDINO, G., H.C. ÖZEN, G.M. NANO, and M. CISERO: Sesquiterpene Coumarin Ethers from the Genus *Heptaptera*. Phytochemistry **31**, 4223 (1992).
12. APPENDINO, G., H.C. ÖZEN, S. TAGLIAPIETRA, and M. CISERO: Coumarins from *Heptaptera anisoptera* Roots. Phytochemistry **31**, 3211 (1992).
13. APPENDINO, G., S. TAGLIAPIETRA, G.M. NANO, and J. JAKUPOVIC: Sesquiterpene Coumarin Ethers from Asafetida. Phytochemistry **35**, 183 (1994).
14. BABA, K., X.Y. QING, M. TANIGUCHI, M. KOZAWA, and E. FUJITA: Studies on Chinese Medicine "Fangfeng", III: Constituents of Shui-Fang-Feng. Shoyakugaku Zasshi **45**, 167 (1991); Chem. Abstr. **116**, 136085 (1992).
15. BABA, K., M. TANIGUTI, Y. YONEDA, and M. KOZAWA: Coumarin Glycosides from *Edgeworthia chrysantha*. Phytochemistry **29**, 247 (1990).
16. BAL-TEMBE, S., D.N. BHEDI, and N.J. DE SOUZA: Improved Synthesis of ($\pm$)-Praeruptorin A and Other Khellactone Esters by Solvent-free DCC Reactions. Heterocycles **29**, 1675 (1989).

17. BANERJI, J., K.P. DHARA, B. DAS, A.K. DAS, and A. CHATTERJEE: Studies on Rutaceae, Part VI: Reactions and Rearrangements of Coumarins. Indian J. Chem. **27B**, 21 (1988).
18. BANTHORPE, D.V., and G.D. BROWN: Two Unexpected Coumarin Derivatives from Tissue Cultures of Compositae Species. Phytochemistry **28**, 3003 (1989).
19. BARTON, D.H.R., D.M.X. DONNELLY, J.P. FINET, and P.J. GUIRY: Application of Aryllead (IV) Derivatives to the Preparation of 3-Aryl-4-hydroxy-1-benzopyran-2-ones. J. Chem. Soc. Perkin 1, 1365 (1992).
20. BASNET, P., S. KADOTA, K. MANANDHAR, M.D. MANNANDHAR, and T. NAMBA: Constituents of *Boenninghausenia albiflora*: Isolation and Identification of some Coumarins. Planta Medica **59**, 384 (1993).
21. BEIER, R.C., G.W. IVIE, and E.H. OERTLI,: Linear Furanocoumarins and Graveolone from the Common Herb Parsley. Phytochemistry **36**, 869 (1994).
22. BHATTACHARJEE, J., and S.K. PAKNIKAR: Synthesis of Coumarins by Transfer of a C_3 Unit of Cinnamic Acids to Phenols Using PPA. Indian J. Chem. **28B**, 205 (1989).
23. BICCHI, C., A. D'AMATO, C. FRATTINI, E.M. CAPPELLETTI, R. CANIATO, and R. FILIPPINI: Chemical Diversity of the Contents from the Secretory Structures of *Heracleum sphondylium* subsp. *sphondylium*. Phytochemistry **29**, 1883 (1990).
24. BILIA, A.R., C. CECCHINI, A. MARSILI, I. MORELLI, and S. MELE: Coumarins and Other Constituents of *Prunus prostrata*. J. Nat. Prod. **56**, 2142 (1993).
25. BILIA, A.R., G. FLAMINI, L. PISTELLI, and I. MORELLI: New Constituents from *Pyracantha coccinea* Leaves. J. Nat. Prod. **55**, 1741 (1992); **56**, 984 (1993).
26. BISWAS, G.K., K. BASU, A.K. BARUA, and P. BHATTACHARYYA:Montmorillonite Clay as Condensing Agent in Pechmann Reaction for the Synthesis of Coumarin Derivatives. Indian J. Chem. **31B**, 628 (1992).
27. BOEYKENS, M., N. DE KIMPE, S.L. DEBENEDETTI, E.L. NADINIC, M.A. GOMEZ, J.D. COUSSIO, A.Z. ABYSHEV, and V.A. GINDIN: Revision of the Structure of Obtusifol. Phytochemistry **36**, 1559 (1994).
28. BORGES, F., and M. PINTO: Synthesis of Coumarins and Derivatives. Biomimetic Synthesis of Esculetin and Halogenated Derivatives. Helv. Chim. Acta **75**, 1061 (1992).
29. BORGES, M.F.M., F.M.F. ROLEIRA, and M.M.M. PINTO: Simultaneous Isocratic HPLC Separation of the Diastereoisomers of Caffeic, Ferulic and Isoferulic Acids and Related Coumarins. J. Liquid Chromatogr. **16**, 149 (1993).
30. BOSE, P., and J. BANERJI: Synthesis of 4-Phenylcoumarins from *Dalbergia volubilis* and *Exostema caribaeum*. Phytochemistry **30**, 2438 (1991).
31. BOURGAUD, F., A. POUTARAUD, and A. GUCKERT: Extraction of Coumarins from Plant Material (Leguminosae). Phytochem. Anal. **5**, 127 (1994); Chem. Abstr. **121**, 129219 (1994).
32. CAIRNS, N., L.M. HARWOOD, and D.P. ASTLES: Syntheses of 6-Acylcoumarins via Highly Regioselective Fries Rearrangements. Total Syntheses of the Linear Coumarins, Geijerin and Dehydrogeijerin. Tetrahedron **48**, 7581 (1992).
33. CAIRNS, N., L.M. HARWOOD, and D.P. ASTLES: Tandem Thermal Claisen-Cope Rearrangements of Coumarate Derivatives. Total Syntheses of the Naturally Occurring Coumarins: Suberosin, Demethylsuberosin, Ostruthin, Balsamiferone and Gravelliferone. J. Chem. Soc. Perkin 1, 3101 (1994).
34. CAPPIELLO, A., G. FAMIGLINI, F. MANGANI, and B. TIRILLINI: Analysis of Coumarins by Micro High-Performance Liquid Chromatography-Mass Spectrometry with a Particle Beam Interface. J. Amer. Soc. Mass Spectrom. **6**, 132 (1995).
35. CARDELLINA, J.H., H.R. BOKESCH, T.C. MCKEE, and M.R. BOYD: HIV Inhibitory Natural Products, Part 25: Resolution and Comparative Anti-HIV Evaluation of the

Enantiomers of Calanolides A and B. Biorg. Med. Chem. Letters **5**, 1011 (1995); Chem. Abstr. **123**, 55525 (1995).
36. CARDONA, L., B. GARCÍA, J.R. PEDRO, and J. PÉREZ: 6-Prenyloxy-7-methoxycoumarin, a Coumarin–Hemiterpene Ether from *Cardus tenuiflorus*. Phytochemistry **31**, 3989 (1992).
37. CAVALEIRO, C., M. CAMPOS, A. PARANHOS, and A. PROENCA DE CUNHA: Furanocoumarin Extraction Assisted by Ultrasound. Bull. Liaison-Groupe Polyphenols **16**, 63 (1992); Chem. Abstr. **123**, 28882 (1995).
38. CECCHERELLI, P., M. CURINI, M.C. MARCOTULLIO, G. MADRUZZA, and A. MENGHINI: Tortuoside, a New Natural Coumarin Glucoside from *Seseli tortuosum*. J. Nat. Prod. **53**, 536 (1990).
39. CHEMESOVA, I.I., T.V. BUKREEVA, and E.V. BOIKO: Phenols of *Artemisia laciniata*. Khim. Prirod. Soedinenii 115 (1990); Chem. Abs. **113**, 37715 (1990).
40. CHEN, C.-T., and S.-J. SHEU: Separation of Coumarins by Micellar Electrokinetic Capillary Chromatography. J. Chromatogr. **710A**, 323 (1995).
41. CHEN, I.-S., C.-T. CHANG, W.-S. SHEEN, C.-M. TENG, I.-L. TSAI, C.-Y. DUH, and F.-N. KO: Coumarins and Anti-Platelet Aggregation Constituents from Formosan *Peucedanum japonicum*. Phytochemistry **41**, 525 (1996).
42. CHEN, I.-S., Y.-C. LIN, I.-L. TSAI, C.-M. TENG, F.-N. KO, T. ISHIKAWA and H. ISHII: Coumarins and Anti-Platelet Aggregation Constituents from *Zanthoxylum schinifolium*. Phytochemistry **39**, 1091 (1995).
43. CHEN, Z.-X., B.-S. HUANG, Q.-L. SHE, and G.F. ZENG,: Study on the Chemical Constituents of the Chinese Medicinal Plant, *Peucedanum praeruptorum* Dunn. Structures of Four New Coumarins. Acta Pharm. Sinica **14**, 486 (1979); Chem. Abstr. **92**, 124903 (1980).
44. CHENERA, B., M.L. WEST, J.A. FINKELSTEIN, and G.B. DREYER,: Total Synthesis of (±)-Calanolide A, a Non-Nucleoside Inhibitor of HIV-1 Reverse Transcriptase. J. Org. Chem. **58**, 5605 (1993).
45. COREY, E.J., and L.I. WU: Enantioselective Total Synthesis of Miroestrol. J. Amer. Chem. Soc. **115**, 9327 (1993).
46. CUCA SUAREZ, L.E., and F. DELLE MONACHE: Constituents of *Murraya exotica* Adapted in Columbia. Rev. Latinoam. Quim. **22**, 38 (1991); Chem. Abstr. **116**, 148161 (1992).
47. D'AGOSTINO, M., F. DE SIMONE, A. DINI, and C. PIZZA: Isolation of 8,3'-Dihydroxy-5,7,4'-trimethoxy-4-phenylcoumarin from *Coutarea hexandra*. J. Nat. Prod. **53**, 161 (1990).
48. DAS, P.C., P.C. JOSHI, S. MANDAL, A. DAS, A. CHATTERJEE, and A. BANERJI: New Coumarino-Lignoids from *Hemidesmus indicus* R. Br. Indian J. Chem. **31B**, 342 (1992).
49. DEBENEDETTI, S.L., E.L. NADINIC, J.D. COUSSIO, N. DE KIMPE, J. FENEAU-DUPONT, and J.P. DECLERCQ: Purpurenol, a Highly Oxygenated Coumarin from *Pterucaulon purpurascens*. Phytochemistry **30**, 2757 (1991).
50. DEBENEDETTI, S.L., E.L. NADINIC, M.A. GOMEZ, J.D. COUSSIO, N. DE KIMPE, and M. BOEYKENS: Purpurasol, a Highly Oxygenated Coumarin from *Pterocaulon purpurascens*. Phytochemistry **31**, 3284 (1992).
51. DEBENEDETTI, S.L., P.S. PALACIOS, E.L. NADINIC, J.D. COUSSIO, N. DE KIMPE, M. BOEYKENS, J. FENEAU-DUPONT, and J.-P. DECLERCQ: 5-(3-Methyl-2-butenyloxy)-6,7-methylenedioxycoumarin, a 5,6,7-Trioxygenated Coumarin from *Pterocaulon virgatum*. J. Nat. Prod. **57**, 1539 (1994).
52. DELLE MONACHE, G., B. BOTTA, V. VINCIGUERRA, and R.M. PINHEIRO: 4-Arylcoumarins from *Coutarea hexandra*. Phytochemistry **29**, 3984 (1990).

53. DESHMUKH, J.G., M.H. JAGDALE, R.B. MANE, and M.M. SALUNKHE: Polymer Supported Hydroxycoumarin Anion. Convenient Method for O-Alkylation of Hydroxycoumarin. J. Indian Chem. Soc. **63**, 442 (1986).
54. DESHPANDE, A.R., H.M. THOMBRE, A.D. NATU, and M.V. PARADKAR: A Convenient Synthesis of 3,7'-Bis (coumarinyl) Ethers: Synthesis of Edgeworin. Indian J. Chem. **31B**, 759 (1992).
55. DESJARDINS, A.E., G.F. SPENCER, and R.D. PLATTNER: Tolerance and Metabolism of Furanocoumarins by the Phytopathogenic Fungus *Gibberella pulicans* (*Fusarium sambucinum*). Phytochemistry **28**, 2963 (1989).
56. DESOKY, E.K: A New Flavonoidal Coumarin from *Murraya exotica* L. Indian J. Chem. **34B**, 747 (1995).
57. DIFAZIO, M.P., and A.T. SNEDEN: A Short Synthesis of (±)-3',6'-Epoxycycloaurapten. J. Nat. Prod. **53**, 1357 (1990).
58. DI PAOLO, E.R., M.O. HAMBURGER, H. STOECKLI-EVANS, C. ROGERS, and K. HOSTETTMANN: New Chromonocoumarin (6H, 7H)-[1]-Benzopyrano [4,3-b] [1]benzopyran-6, 7-dione Derivatives from *Polygala fruticosa* Berg. Helv. Chim. Acta **72**, 1455 (1989).
59. DÖPKE, W., D. ZAIGAN, P.T. SON, V.N. HUONG, and N.T. MINH: Isolation of a New Coumarin Derivative from *Artemisia caruifolia*. Z. Chem. **30**, 375 (1990); Chem. Abstr. **114**, 81324 (1991).
60. DÖPKE, W., D. ZAIGAN, P.T. SON, V.N. HUONG, and N.T. MINH: Isolation of a New Coumarin Derivative from *Artemisia caruifolia*. Pharmazie **45**, 696 (1990).
61. DUBOIS, M. -A., M. WIERER, and H. WAGNER: Palustroside, a New Coumarin Glucoside Ester from *Ledum palustre*. Phytochemistry **29**, 3369 (1990).
62. DUH, C.-Y., S.-K. WANG, and Y.-C. WU: Cytotoxic Pyranocoumarins from the Aerial parts of *Peucedanum japonicum*. Phytochemistry **30**, 2812 (1991).
63. DUH, C.-Y., S.-K. WANG, and Y.-C. WU: Cytotoxic Pyranocoumarins from Roots of *Peucedanum japonicum*. Phytochemistry **31**, 1829 (1992).
64. ELGAMAL, M.H.A., N.M.M. SHALABY, H. DUDDECK, and M. HIEGEMANN: Coumarins and Coumarin Glucosides from the Fruits of *Ammi majus*. Phytochemistry **34**, 819 (1993).
65. EL MODAFAR, C., A. CLERIVET, A. FLEURIET, and J.J. MACHEIX: Inoculation of *Platanus acerifolia* with *Ceratocystis fimbriata* F. Sp. *platani* Induces Scopoletin and Umbelliferone Accumulation. Phytochemistry **34**, 1271 (1993).
66. EL-TURBI, J.A., J.A. ALEXANDER, A.I. GRAY, and P.G. WATERMAN: Further Novel 6, 7-Dimethoxy-8-Prenylated Coumarins from Aerial Parts of *Phebalium elastius* ssp. *beckleri*. Z. Naturforsch. **45C**, 927 (1990).
67. EL-TURBI, J.A., J.A. ARMSTRONG, A.I. GRAY, and P.G. WATERMAN: 6-Methoxymurranganon Derivatives from the Aerial Parts of *Phebalium elastius* subsp. *beckleri*. Phytochemistry **29**, 3982 (1990).
68. FOO, L.Y.: Flavanocoumarins and Flavanophenylpropanoids from *Phyllocladus trichomanoides*. Phytochemistry **28**, 2477 (1989).
69. FUKAI, T., H. KATO, and T. NOMURA: Gancaonin W, a New Prenylated 3-Arylcoumarin, from Licorice. Shoyakugaku Zasshi, **47**, 326 (1993); Chem. Abstr. **120**, 294089 (1994).
70. FUKAI, T., L. ZENG, J. NISHIZAWA, Y.-H. WANG, and T. NOMURA: Four Isoprenoid-Substituted Flavonoids from *Glycyrrhiza aspera*. Phytochemistry **36**, 233 (1994).
71. FULLER, R.W., H.R. BOKESCH, K.R. GUSTAFSON, T.C. MCKEE, J.H. CARDELLINA II, J.B. MCMAHON, G.M. CRAGG, D.D. SOEJARTO, and M.R. BOYD: HIV-Inhibitory

Coumarins from Latex of the Tropical Rain Forest Tree *Calophyllum teysmannii* var. *inophylloide*. Bioorg. Med. Chem. Letters **4**, 1961 (1994).
72. FURUKAWA, H., C. ITO, T. MIZUNO, M. JU-ICHI, M. INOUE, I. KAJIURA, and M. OMURA: Spectrophotometric Elucidation of Acrimarines, the First Naturally Occurring Acridone-Coumarin Dimers. J. Chem. Soc. Perkin 1, 1593 (1990).
73. GRAY, A.I., M.A. RASHID, and P.G. WATERMAN: NMR Assignments for the Pentacyclic Coumarins Bruceol and Deoxybruceol. J. Nat. Prod. **55**, 681 (1992).
74. GU, L.H., X. LI, S.Q. YAN, and T.R. ZHU: Studies on Antibacterial Constituents from *Gerbera anandria*. Acta Pharm. Sinica **24**, 744 (1989); Chem. Abstr. **112**, 73872 (1990).
75. GU, Z., D. ZHANG, X. YANG, M. HATTORI, and T. NAMBA: Two New Coumarin Glycosides from *Notopterygium forbesii*. Chem. Pharm. Bull. **38**, 2498 (1990).
76. GUPTA, S., B.N. JHA, G.K. GUPTA, B.K. GUPTA, and K.L. DHAR: Coumestans from Seeds of *Psoralea corylifolia*. Phytochemistry **29**, 2371 (1990).
77. GUSTAFSON, K.R., H.R. BOKESCH, R.W. FULLER, J.H. CARDELLINA II, M.R. KADUSHIN, D.D. SOEJARTO, and M.R. BOYD: Calanone, a Novel Coumarin from *Calophyllum teysmannii*. Tetrahedron Letters **35**, 5821 (1994).
78. HALIM, A.F., H.-E.A. SAAD, M.F. LAHLOUB, and A.F. AHMED: Pituranthoside from *Pituranthos triradiatus*. Phytochemistry **40**, 927 (1995).
79. HARAYAMA, T., K. KATSUNO, H. NISHIOKA, M. FUJII, Y. NISHITA, H. ISHII, and Y. KANEKO: A Convenient Synthesis of Simple Coumarin from Salicylaldehyde and Wittig Reagent (I): A Synthesis of Methoxy- and Hydroxycoumarins. Heterocycles **39**, 613 (1994).
80. HARAYAMA, T., K. KATSUNO, Y. NISHITA, and M. FUJII: Revision of Structure of a New Coumarin Isolated from *Artemisia caruifolia* Wall. Chem. Pharm. Bull. **42**, 1550 (1994).
81. HÄRMÄLÄ, P., L. BOTZ, O. STICHER, and R. HILTUNEN: Two-Dimensional Planar Chromatographic Separation of a Complex Mixture of Closely Related Coumarins from the Genus *Angelica*. J. Planar chromatogr. -Mod. TLC **3**, 515 (1990); Chem. Abstr. **114**, 171400 (1991).
82. HÄRMÄLÄ, P., S. KALTIA, H. VUORELA, and R. HILTUNEN: A Furanocoumarin from *Angelica archangelica*. Planta Medica **58**, 287 (1992).
83. HÄRMÄLÄ, P., H. VUORELA, P. LEHTONEN, and R. HILTUNEN: Optimisation of HPLC of Coumarins in *Angelica archangelica* with Reference to Molecular Structure. J. Chromatogr. **507**, 367 (1990).
84. HASAN, C.M., D.-Y. KONG, A.I. GRAY, P.G. WATERMAN, and J.A. ARMSTRONG: Microcybin: A Novel Dimeric Coumarin from *Microcybe multiflorus* and *Nematolepis phebalioides*. J. Nat. Prod. **56**, 1839 (1993).
85. HATANO, T., T. YASUHARA, T. FUKUDA, T. NORA, and T. OKUDA: Phenolic Constituents of Licorice, II: Structures of Licopyranocoumarins, Licoarylcoumarin and Glisoflavone, and Inhibitory Effects of Licorice Phenolics on Xanthine Oxidase. Chem. Pharm. Bull. **37**, 3005 (1989).
86. HERNANDEZ-GALAN, R., J. SALVA, G.M. MASSANET, and I.G. COLLADO: An Improved Synthesis of 3-(1, 1-Dimethylallyl) coumarins. Tetrahedron **49**, 1701 (1993).
87. HOEFNAGEL, A.J., E.A. GUNNEWEGH, R. DOWNING, and H. VAN BEKKUM: Synthesis of 7-Hydroxycoumarins Catalysed by Solid Acid Catalysts. J. Chem. Soc., Chem. Commun. 225 (1995).
88. IKESHIRO, Y., I. MASE, and Y. TOMITA: Dihydropyranocoumarins from Roots of *Peucedanum japonicum*. Phytochemistry **31**, 4303 (1992).
89. IKESHIRO, Y., I. MASE, and Y. TOMITA: Dihydropyranocoumarins from *Peucedanum japonicum*. Phytochemistry **33**, 1543 (1993).

90. IKESHIRO, Y., I. MASE, and Y. TOMITA: Coumarin Glycosides from *Peucedanum japonicum*. Phytochemistry **35**, 1339 (1994).
91. INOUE, T., T. TOYONAGA, S. NAGUMO, and M. NAGAI: Biosynthesis of 4-Hydroxy-5-methylcoumarin in a *Gerbera jamesonii* Hybrid. Phytochemistry **28**, 2329 (1989).
92. ISHII, H., T. ISHIKAWA, H. WADA, H. MIYAZAKI, Y. KANEKO, and T. HARAYAMA: Synthetic Studies on Naturally Occurring Coumarins, II: Synthesis of 6,7-Dimethoxy- and 7,8-Dimethoxy-5-[(*E*)-3-oxo-1-butenyl] coumarins. Chem. Pharm. Bull. **40**, 2614 (1992).
93. ISHII, H., Y. KANEKO, H. MIYAZAKI, and T. HARAYAMA: A Convenient Synthesis of a Simple Coumarin. Chem. Pharm. Bull. **39**, 3100 (1991).
94. ISHII, H., K. KENMOTSU, W. DÖPKE, and T. HARAYAMA: Synthetic Studies on Naturally Occurring Coumarins, I: A Convenient Synthesis of 5,8-Dimethoxy- and 7,8-Dimethoxycoumarin. Chem. Pharm. Bull. **40**, 1770 (1992).
95. ISHII, H., J. KOBAYASHI, M. ISHIKAWA, J. HAGINIWA, and T. ISHIKAWA: Studies on the Chemical Constituents of Rutaceous Plants LXVI: The Chemical Constituents of *Toddalia asiatica* (L.) Lam. (*T. aculeata* Pers.), 1: Chemical Constituent of the Root Bark. Yakugaku Zasshi **111**, 365 (1991); Chem. Abstr. **116**, 3546 (1992).
96. ISHII, H., J. KOBAYASHI, and T. ISHIKAWA: Studies on the Chemical Constituents of Rutaceous Plants, 72: Toddacoumalone, a Novel Mixed Dimer of Coumarin and Quinolone from *Toddalia asiatica* (L.) Lam. (*T. aculeata* Pers.). Tetrahedron Letters **32**, 6907 (1991).
97. ISHII, H., J.-I. KOBAYASHI, E. SAKURADA, and T. ISHIKAWA: The Absolute Stereochemistries of (+)-Toddalolactone and Its Related Chiral Coumarins from *Toddalia asiatica* (L.) Lam. (*T. aculeata* Pers.) and Their Optical Purities. J. Chem. Soc. Perkin 1, 1681 (1992).
98. ISHII, H., J.-I. KOBAYASHI, H. SEKI, and T. ISHIKAWA: Toddacoumaquinone, a Unique Coumarin-Naphthoquinone Dimer from *Toddalia asiatica* (L.) Lam. (*T. aculeata* Pers.). Chem. Pharm. Bull. **40**, 1358 (1992).
99. ISHII, H., J. KOBAYASHI, K. YAMAGUCHI, and T. ISHIKAWA: Toddalosin, a New Biscoumarin, from *Toddalia asiatica* (L.) Lam. (*T. aculeata* Pers.). Chem. Pharm. Bull. **41**, 1655 (1993).
100. ISHII, H., S. OHTA, H. NISHIOKA, N. HAYASHIDA, and T. HARAYAMA: A Convenient Preparation of Salicylaldehydes from 2-Methylbenzofurans by Ozonolysis. Chem. Pharm. Bull. **41**, 1166 (1993).
101. ISHII, H., S. TAN, J.P. WANG, I.-S. CHEN, and T. ISHIKAWA: Examination of Coumarins Using Supercritical Fluid and Soxhlet Extraction. Is Toddalolactone a Genuine Natural Coumarin? Yakugaku Zasshi **111**, 376 (1991); Chem. Abstr. **115**, 252094 (1991).
102. ISHIKAWA, T., K.-I. KOTAKE, and H. ISHII: Synthesis of Toddacoumaquinone, a Coumarin-Naphthoquinone Dimer, and Its Antiviral Activities. Chem. Pharm. Bull. **43**, 1039 (1995).
103. ITO, C., K. FUJIWARA, M. KAJITA, M. JU-ICHI, Y. TAKEMURA, Y. SUZUKI, K. TANAKA, M. OMURA, and H. FURUKAWA: New Coumarins from *Citrus* Plants. Chem. Pharm. Bull. **39**, 2509 (1991).
104. ITO, C., H. FURUKAWA. H. ISHII, T. ISHIKAWA, and J. HAGINIWA: The Chemical Composition of *Murraya paniculata*. The Structure of Five New Coumarins and One New Alkaloid and the Stereochemistry of Murrangatin and Related Coumarins. J. Chem. Soc. Perkin 1, 2047 (1990).
105. ITO, C., M. MATSUOKA, T. OKA, M. JU-ICHI, M. NIWA, M. OMURA, and H. FURUKAWA: New Binary Coumarins from *Citrus* Plants. Chem. Pharm. Bull. **38**, 1230 (1990).

106. Ito, C., T. Mizuno, S. Tanahashi, H. Furukawa, M. Ju-ichi, M. Inoue, M. Muraguchi, M. Omura, D.R. McPhail, and A.T. McPhail: Structure of Bisosthenon, a Novel Dimeric Coumarin from *Citrus* Plants. Chem. Pharm. Bull. **38**, 2102 (1990).
107. Ito, C., M. Nakagawa, M. Inoue, Y. Takemura, M. Ju-ichi, M. Omura, and H. Furukawa: A New Biscoumarin from *Citrus* Plants. Chem. Pharm. Bull. **41**, 1657 (1993).
108. Ito, C., T. Ono, Y. Takemura, Y. Nakata, H. Ten, M. Ju-ichi, M. Okano, N. Fukamiya, and H. Furukawa: Structure of Hassmarin, a Novel Biscoumarin from a *Citrus* Plant. Chem. Pharm. Bull. **41**, 1302 (1993).
109. Ito, C., T. Ono, E. Tanaka, Y. Takemura, T. Nakata, H. Uchida, M. Ju-ichi, M. Omura, and H. Furukawa: Structure of Pummeloquinone, a New Coumarin-Nathoquinone Dimer Isolated from *Citrus* Plants. Chem. Pharm. Bull. **41**, 205 (1993).
110. Ito, C., S. Tanahashi, Y. Tani, M. Ju-ichi, M. Omura, and H. Furukawa: Isolation of New Constituents from *Citrus* Plants. Chem. Pharm. Bull. **38**, 2586 (1990).
111. Jackson, Y.A: Improved Synthesis of Esculetin. Hetrocycles **41**, 1979 (1995).
112. Jain, N., M.S. Alam, M. Kamil, M. Ilyas, and M. Ali: A Coumarin from *Setaria italica*. Phytochemistry **30**, 3826 (1991).
113. Jiang, T., M. Hong, J. Pan, and X. Yang: NMR Study of Two New Compounds from the Roots of *Zanthoxylum schinifolium* Sieb. et Zucc. Chin. J. Mag. Reson. **9**, 413 (1992); Chem. Abstr. **119**, 245523 (1993).
114. Jong, T.-T., H.-C. Hwang, M.-Y. Jean, T.-S. Wu, and C.-M. Teng: An Antiplatelet Aggregation Principle and X-Ray Structural Analysis of *cis*-Khellactone Diester from *Peucedanum japonicum*. J. Nat. Prod. **55**, 1396 (1992).
115. Joshi, P.C., S. Mandal, P.C. Das, and A. Chatterjee: Albiflorin-1, a Coumarin from *Boenninghausenia albiflora*. Phytochemistry **30**, 2094 (1991).
116. Joshi, P.C., S. Mandal, P.C. Das, and A. Chatterjee: Two Minor Coumarins of *Boenninghausenia albiflora*. Phytochemistry **32**, 481 (1993).
117. Ju-ichi, M., Y. Takemura, M. Azuma, K. Tanaka, M. Okano, N. Fukamiya, C. Ito, and H. Furukawa: New Coumarins from *Citrus hassaku*. Chem. Pharm. Bull. **39**, 2252 (1991).
118. Ju-ichi, M., Y. Takemura, M. Okano, N. Fukamiya, C. Ito, and H. Furukawa: Structure of Bisclausarin, a New Bicoumarin from Roots of *Citrus hassaku*. Heterocycles **32**, 1189 (1991).
119. Jung, D.J., A. Porzel, and S. Huneck: Gigasol and Other Coumarins from *Angelica gigas*. Phytochemistry **30**, 710 (1991).
120. Jung, M., H. Geiger, and H.D. Zinsmeister: Tri- and Tetrahydroxycoumarin Derivatives from *Tetraphis pellucida*. Phytochemistry **39**, 379 (1995).
121. Jung, M., H.D. Zinsmeister, and H. Geiger: New Three- and Tetraaoxygenated Coumarin Glucosides from the Mosses *Atrichum undulatum* and *Polytrichum formosum*. Z. Naturforsch. **49C**, 697 (1994).
122. Kashman, Y., K.R. Gustafson, R.W. Fuller, J.H. Cardellina II, J.B. McMahon, M.J. Currens, R.W. Buckheit, Jr., S.H. Hughes, G.M. Cragg, and M.R. Boyd: HIV Inhibitory Natural Products, Part 7: The Calanolides, a Novel HIV-Inhibitory Class of Coumarin Derivatives from the Tropical Rainforest Tree, *Calophyllum lanigerum*. J. Med. Chem. **35**, 2735 (1992); **36**, 1110 (1993).
123. Kayser, O., and H. Kolodziej: Highly Oxygenated Coumarins from *Pelargonium sidoides*. Phytochemistry **39**, 1181 (1995).
124. Kondo, Y., A. Kato, Y. Kubota, and S. Nozoe: Bakuchicin from *Psoralea corylifolia* Seeds. Heterocycles **31**, 187 (1990).

125. KONG, L.Y., X. LI, Y.H. PEI, Z.D. MIN, and T.R. ZHU: Structure Elucidation of Qianhucoumarin F by 2D NMR. Chin. J. Mag. Reson. **11**, 245 (1994); Chem. Abstr. **121**, 276673 (1994).
126. KONG, L.Y., X. LI, Y.H. PEI, and T.R. ZHU: Isolation and Structural Elucidation of Qianhucoumarin D and Qianucoumarin E from *Peucedanum praeruptorum*. Acta Pharm. Sinica **29**, 49 (1994); Chem. Abstr. **121**, 129871 (1994).
127. KONG, L.Y., Y.H. PEI, X. LI, and T.R. ZHU: New Coumarins from *Peucedanum praeruptorum*. Chin. Chem. Letters **4**, 35 (1993); Chem. Abstr. **121**, 31072 (1994).
128. KONG, L.Y., Y.H. PEI, X. LI, and T.R. ZHU: New Compounds from *Peucedanum praeruptorum*. Chin. Chem. Letters **4**, 37 (1993); Chem. Abstr. **121**, 31073 (1994).
129. KONG, L.Y., Y.H. PEI, X. LI, T.R. ZHU, Y. LU, Z. TIAN, and Q. ZHENG: 2D NMR Studies on Relative Configuration of Acylated Khellactones. Chin. J. Mag. Reson. **10**, 433 (1993); Chem. Abstr. **121**, 35058 (1994).
130. KONG, L.Y., Y.H. PEI, X. LI, T.R. ZHU, and T. OKUYAMA: Isolation and Structure Elucidation of Qianhucoumarin A. Acta Pharm. Sinica **28**, 432 (1993); Chem. Abstr. **119**, 156299 (1993).
131. KREHER, B., A. NESZMÉLYI, and H. WAGNER: Triumbellin, a Tricoumarin Rhamnopyranoside from *Daphne mezereum*. Phytochemistry **29**, 3633 (1990).
132. KUMAR, D., and D.K. MUKHARYA: Anthraquinone and Coumarin Glycoside of *Hesperethusa crenulata* (Roxb.). Acta Cienc. Indica, Chem. **16C**, 411 (1992); Chem. Abstr **116**, 170107 (1992).
133. KUMAR, V., N.M.M. NIYAZ, and D.B.M. WICKRAMARATNE: Coumarins from Stem Bark of *Paramignya monophylla*. Phytochemistry **38**, 805 (1995).
134. KUSTER, R.M., R.R. BERNARDO, A.J.R. DA SILVA, J.P. PARENTE, and W.B. MORS: Furocoumarins from the Rhizomes of *Dorstenia brasiliensis*. Phytochemistry **36**, 221 (1994).
135. KUWAJIMA, H., M. MORITA, K. TAKAISHI, K. INOUE, T. FUJITA, Z.-D. HE, and C.-R. YANG: Secoiridoid, Coumarin and Secoiridoid-Coumarin Glucosides from *Fraxinus chinensis*. Phytochemistry **31**, 1277 (1992).
136. KWON, Y.S., and C.M. KIM: Coumarin Glycosides from the Roots of *Angelica dahurica*. Kor. J. Pharmacogn. **23**, 221 (1992); Chem. Abstr. **119**, 146436 (1993).
137. LAASKO, J.A., E.D. NARSKE, J.B. GLOER, D.T. WICKLOW, and P.F. DOWD: Isokotanins A.-C: New Bicoumarins from the Sclerotia of *Aspergillus alliaceus*. J. Nat. Prod. **57**, 128 (1994).
138. LAMNAOUER, D., O. FRAIGUI, M.-T. MARTIN, and B. BODO: Structure of Ferulenol Derivatives from *Ferula communis* var. *genuina*. Phytochemistry **30**, 2383 (1991).
139. LAMNAOUER, D., O. FRAIGUI, M.-T. MARTIN, J.-F. GALLARD, and B. BODO: Structure of Isoferprenin, a 4-Hydroxycoumarin Derivative from *Ferula communis* var. *genuina*. J. Nat. Prod. **54**, 576 (1991).
140. LARSEN, I.K., E. LEMMICH, S.T. THIILBORG, H.M. NIELSEN, F.M. PIEDADE, M.M. KADY, and S. BROEGGER CHRISTENSEN: The Absolute and Relative Configuration of the Molluscicides Ethuliacoumarin A and Isoethuliacoumarin A. Acta Chem. Scand. **46**, 750 (1992).
141. LASCHOBER, R., and T. KAPPE: New and Efficient Synthesis of Coumestan and Coumestrol. Synthesis 387 (1990).
142. LEMMICH, J.: Monoterpene, Chromone and Coumarin Glucosides of *Diplolophium buchananii*. Phytochemistry **38**, 427 (1995).
143. LIN, J.K., and T.-S. WU: Constituents of Flowers of *Murraya paniculata*. J. Chin. Chem. Soc. (Taipei) **41**, 213 (1994); Chem. Abstr. **121**, 5109 (1994).

144. LIN, L.-J., L.-Z. LIN, N. RUANGRUNGSI, and G.A. CORDELL: 3-Hydroxycoumarin Glycosides from *Alyxia reinwardti* var. *lucida*. Phytochemistry **34**, 825 (1993).
145. LIU, J.-H., Y.-J. GUO, S.-X. XU, and X.-S. YAO: Two New Coumarins from *Angelica pubescens* f. *biserrata*. Chin. Chem. Letters **4**, 885 (1993); Chem. Abstr. **121**, 31036 (1994).
146. LIU, J.-H., S.-X. XU, X.-S. YAO, and H. KOBAYASHI: Angelol-Type Coumarins from *Angelica pubescens* f. *biserrata* and Their Inhibitory Effect on Platelet Aggregation. Phytochemistry **39**, 1099 (1995).
147. LIU, J.-H., S.-X. XU, X.-S. YAO, and H. KOBAYASHI: Two New 6-Alkylcoumarins from *Angelica pubescens* f. *biserrata*. Planta Medica **61**, 482 (1995).
148. LLABRES, G., M. BAIWIR, W. VILEGAS, G.L. POZETTI, and J.H.Y. VILEGAS: A Proton and Carbon-13 NMR Study of a Novel Naturally Occurring Furanocoumarin from *Dorstenia cayapiaa*. Spectrochim. Acta **48A**, 1347 (1992).
149. MAGALHÃES, A.F., E.G. MAGALHÃES, H.F. LEITÃO FILHO, R.T.S. FRIGHETTO, and S.M.G. BARROS: Coumarins from *Pterocaulon balansae* and *P. lanatum*. Phytochemistry **20**, 1369 (1981).
150. MAGALHÃES, A.F., B.H.L.N. SALES, E.G. MAGALHÃES, and I.F.M. VALIO: Flavonoids and 3-Phenylcoumarins from the Seeds of *Pachyrhizus tuberosus*. Phytochemistry **31**, 1831 (1992).
151. MAHMOUD, Z.F., M.E. AMER, M.S. ABDEL KADER, and N.A. ABDEL-SALAM: A Coumestan from *Lotus creticus*. Phytochemistry **29**, 355 (1990).
152. MALDONADO, E., E. HERNÁNDEZ, and A. ORTEGO: Amides, Coumarins and Other Constituents from *Simsia cronquistii*. Phytochemistry **31**, 1413 (1992).
153. MALI, R.S., N.A. PANDHARE, and M.D. SINDKHEDKAR: Convenient Two-Step Syntheses of Seselin and Angelicin Derivatives. Tetrahedron Letters **36**, 7109 (1995).
154. MALI, R.S., P.K. SANDHU, and A. MANEKAR-TILVE: Efficient Synthesis of 6-Prenylcoumarins; Total Syntheses of Suberosin, Toddaculin, O-Methylapigravin (O-Methylbrosiparin) and O-Methylbalsamiferone. J. Chem. Soc., Chem. Commun. 251 (1994).
155. MANDAL, S., P.C. DAS, P.C. JOSHI, and A. CHATTERJEE: Hemidesminine, a New Coumarinolignoid from *Hemidesmus indicus* R. Br. Indian J. Chem. **30B**, 712 (1991).
156. MARCO, J.A., J.F. SANZ, A. YUSTE, and A. RUSTAIYAN: New Umbelliferone Sesquiterpene Ethers from Roots of *Ligularia persica*. Liebigs Ann. Chem. 929 (1991).
157. MARSTON, A., K. HOSTETTMANN, and J.D. MSONTHI: Isolation of Antifungal and Larvicidal Constituents of *Diplolophium buchananii* by Centrifugal Partition Chromatography. J. Nat. Prod. **58**, 128 (1995).
158. MASUDA, T., Y. MUROYA, and N. NAKATANI: 7-Hydroxycoumarin Derivatives from the Juice Oil of *Citrus hassaku*. Phytochemistry **31**, 1363 (1992).
159. MATA, R., M.R. CAMACHO, E. CERVERA, R. BYE, and E. LINARES: Secondary Metabolites from *Hintonia latiflora*. Phytochemistry **29**, 2037 (1990).
160. MATA, R., M.d.R. CAMACHO, S. MENDOZA, and M.d.C. CRUZ: A Phenylstyrene from *Hintonia latiflora*. Phytochemistry **31**, 3199 (1992).
161. MATSUDA, N., and M. KIKUCHI: A Coumarin Glycoside from *Lonicera gracilipes* var. *glandulosa*. Phytochemistry **38**, 803 (1995).
162. MAZUR, J., M. GRABOWSKA, and T. ZAWADOWSKI: Synthesis of Novel Sphondin (Angular Furocoumarin) Derivatives. Acta Pol. Pharm., **51**, 343 (1994); Chem. Abstr. **123**, 143676 (1995).
163. MCKEE, T.C., J.H. CARDELLINA II, G.B. DREYER, and M.R. BOYD: The Pseudocalanolides: Structure Revision of Calanolides C and D. J. Nat. Prod. **58**, 916 (1995).
164. MCNAB, H: A Reassignment of the Structure of "Bakuchicin". J. Chem. Res., Synop. 116 (1995).

165. MISKI, M., and J. JAKUPOVIC: Cyclic Farnesyl-Coumarin and Farnesyl-Chromone Derivatives from *Ferula communis* subsp. *communis*. Phytochemistry **29**, 1995 (1990).
166. MIZUNO, A., C. KAWASAKI, Y. OKADA, and T. OKUYAMA: Studies on the Chemical Structures and Pharmacological Activities of Umbelliferous Plants, XXIII: Structures of Coumarin Glycosides Isolated from *Angelica edulis*. Chin. Pharm. J. (Taipei) **46**, 249 (1994); Chem. Abstr. **122**, 209702 (1995).
167. MIZUNO, A., M. TAKATA, Y. OKADA, T. OKUYAMA, H. NISHINO, A. NISHINO, J. TAKAYASU, and A. IWASHIMA: Structures of New Coumarins and Antitumor-Promoting Activity of Coumarins from *Angelica edulis*. Planta Medica **60**, 333 (1994).
168. MIZUNO, M., H. KOJIMA, M. IINUMA, and T. TANAKA: Chemical Constituents and Their Variations Among *Coptis* Species in Japan. Shoyakugaku Zasshi **46**, 42 (1992); Chem. Abstr. **118**, 87473 (1993).
169. MIZUNO, M., H. KOJIMA, M. IINUMA, T. TANAKA, and K. GOTO: Coumarin Derivatives in *Coptis trifolia*. Phytochemistry **31**, 717 (1992).
170. MÜLLER, A.H., L.R.O. DEGÁSPARI, P.C. VIEIRA, M.F. DAS G.F. DA SILVA, J.B. FERNANDES, and J.R. PIRANI: 3-Methoxyfurocoumarins from *Pilocarpus riedelianus*. Phytochemistry **34**, 585 (1993).
171. MURRAY, R.D.H.: Naturally Occurring Plant Coumarins. Fortschr. Chem. Org. Naturstoffe **35**, 199 (1978).
172. MURRAY, R.D.H.: Naturally Occurring Plant Coumarins. Fortschr. Chem. Org. Naturstoffe **58**, 83 (1991).
173. NAGASAWA, K., Y. HIGUCHI, K. ITO, M. IMANARI, and J. FUJII: An Easy and Absolute Diagnosis for the Coumarin/Chromone Discrimination by Using Oxygen-17 NMR. Chem. Pharm. Bull. **41**, 211 (1993).
174. NARANTUYAA, S., D. BATSUREN, Y.V. RASHKES, and E.G. MIL'GROM: Chemical Study of Plants of Mongolia. Coumarins of *Stellera chamaejasme*. Structure of Chamaejasmoside, a New Bicoumarian Glycoside. Khim. prirod. Soedinenii 216 (1994); Chem. Abstr. **122**, 310690 (1995).
175. NEMA, D., and S.K. SRIVASTAVA: New Pigments from the Stem Bark of *Aegle marmelos*. Proc. Natl. Acad. Sci., India **61A**, 465 (1991); Chem. Abstr. **117**, 44563 (1992).
176. NGADJUI, B.T., J.F. AYAFOR, B.L. SONDENGAM, and J.D. CONNOLLY: Prenylated Coumarins from the Leaves of *Clausena anisata*. J. Nat. Prod. **52**, 243 (1989).
177. NGADJUI, B.T., S.M. MOUNCHEROU, J.F. AYAFOR, B.L. SONDENGAM, and F. TILLEQUIN: Geranyl Coumarins from *Clausena anisata*. Phytochemistry **30**, 2809 (1991).
178. NIWA, M., H. SUGINO, S. TAKASHIMA, T. SAKAI, Y.C. WU, T.-S. WU, and C.S. KUOH: New Coumarin Glucoside from *Daphne arisanensis*. Chem. Pharm. Bull. **39**, 2422 (1991).
179. NKENGFACK, A.E., J. KOUAM, T.W. VOUFFO, M. MEYER, M.S. TEMPESTA, and Z.T. FOMUM: An Isoflavanone and a Coumestan from *Erythrina sigmoidea*. Phytochemistry **35**, 521 (1994).
180. NOZAWA, K., S. NAKAJIMA, K. KAWAI, S. UDAGAWA, and M. MIYAJI: Bicoumarins from Ascostromata of *Petromyces alliaceus*. Phytochemistry **35**, 1049 (1994).
181. NYKOLOV, N., T. IOSSIFOVA, E. VASSILEVA, I. KOSTOVA, and G. STOEV: Reverse-Phase High Pressure Liquid Chromatographic Analysis of Hydroxycoumarins in Plant Extracts. Quantitative Determination of Hydroxycoumarins in *Fraxinus ornus*. Phytochem. Anal. **4**, 86 (1993).
182. OCHOCKA, R.J., D. RAJZER, P. KOWALSKI, and H. LAMPARCZYK: Determination of Coumarins from *Crysanthemum segetum* L. by Capillary Electrophoresis. J. Chromatogr. **709A**, 197 (1995).

183. OHASHI, K., H. WATANABE, K. OHI, H. ARIMOTO, and Y. OKUMURA: Two New 7-Geranyloxycoumarins from the Bark of *Aegle marmelos*, an Indonesian Medicinal Plant. Chem. Letters 881 (1995).
184. OKADA, Y., N. MIYAUCHI, K. SUZUKI, T. KOBAYASHI, C. TSUTSUI, K. MAYUZUMI, S. NISHIKE, and T. OKUYAMA: Search for Naturally Occurring Substances to Prevent the Complication of Diabetes, II: Inhibitory Effect of Coumarin and Flavonoid Derivatives on Bovine Lens Aldose Reductase and Rabbit Platelet Aggregation. Chem. Pharm. Bull. **43**, 1385 (1995).
185. OKUYAMA, T., and S. SHIBATA: Studies on Coumarins of a Chinese Drug "Qian-Hu". Planta Medica **42**, 89 (1981).
186. OLIVEIRA, F.M., A.E.G. SANT'ANA, L.M. CONSERVA, J.G.S. MAIA, and G.M.P. GUILHON: Alkaloids and Coumarins from *Esenbeckia* Species. Phytochemistry **41**, 647 (1996).
187. OSOBA, O.A., and M.R. ROBERTS: Methyltransferase Activity in *Ailanthus altissima* Cell Suspension Cultures. Plant Cell Rep. **13**, 277 (1994); Chem. Abstr. **121**, 104070 (1994).
188. PADHA, N., K.N. GOSWAMI, and V.S. YADAVA: Crystal and Molecular Structure of 5-(3-Methylbut-2-enyloxy)-7-methoxy-8-(3-methylbut-2-enyl)- 2H-[1]benzopyran-2-one (Sesibiricin). Z. Kristallogr. **209**, 878 (1994); Chem. Abstr. **122**, 265086 (1995)
189. PAKNIKAR, S.K., J. BHATTACHARJEE, and K.K. NADKARNI: A Single-Step Synthesis of Ayapin, Limettin, Scoparone, Leptodactylone, Fraxinol, Isoscopoletin, 5, 6, 7-Trimethoxycoumarin and 7-Methoxy-8-hydroxycoumarin. J. Indian Inst. Sci. **74**, 277 (1994); Chem. Abstr. **122**, 290496 (1995).
190. PAKNIKAR, S.K., K.P.P. FONDEKAR, J.K. KIRTANY, and S. NATORI: 4-Hydroxy-5-Methylcoumarin Derivatives from *Diospyros kaki* Thunb. and *D. kaki* var. *sylvestris* Makino; Structure and Synthesis of 11-Methylgerberinol. Phytochemistry **41**, 931 (1996).
191. PALMER, C.J., and J.L. JOSEPHS: Synthesis of the *Calophyllum* Coumarins. Tetrahedron Letters **35**, 5363 (1994).
192. PARADKAR, M.V., M.S. KULKARNI, S.A. KULKARNI, and H.M. GODBOLE: Synthesis of Naturally occurring 6-Acyl-7-methoxycoumarins. J. Chem. Res., Synop. 262 (1995).
193. PRAKASH, D., V. LAKSHMI, K. RAJ, and R.S. KAPIL: Constituents of *Clausena indica* Root. Fitoterapia **60**, 347 (1989); Chem. Abstr. **112**, 73862 (1990).
194. QIN, B., Q.-P. CHEN, and Z.-C. LOU: Active Constituents of *Viola prionantha* Bge. J. Chin. Pharm. Sci. **3**, 91 (1994); Chem. Abstr. **122**, 209705 (1995).
195. QIN, B., Q.-P. CHEN, L.-W. SHI and Z.-C. LOU: Separation and Quantitative Determination of Three Coumarins in the Chinese Traditional Drug Zihuadiding, Herba violae, by High-Performance Liquid Chromatography. J. Chin. Pharm. Sci. **3**, 157 (1994); Chem. Abs. **122**, 222980 (1995).
196. QUADER, M.A., J.A. EL-TURBI, J.A. ARMSTRONG, A.I. GRAY, and P.G. WATERMAN: Coumarins and Their Taxonomic Value in the Genus *Phebalium*. Phytochemistry **31**, 3083 (1992).
197. RAHMANI, M., T.Y.Y. HIN, H.B.M. ISMAIL, M.A. SUKARI, and A.R. MANAS: Microminutinin: a Novel Coumarin from *Micromelum minutum*. Planta Medica **59**, 93 (1993).
198. RAHMANI, M., Y.H. TAUFIQ-YAP, H.B.M. ISMAIL, A. SUKARI, and P.G. WATERMAN: New Coumarin and Dihydrocinnamic Acid Derivatives from Two Malaysian Populations of *Micromelum minutum*. Phytochemistry **37**, 561 (1994).
199. RAJNIKANT, G.K.N., V.S. YADAVA, and V.M. PADAMANABHAN: Seselinal, a Simple Coumarin Structure. Mol. Cryst. Liq. Cryst. Sci. Technol. **3C**, 61 (1993); Chem. Abstr. **120**, 191380 (1994).

200. RAO, G., Y. DAI, F. ZHANG, F. CAI, and H. SUN: Chemical Constituents of the Chinese Drug Han-Qian-Hu (*Ligusticum daucoides*) Acta Bot. Yunnanica **13**, 209 (1991); Chem. Abstr. **117**, 118295 (1992).
201. RAO, G., F. NIU, and H. SUN: Study on the Chemical Constituents of the Chinese Drug "Zhong-Dian Qian-Hu" (*Peucedanum* sp.). Acta Bot. Yunnanica **12**, 434 (1990); Chem. Abstr. **114**, 234923 (1991).
202. RAO, G., H. SUN, Z. LIN, and F. NIE: Chemical Studies of Traditional Chinese Medicine Qian-Hu. Nat. Prod. Res. Development **5**, 1 (1993); Chem. Abstr. **121**, 53896 (1994).
203. RAO, M. N., G.L.D. KRUPADANAM, and G. SRIMANNARAYAMA: Four Isoflavones and Two 3-Aryl Coumarins from Stems of *Derris scandens*. Phytochemistry **37**, 267 (1994).
204. RASHID, M.A., J.A. ARMSTRONG, A.I. GRAY, and P.G. WATERMAN: Protobruceols: New 6-*C*-Monoterpenyl-5,7-oxygenated Coumarins from *Eriostemon brucei*. Nat. Prod. Letters **1**, 79 (1992).
205. RASHID, M.A., J.A. ARMSTRONG, A.I. GRAY, and P.G. WATERMAN: Pyranocoumarins as Chemotaxonomic Markers in *Eriostemon coccineus* and *Philotheca citrina*. Phytochemistry **30**, 4033 (1991).
206. RASHID, M.A., J.A. ARMSTRONG, A.I. GRAY, and P.G. WATERMAN: Alkaloids, Flavonols and Coumarins from *Drummondita hassellii* and *D. calida*. Phytochemistry **31**, 1265 (1992).
207. RASHID, M.A., J.A. ARMSTRONG, A.I. GRAY, and P.G. WATERMAN: Tetra- and Pentacyclic 6-*C*-Monoterpenyl-5,7-dioxycoumarins from *Eriostemon brucei*. Phytochemistry **31**, 3583 (1992).
208. RASHID, M.A., J.A. ARMSTRONG, A.I. GRAY, and P.G. WATERMAN: Novel *C*-Geranyl 7-Hydroxycoumarins from the Aerial Parts of *Eriostemon tomentellus*. Z. Naturforsch. **47B**, 284 (1992).
209. RASHID, M.A., A.I. GRAY, P.G. WATERMAN, and J.A. ARMSTRONG: Coumarins from *Eriostemon spicatus*. J. Nat. Prod. **55**, 685 (1992).
210. RASHID, M.A., A.I. GRAY, P.G. WATERMAN, and J.A. ARMSTRONG: Coumarins from *Phebalium tuberculosum* ssp. *megaphyllum* and *P. filifolium*. J. Nat. Prod. **55**, 851 (1992).
211. RASOOL, N., A.Q. KHAN, V.U. AHMAD, and A. MALIK: A Benzoquinone and a Coumestan from *Psoralea plicata*. Phytochemistry **30**, 2800 (1991).
212. REISCH, J., and S.H. ACHENBACH: A Furanocoumarin Glucoside from Stembark of *Skimmia japonica*. Phytochemistry **31**, 4376 (1992).
213. REISCH, J., H.M.T.B. HERATH, and N.S. KUMAR: Synthesis of the Natural Coumarins (*E*)-Suberenol, Cyclobisuberodiene and Two Other Related New Coumarins. Liebigs Ann. Chem. 931 (1990).
214. REISCH, J., H.M.T.B. HERATH, and N.S. KUMAR: Synthesis of *rac*-Acridone-Coumarin Dimers (Acrimarins) Liebigs Ann. Chem. 839 (1991).
215. REISCH, J. and A.A.W. VOERSTE: Natural Product Chemistry, Part 181: Investigations on the Synthesis of Dihydropyrano- and Dihydrofurano-Coumarins by Application of Catalytic Enantioselective *cis*-Dihydroxylation. J. Chem. Soc. Perkin 1, 3251 (1994).
216. REISCH, J., A. WICKRAMASINGHE, and V. KUMAR: Synthesis of Dicoumarinyl Ethers with Structures Possible for Oreojasmin: Coumarins of *Ruta oreojasme* Fruits. J. Nat. Prod. **52**, 1379 (1989).
217. REISCH, J., A. WICKRAMASINGHE, and D.B.M. WICKREMARATNE: Synthesis of the Natural Coumarins Gleinene and Gleinadiene. Liebigs Ann. Chem. 209 (1990).
218. ROUESSAC, F., and A. LECLERC: An Efficient Synthesis of Isofraxidin. Synth. Commun. **23**, 1147 (1993).
219. ROUESSAC, F., and A. LECLERC: An Efficient Preparation of Coumarins. Synth. Commun. **23**, 2709 (1993).

220. SALVÁ, J., F.R. LUIS, E. PANDO, G.M. MASSANET, and R.H. GALÁN: Synthesis of C-8 Prenylated and Angular 3-(1',1'-Dimethylallyl) Coumarins. Heterocycles 31, 255 (1990).
221. SARKER, S.D., J.A. ARMSTRONG, A.I. GRAY, and P.G. WATERMAN: Sesquiterpenyl Coumarins and Geranyl Benzaldehyde Derivatives from the Aerial Parts of *Eriostemon myoporoides*. Phytochemistry 37, 1287 (1994).
222. SARKER, S.D., J.A. ARMSTRONG, and P.G. WATERMAN: An Alkaloid, Coumarins and a Triterpene from *Boronia algida*. Phytochemistry 39, 801 (1995).
223. SARKER, S.D., A.I. GRAY, P.G. WATERMAN, and J.A. ARMSTRONG: Coumarins from Two *Asterolasia* Species. J. Nat. Prod. 57, 324 (1994).
224. SARKER, S.D., A.I. GRAY, P.G. WATERMAN, and J.A. ARMSTRONG: Coumarins from *Asterolasia trymalioides*. J. Nat. Prod. 57, 1549 (1994).
225. SARKER, S.D., P.G. WATERMAN, and J.A. ARMSTRONG: Coumarin Glycosides from Two Species of *Eriostemon*. J. Nat. Prod. 58, 1109 (1995).
226. SCHULZ, H., G. ALBROSCHEIT, and D. NOWAK: Characterisation of Grapefruit Oil and Juice by HPLC. Lebensm. -Unters. Forsch. 195, 254 (1992); Chem. Abstr. 118, 58289 (1993).
227. SCHUSTER, N., C. CHRISTIANSEN, J. JAKUPOVIC, and M. MUNGAI: An Unusual [2 + 2] Cycloadduct of Terpenoid Coumarin from *Ethulia vernonioides*. Phytochemistry 34, 1179 (1993).
228. SEONG, B.W., C.S. YOOK, H.S. CHUNG, and W.S. WOO: New *cis*-Khellactone Esters from *Angelica flaccida*. Planta Medica 57, 496 (1991).
229. SHARIFI, S., H.C. MICHAELIS, E. LOTTERER, and J. BIRCHER: A Selective and Sensitive Method for the Determination of Coumarin and Its Main Metabolites 7-Hydroxycoumarin, 7-Hydroxycoumarin Glucuronide and 3-Hydroxycoumarin in Human Plasma by HPLC. J. Liq. Chromatogr. 16, 1263 (1993).
230. SHEN, X.-W., S.-Z. ZHENG, Z.-D. YIN, Z.-W. SONG, and L. WANG: Five New Compounds from *Elsholtzia densa*. Chem. J. Chin. Univ. 15, 540 (1994); Chem. Abstr. 121, 276671 (1994).
231. SHKARENDA, V.V., and P.V. KUZNETSOV: Present Status of Liquid Column Chromatography of Coumarins, II: High-Performance Liquid Chromatography of Coumarin Derivatives. Khim. prirod. Soedinenii 71, (1993); Chem. Abstr. 123, 221992 (1995).
232. SINGH, R.P., and V.B. PANDEY: Nivetin, a Neoflavonoid from *Echinops niveus*. Phytochemistry 29, 680 (1990).
233. SINGH, S., M.K. PALIWAL, I.R. SIDDIQUI, and J. SINGH: Two New Coumarin Glycosides from *Pterocarpus santalinus*. Fitoterapia 63, 555 (1992); Chem. Abstr. 119, 135568 (1993).
234. SKALTSOUNIS, A.L., S. MITAKU, G. GAUDEL, F. TILLEQUIN, and M. KOCH: Synthesis, Glycosidation and Resolution of ($\pm$)-Lomatin. Heterocycles 34, 121 (1992).
235. SORIANO-GARCÍA, M., R. VILLENA IRIBE, S. MENDOZA-DÍAZ, M. DEL RAYO COMACHO, and R. MATA: Structure of 4-(3,4-Dihydroxyphenyl)-5-(O-β-D-galactopyranosyl)-7-methoxycoumarin Trihydrate. Acta Cryst. 49C, 329 (1993).
236. SPENCER, G.F., A.E. DESJARDINS, and R.D. PLATTNER: 5-(2-Carboxyethyl)-6-hydroxy-7-methoxybenzofuran, a Fungal Metabolite of Xanthotoxin. Phytochemistry 29, 2495 (1990).
237. SRIPATHI, S.K., R. GANDHIDASAN, P.V. RAMAN, N.R. KRISHNASWAMY, and S. NANDURI: First Isolation of Xanthone and Isolation of a 6-Ketodehydrorotenoid from *Dalbergia sissoides*. Phytochemistry 37, 911 (1994).

238. SRIVASTAVA, S.K., and S.D. SRIVASTAVA: Two New Coumarins and a New Saponin from *Ruta graveolens*. Fitoterapia **65**, 301 (1994); Chem. Abstr. **122**, 128534 (1995).
239. SULISTYUWATI, L., P.J. KEANE, and J.W. ANDERSON: 6,7-Dimethoxycoumarin is a Phytoalexin Which is Produced Enzymatically in Living Cells in the Plant in Response to Invading Fungus-*Citrus* Seedlings and *Phytophthora citrophora*. Physiol. Mol. Plant Pathol. **37**, 451 (1990); Chem. Abstr. **114**, 244403 (1991).
240. SWAIN, L.A., J.M.E. QUIRKE, S.A. WINKLE, and K.R. DOWNUM: A Furanocoumarin from *Dorstenia contrajerva*. Phytochemistry **30**, 4196 (1991).
241. TAKATA, M., S. SHIBATA, and T. OKUYAMA: Structures of Angular Pyranocoumarins of Bai-Hua Qian-Hu, the Root of *Peucedanum praeruptorum*. Planta Medica **56**, 307, (1990).
242. TAKEMURA, Y., M. INOUE, H. KAWAGUCHI, M. JU-ICHI, C. ITO, H. FURUKAWA, and M. OMURA: New Acrimarines from *Citrus* Plants. Heterocycles **34**, 2363 (1992).
243. TAKEMURA, Y., M. JU-ICHI, K. HATANO, C. ITO, and H. FURUKAWA: Srtucture of Furobinordentatin, a Novel Bicoumarin from *Citrus yuko*. Chem. Pharm. Bull. **42**, 997 (1994).
244. TAKEMURA, Y., M. JU-ICHI, K. HATANO, C. ITO, and H. FURUKAWA: Structures of Furobinordentatin and Furobiclausarin, Two Novel Bicoumarins from *Citrus* Plants. Chem. Pharm. Bull. **42**, 2436 (1994).
245. TAKEMURA, Y., M. JU-ICHI, T. KUROZUMI, M. AZUMA, C. ITO, K. NAKAGAWA, M. OMURA, and H. FURUKAWA: Structural Elucidation of Citrumarins: Four Novel Binary Coumarins Isolated from *Citrus* Plants. Chem. Pharm. Bull. **41**, 73 (1993).
246. TAKEMURA, Y., M. JU-ICHI, M. OMURA, M. HARUNA, C. ITO, and H. FURUKAWA: Three New Acridone-Coumarin Dimers from a *Citrus* Plant. Heterocycles **38**, 1937 (1994).
247. TAKEMURA, Y., T. KUROZUMI, M. JU-ICHI, M. OKANO, N. FUKAMIYA, C. ITO, T. ONO, and H. FURUKAWA: The Structures of Neoacrimarines-C and -D, Two New Acridone-Coumarin Dimers from *Citrus hassaku*. Chem. Pharm. Bull. **41**, 1757 (1993).
248. TAKEMURA, Y., S. MAKI, M. JU-ICHI, M. OMURA, C. ITO, and H. FURUKAWA: Two Novel Acridone-Coumarin Dimers, Neoacrimarines-A and -B, from *Citrus* Plants. Heterocycles **36**, 675 (1993).
249. TAKEMURA, Y., Y. NAKATA, M. AZUMA, M. JU-ICHI, M. OKANO, N. FUKAMIYA, M. OMURA, C. ITO, K. NAKAGAWA, and H. FURUKAWA: Five New Pyranocoumarins from Some *Citrus* Plants. Chem. Pharm. Bull. **41**, 1530 (1993).
250. TAKEMURA, Y., T. NAKATA, M. JU-ICHI, M. OKANO, N. FUKAMIYA, C. ITO, and H. FURUKAWA: Three New Bicoumarins from *Citrus hassaku*. Chem. Pharm. Bull. **42**, 1213 (1994).
251. TAKEMURA, Y., T. NAKATA, H. UCHIDA M. JU-ICHI, K. HATANO, C. ITO, and H. FURUKAWA: Structure of Claudimerin-A, a Novel Dimeric Coumarin from *Citrus hassaku*. Chem. Pharm. Bull. **41**, 2061 (1993).
252. TE PASKE, M.R., J.B. GLOER, D.T. WICKLOW, and P.F. DOWD: Aflavarin and β-Aflatrem: New Anti-Insectan Metabolites from the Sclerotia of *Aspergillus flavus*. J. Nat. Prod. **55**, 1080 (1992).
253. TERREAUX, C., M. MAILLARD, H. STOECKLI-EVANS, M.P. GUPTA, K.R. DOWNUM, J.M.E. QUIRKE, and K. HOSTETTMANN: Structure Revision of a Furanocoumarin from *Dorstenia contrajerva*. Phytochemistry **39**, 645 (1995).
254. ULUBELEN, A., A.H. MERICLI, F. MERICLI, U. SONMEZ, and R. ILARSLAN: Alkaloids and Coumarins from *Haplophyllum thesioides*. Nat. Prod. Letters **1**, 269 (1993).
255. ULUBELEN, A., A.H. MERICLI, F. MERICLI, and N. TAN: Two New Coumarins from *Haplophyllum ptilostylum*. J. Nat. Prod. **56**, 1184 (1993).

256. ULUBELEN, A., A.H. MERICLI, F. MERICLI, and N. TAN: Further Compounds from *Haplophyllum ptilostylum.* Nat. Prod. Letters **3**, 145 (1993).
257. VILEGAS, J.H.Y., O.R. GOTTLIEB, and H.E. GOTTLIEB: A Cinnamoylglucose from *Gomortega keule.* Phytochemistry **30**, 4200 (1991).
258. VILEGAS, W., N. BORALLE, A. CABRERA, A.C. BERNARDI, G.L. POZETTI, and S.F. ARANTES: Coumarins and a Flavonoid from *Pterocaulon alopecuroides.* Phytochemistry **38**, 1017 (1995).
259. VILEGAS, W., G.L. POZETTI, and J.H.Y. VILEGAS: Coumarins from *Brosimum guadichaudi.* J. Nat. Prod. **56**, 416, (1993).
260. WAIGH, R.D., B.M. ZERIHUN, and D.J. MAITLAND: Ten 5-Methylcoumarins from *Clutia abyssinica.* Phytochemistry **30**, 333 (1991).
261. WANG, C.L., R.H. ZHANG, Y.S. HAN, X.G. DONG, and W.B. LIU: Chemical Studies of Coumarins from *Glycyrrhiza uralensis.* Acta Pharm. Sinica **26**, 147 (1991); Chem. Abstr. **115**, 68400 (1991).
262. WU, T.-S., S.-C. HUANG, and J.-S. LAI: Stem Bark Coumarins of *Citrus grandis.* Phytochemistry **36**, 217 (1994).
263. XIAO, Y.Q., K. BABA, M. TANIGUCHI, X.H. LIU, Y.F. SUN, and M. KOZAWA: Coumarins from *Notopterygium incisum* Ting. Acta Pharm. Sinica **30**, 274 (1995); Chem. Abstr. **123**, 107848 (1995).
264. XIAO, Y.Q., X.H. LIU, Y.F. SUN, K. BABA, M. TANIGUCHI, and M. KOZAWA: Three New Furocoumarins from *Notopterygium incisum* Ting. Chin. Chem. Letters **5**, 593 (1994); Chem. Abstr. **121**, 226435 (1994).
265. YANG, L., H. CHEN, and J.J. ZHONG: Lignan and a Coumarin from *Cremanthodium ellisii.* Indian J. Chem. **34B**, 975 (1995).
266. ZDERO, C., F. BOHLMANN, and H.M. NIEMEYER: A Heliangolide, 3-Hydroxyumbelliferone Derivatives and Diterpenes from *Bahia ambrosioides.* Phytochemistry **29**, 205 (1990).
267. ZENG, L., R.-Y. ZHANG, T. MENG, and Z. LOU: Coumarins in Licorice Root by HPLC. J. Chromatogr. **513**, 247 (1990).
268. ZENG, L., R. ZHANG, D. WANG, C. GAO, and Z. LOU: The Chemical Constituents of *Glycyrrhiza aspera* Root. Acta Bot. Sinica **33**, 124 (1991); Chem. Abstr. **115**, 228388 (1991).
269. ZHANG, D., X. LI, L. WU, S. SU, and T. ZHU: Studies on Coumarin Analogues from Wannianthao (*Artemisia sacrorum*). Zhongcaoyao **20**, 487 (1989); Chem. Abstr. **112**, 175636 (1990).
270. ZHANG, X., A. ARCHELAS, A. MEOU, and R. FURSTOSS: Microbiological Transformations 21: An Expedient Route to Both Enantiomers of Marmin and Epoxyauraptens via Microbiological Dihydroxylation of 7-Geranyloxycoumarin. Tetrahedron: Asymmetry **2**, 247 (1991).
271. ZHOU, C., and J. HU: Synthesis of Coumarin. Nat. Sci. J. Xiangtan Univ. **12**, 79 (1990); Chem. Abstr. **114**, 81502 (1991).
272. ZHOU, C., and G. XIE: Synthesis of Coumarin Catalysed by Crown Ether. Nat. Sci. J. Xiangtan Univ. **15**, 87 (1993); Chem. Abstr. **121**, 9105 (1994).
273. ZOU, K., R.Y. ZHANG, and X.B. YANG: Structure Determination of Inflacoumarin A from *Glycyrrhiza inflata.* Acta Pharm. Sinica **29**, 397 (1994); Chem. Abstr. **121**, 226362 (1994).
274. ZUBIA, E., F.R. LUIS, G.M. MASSANET, and I.G. COLLADO: An Efficient Synthesis of Furanocoumarins. Tetrahedron **48**, 4239 (1992).

(*Received April 23, 1996*)

Artemisinin: An Endoperoxidic Antimalarial from *Artemisia annua* L.

H. ZIFFER[1], R. J. HIGHET[2], and D. L. KLAYMAN[3,#]

[1] Laboratory of Chemical Physics, National Institute of Diabetes, Digestive and Kidney Diseases, National Institutes of Health, Bethesda, MD 20892, U.S.A.

[2] Laboratory of Biophysical Chemistry, National Heart, Lung and Blood Institute, National Institutes of Health, Bethesda, MD 20892, U.S.A.

[3] Walter Reed Army Institute of Research, Division of Experimental Therapeutics, Washington, DC 20307, U.S.A.

Contents

1. Introduction 123
   1.1. Historical 123
   1.2. History of the Qinghao Plant 124
   1.3. Modern History of *Artemisia annua* L. in Chinese Medicine 125
2. *Artemisia annua* L. and its Constituents 126
   2.1. Taxonomy 126
   2.2. Geographic Distribution 126
   2.3. Cultivation 126
   2.4. Cell Culture 127
   2.5. Isolation of Artemisinin from *A. annua* 127
   2.6. Other Constituents of *A. annua* 127
3. Artemisinin Structure Determination 128
4. Artemisinin Syntheses 129
   4.1. From (−)-Isopulegol 129
   4.2. From (R)-(+)-Hydroxymenthol 131
   4.3. From 3(R)-Methyl-6-phenylsulfinyl-cyclohexanone 133
   4.4. From (+)-Isolimonene 133
   4.5. From Artemisinic Acid 133
   4.6. From Arteannuin B 136

[#] Deceased October 29, 1992.

5. Physical Measurements and Analyses 137
   5.1. NMR 137
      5.1.1. ^{1}H 137
      5.1.2. ^{13}C 139
   5.2. Circular Dichroism 139
   5.3. Infrared 139
   5.4. Mass Spectroscopy 140
   5.5. X-Ray Crystallography 141
   5.6. Quantitative TLC 141
   5.7. Titrimetric 141
   5.8. HPLC 141
      5.8.1. Electrochemical Detection 141
      5.8.2. UV Detection Methods 142
      5.8.3. Capillary Gas Chromatography 143
      5.8.4. Diverse Analytical Methods 143
      5.8.5. Radiolabelling 143
      5.8.6. Radioimmuno Assay 143
6. Reactions of Artemisinin and its Derivatives 144
   6.1. Thermolysis 144
   6.2. Chemical 145
      6.2.1. Reactions with Alkali 145
      6.2.2. Reactions with Ammonia and Amines 146
      6.2.3. Reactions with Acid 148
         6.2.3.1. Arteether 148
         6.2.3.2. Dihydroartemisinin 149
         6.2.3.3. Acid-Catalyzed Additions to Anhydro-
                  dihydroartemisinin 150
            6.2.3.3.1. Triphenylphosphine Hydrobromide 150
            6.2.3.3.2. p-Toluenesulfonic Acid 150
         6.2.3.4. Acid-Catalyzed Rearrangements of Artemisinin
                  Derivatives 150
            6.2.3.4.1. Lewis Acids 150
            6.2.3.4.2. Silica Gel-Catalyzed Rearrangements ... 150
      6.2.4. Reaction of Artemisinin with Reducing Agents 153
         6.2.4.1. Lithium Aluminum Hydride 153
         6.2.4.2. Sodium Borohydride 153
         6.2.4.3. A Mixture of Sodium Borohydride and
                  Boron Trifluoride 153
         6.2.4.4. Hydrogenation 154
      6.2.5. Bromination 155
      6.2.6. Fluorinated Artemisinin Derivatives 155
      6.2.7. Epoxidation of Anhydrodihydroartemisinin 156
      6.2.8. Osmium Tetroxide Oxidation 157
7. Dihydroartemisinin Derivatives 158
   7.1. Derivatives with Enhanced Oil Solubility 158
      7.1.1. Ethers 158
      7.1.2. Esters 159
      7.1.3. Carbonates 159
   7.2. Derivatives with Enhanced Water Solubility 160
      7.2.1. Sodium Artesunate 160

References, pp. 202–214

        7.2.2. Sodium Artelinate and Related Derivatives 160
   7.3. Artemisinin Derivatives 161
        7.3.1. (+)-Deoxoartemisinin 161
        7.3.2. (+)-Homodeoxoartemisinin 162
        7.3.3. (+)-10-Alkyldeoxoartemisinin 163
        7.3.4. (+)-10β-Allyldeoxoartemisinin 164
        7.3.5. C-3 and C-9 Substituted 10-Deoxoartemisinins 164
        7.3.6. C-14 Modified Deoxoartemisinins 165
8. Simplified Artemisinin Derivatives 166
   8.1. 9-Desmethylartemisinin 166
   8.2. 6,9-Bisnorartemisinin 166
   8.3. (+)-8a,9-Secoartemisinin 167
   8.4. (+)-4,5-Secoartemisinin 168
   8.5. (+)-Hexahydroisochroman-3-one 170
   8.6. 4,5-Desethanoartemisinin 170
   8.7. 9-Alkyl-9-desmethylartemisinin 170
   8.8. C-3 and C-9 Modified Artemisinin Derivatives 171
   8.9. Carba-Analogs of Artemisinin 172
9. Quantitative Structure-Activity Analyses 173
10. Tricyclic 1,2,4-Trioxane Analogs 174
11. Metabolism 175
    11.1. Microbial Metabolites of Artemisinin and its Derivatives 175
    11.2. Mammalian Metabolites 177
12. Test Data of Artemisinin Derivatives 178
    12.1. *In Vitro* 178
    12.2. *In Vivo* 188
13. Toxicity 193
14. Pharmacology and Pharmacokinetics 194
15. Clinical Evaluation of Artemisinin and Derivatives 195
    15.1. Dihydroartemisinin 196
16. Mechanisms of Action 196
17. Other Peroxides 198
    17.1. Naturally Occurring Peroxides 198
    17.2. Synthetic Peroxides 199
18. Conclusion 201
References 202

1. Introduction

1.1. Historical

For thousands of years, physicians in China have treated fever with a decoction of the plant qinghao (*1*). In 1972 Chinese chemists isolated the active febrifuge from this plant, determining its structure, **1**, by single crystal x-ray crystallography (*2, 3*). The discovery was timely, for the world sorely needs a better treatment for malaria. More than 270 million people suffer from the disease, two to three million dying each year. The majority of the deaths are of children under 5 years of age, who are especially sensitive because of their lack of immunity to the disease.

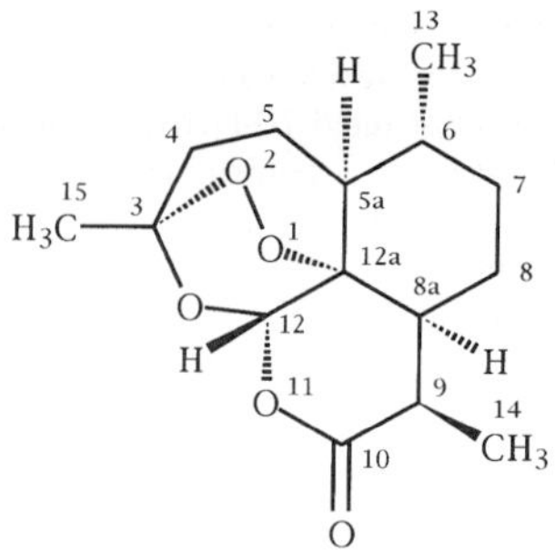

1

Malaria is caused by protozoa of the genus *Plasmodium* which are injected into the blood stream by the bite of an infected female *Anopheles* mosquito seeking a blood meal during the reproductive phase of her life. Proliferation of the protozoan in the blood stream causes fever, chills, sweating, dizziness, headache, and diarrhea. Of the four species of *Plasmodium* that infect humans, *Plasmodium vivax, P. malariae, P. ovale*, and *P. falciparum*, the last is responsible for cerebral malaria, which can cause the patient to lapse into a coma and ultimately leads to death. In many parts of the world, strains of *P. falciparum* have emerged which are resistant to chloroquine, otherwise the drug of choice. Its chemical structure, like that of many of the synthetic antimalarial agents – chloroquine, primaquine, amodiaquine, and mefloquine – is patterned after the cinchona alkaloid quinine. Novel classes of antimalarial agents are needed to overcome the resistant strains.

1.2. History of the Qinghao Plant

Artemisinin is not the first Chinese natural product to offer promise against malaria. Prior to World War II, the plant called "chang-shan", *Dichroa febrifuga*, a member of the Hydrangea family, was thought to offer a therapeutic solution. Its main antimalarial alkaloid, febrifugine, has proven to be too toxic for human use and structural modifications have been unable to overcome the poor activity/toxicity ratio.

The older Chinese medical literature on the uses of Qinghaosu was summarized by LUO and SHEN in their review of the chemistry,

References, pp. 202–214

pharmacology, and clinical applications of qinghaosu and its derivatives (4) with references to the plant's use dating back 2,000 years. The qinghao plant is an article of commerce available from virtually all warehouses for Chinese herbs (5). It was recommended that a feverish patient take a handful of sweet wormwood, soak it in a *sheng* (ca. 1 liter) of water, squeeze out the juice, and drink it all (6).

1.3. Modern History of *Artemisia annua* L. in Chinese Medicine

In the 1960's a program in the P. R. of China of re-examining traditional herbal remedies by modern standards included the qinghao plant, known for its antipyretic properties. Early efforts to isolate the active principle from the plant were disappointing. In 1971, according to the popular Chinese press, an anonymous female pharmaceutical research institute researcher (possibly Professor YOU-YOU TU of the Institute of Chinese Materia Medica, Beijing) made a low temperature extraction with diethyl ether to obtain a material which exhibited positive antimalarial activity in mice infected with *Plasmodium berghei* as well as in infected monkeys. The active ingredient, **1**, isolated in 1972 (5, 7), acquired several names: qinghaosu, arteannuin, and artemisinin. [Artemisinin was also called artemisinine; however, since a final "e" suggests a nitrogen-containing compound the name is not favored by Chemical Abstracts.]

Although Chinese workers believed they were the first to isolate artemisinin, JEREMIC and co-workers (8) in Belgrade, Yugoslavia may have isolated the compound earlier; however, they assigned an incorrect ozonide structure for the compound and did not follow up their discovery. After some animal testing, artemisinin was administered to humans infected with malaria and found to be an efficacious schiztonocide, with little or no toxicity. By 1972, artemisinin and derivatives had been used in ten regions of China and administered to some 6,000 patients.

Chemically, **1** is unlike previous antimalarials in that it does not contain a nitrogen atom but is instead a sesquiterpene lactone with an endoperoxide moiety, an unusual functional group in natural products. The discovery of this new and unusual natural product with antimalarial activity has stimulated the search for new, more active and longer lasting antimalarial drugs.

The chemistry and pharmacology of artemisinin has been reviewed by KLAYMAN (*1*), LUO and SHEN (*4*), WOERDENBAG et al. (*9*), ZAMAN (*10*), BUTLER (*11*), JUNG (*12*), and MESHNICK (*13*).

2. *Artemisia annua* L. and its Constituents

2.1. Taxonomy

The genus *Artemisia* (Compositae) comprises over 300 species, many of which have been used as spices, insect repellants and as a source for essential oils. *A. vulgaris* (common mugwort, motherwort, sailor's tobacco) is used in folk medicine for the treatment of stomach ache, headache, diarrhea, fever, rheumatism, bronchitis, poison oak, and to heal wounds; *A. dracunculus* is familiar in western cuisine as tarragon; *A. absinthium* (wormwood, mugwort, absinthe, mingwort, old woman) provides absinth, a narcotic and now illegal drink and a source of volatile oils for massage and stomach disorders; *A. tridentata* is sagebrush.

2.2. Geographic Distribution

A. annua, an annual, flourishes in many places in the temperate zone as well as throughout much of China. In the United States the weed grows primarily along rivers and has been reported in the states of New York, New Jersey, Maryland, Virginia, and West Virginia; however, it is not believed to be native to the United States (*14*), but may have been introduced inadvertently. The plant is known by the common names of annual wormwood, sweet wormwood, and sweet Annie. It also grows wild in countries of central Europe, such as the former Yugoslavia, Hungary, Bulgaria, and Romania as well as Italy, France, Spain, Turkey, the former Soviet Union, and Argentina.

2.3. Cultivation

Studies in China (*15*), India (*16, 17*), Turkey (*18*) and Australia (*19*) are currently in progress with the objective of increasing the yields of artemisinin from the plant. Although Chinese plants have been reported to yield as much as 0.9% of artemisinin (*20, 21*), elsewhere plants yield about 0.1% (*22*), with the highest content found in the leaves of the top 50 cm of the plant (*23*), the highest yields occuring just before flowering (*24*). Since artemisinin occurs in *A. annua* to only a small extent (*25*), economic and practical considerations dictate that plants with maximum contents of artemisinin be found and ways to increase their content be sought. Thus far, the *de novo* syntheses of **1** are too complex and lengthy to be practical or afford a useful source of the drug, but the use

References, pp. 202–214

of other *A. annua* constituents (Scheme 1) offers promise (see Section 4.5. on the use of artemisinic acid, **2**, as a starting material for the synthesis of artemisinin).

2.4. Cell Culture

Several groups have investigated the possibility of growing tissue cultures of *A. annua* in order to obtain artemisinin (*26, 27*). KUDAKASSERIL *et al.* have shown that cultures of shoots and roots incorporate labelled isopentenyl acetate (*28, 29*). WEATHERS *et al.* have grown transformed root cultures infected with *Agrobacterium* (*30*) and shown that artemisinin is produced up to 0.42% dry weight yield.

2.5. Isolation of Artemisinin from *A. annua*

Although artemisinin was isolated first in China or perhaps Yugoslavia, the first laboratory procedure for its preparation was published by KLAYMAN and co-workers (*31*). The air-dried leaves of *A. annua* were extracted with 30–60° petroleum ether, and the extract concentrated and redissolved in chloroform, to which acetonitrile was then added to precipitate inert plant components such as sugars and waxes. Chromatography of the concentrate on silica gel by eluting with chloroform-ethyl acetate was monitored by TLC on silica gel plates with iodine vapor; artemisinin was identified by a sharp singlet at δ 5.80 corresponding to H-12 in the nmr spectrum. Fractions containing high concentrations of artemisinin could be crystallized from cyclohexane or 50% ethanol. Artemisinin has also been isolated using the ITO multilayer coil separator-extractor (*32*). ELSOHLY and co-workers (*33*) described a large scale extraction technique for the purification of artemisinin.

2.6. Other Constituents of *A. annua*

The value of artemisinin as an antimalarial has prompted studies of other constituents, shown in Scheme 1, in *A. annua*. Artemisitene, **5**, first detected among HPLC-EC (electrochemical detection) positive compounds (*v. i.*), was isolated from *A. annua* in very low yield (*34*). It was later synthesized from artemisinic acid (Section 4.5.). Its *in vitro* antimalarial activity is somewhat lower than that of artemisinin (*35*).

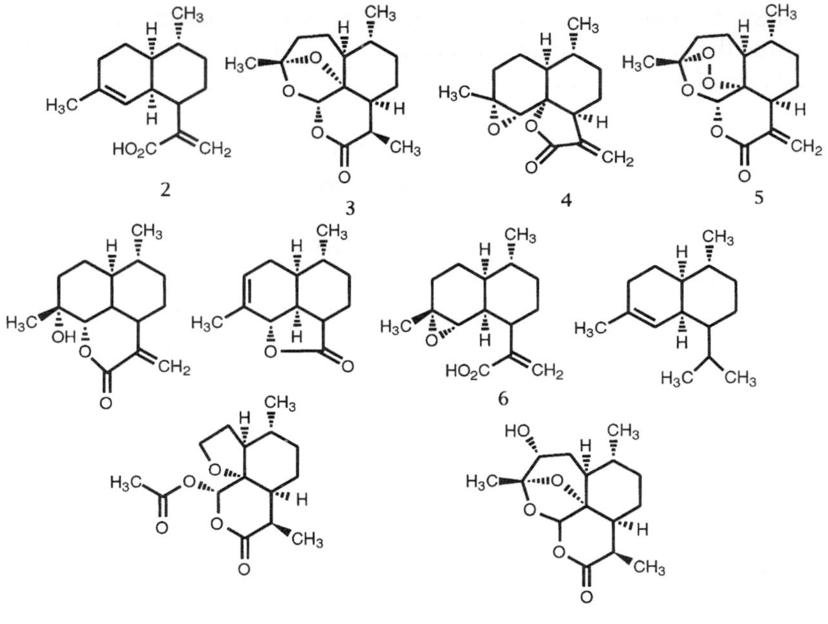

Scheme 1

A new sesquiterpene, artemisinin G, **6**, was recently isolated from *A. annua* by WEI *et al.* (*36*).

3. Artemisinin Structure Determination

Artemisinin is a colorless sesquiterpene, mp 156–157°, $[\alpha]^{25}_D +66.3$ (c 1.64 CHCl$_3$), with a mass spectrum and elementary analysis conforming to $C_{15}H_{22}O_5$. Absorption in the infrared at 1745 cm^{-1} corresponds to a δ-lactone and at 831, 881, and 1115 cm^{-1} to a peroxide group. LIU *et al.* confirmed the presence of a δ-lactone by opening the lactone with base and reforming it with dilute acid (*3*).

Corresponding to the crystal structure for **1** the ^{1}H NMR spectrum shows the presence of two secondary and one tertiary methyl group, and an acetal proton. Besides confirming the presence of a carbonyl carbon and three methyl groups, the ^{13}C-nmr spectrum shows four carbons bearing two hydrogens, five bearing one hydrogen, and two quaternary carbons (*37*). Its systematic name is 3,6,9-trimethyl-9,10b-epidioxyperhydropyrano[4.3.2-jk]benzoxepin-2-one.

4. Artemisinin Syntheses

The potential medicinal value of a new antimalarial drug and its novel structure stimulated vigorous synthetic studies of the natural material and analogs. Isopulegol, **7**, with three of the asymmetric centers needed, is a convenient starting material, but an even better one is available from the striking observation of SCHULTE and OHLOFF that treatment of **7** with diborane and subsequent alkaline peroxide oxidation produced the hydroxymenthol, **8**, with 95% stereospecificity (Scheme 2) (*38*).

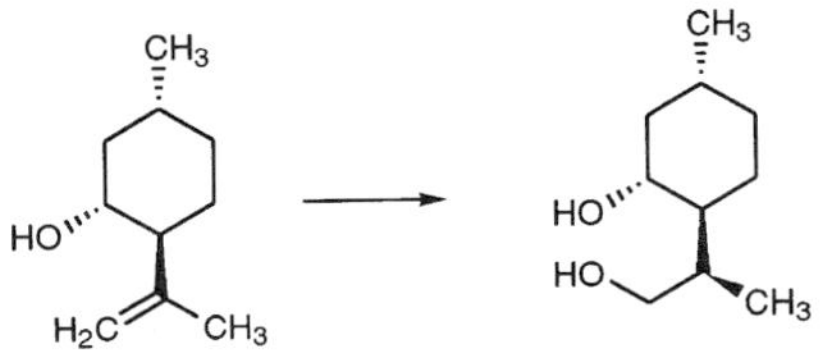

Scheme 2

The introduction of a peroxide function poses a problem. Most workers have employed photochemical oxidation to introduce the group, frequently in low yield. A somewhat more attractive alternative is based on the observation by BÜCHI and WÜEST that ozonolysis of vinyl silanes leads to hydroperoxides (Scheme 3) (*39*).

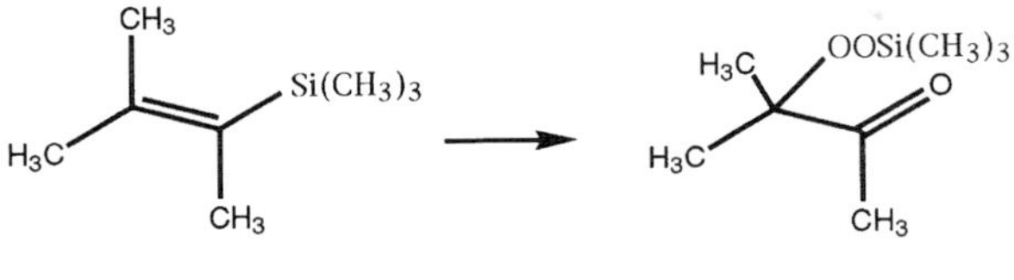

Scheme 3

4.1. From (−)-Isopulegol

SCHMID and HOFHEINZ (*40*) prepared artemisinin in 1982 from (−)-isopulegol, **7** (Scheme 4). The hydroxyl group of **7** was protected as the methoxymethyl ether before hydroboration of the double bond. Following benzylation of the newly formed hydroxyl group, the methoxymethyl ether was cleaved with acid and oxidized to yield the menthone, **9**. Kinetic deprotonation and treatment with a tenfold excess of the alkylating agent produced **10** as the major product, convertible by lithium methoxy(trimethylsilyl)-methylide to **11**. After reductive clea-

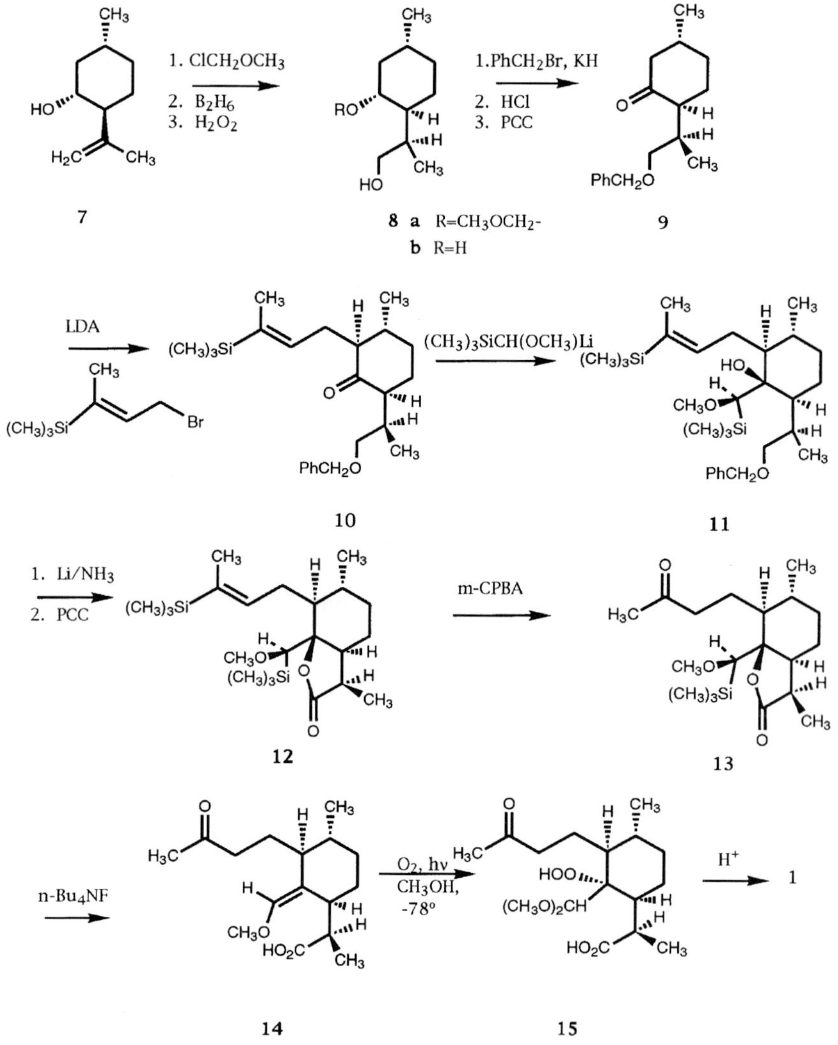

Scheme 4

vage of the benzyl ether followed by oxidation, acidification yielded the lactone **12**. Treatment of the ene-silyl function by m-CPBA produced the ketone **13**, which was desilylated with fluoride ion to the enol ether **14**. It was this material which was irradiated with oxygen and methylene blue to yield a mixture of peroxides, including **15**, which acid treatment converted into artemisinin, **1**, in small yield.

4.2. From (R)-(+)-Hydroxymenthol, 8b

Chinese workers have exploited the availability of the hydroxymenthol **8b** in a somewhat similar synthesis given in Scheme 5 (*41*, *42*). Oxidation of the benzyl ether produced ketone **9**, and selective deprotonation and a Michael addition of 3-trimethylsilyl-butenone yielded a diketone which was cyclized to decalenone, **16**. Borohydride reduction and Jones oxidation provided a saturated ketone which was treated with methyl magnesium iodide and dehydrated to a mixture of

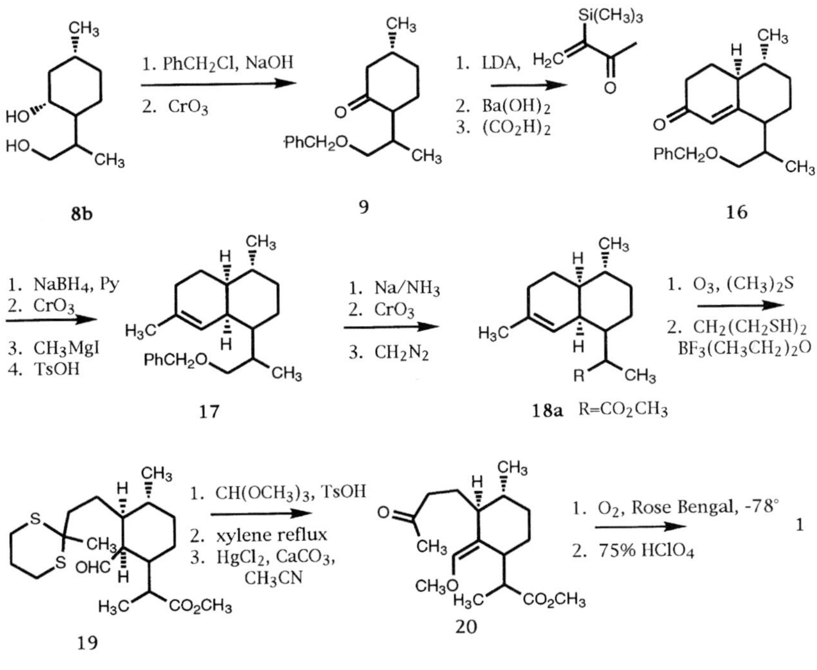

Scheme 5

olefins including **17** which could be separated by chromatography. The benzyl group was removed using sodium in liquid ammonia to provide an alcohol which was oxidized to a carboxyl group and esterified by diazomethane to give **18a**. The double bond was now cleaved by ozone and the ketone selectively transformed into the dithiane **19**. Alkylation by methyl orthoformate provided the acetal, which was converted to the olefin by xylene reflux; removal of the dithiane by mercuric chloride provided the ketone **20**. Photolysis in the presence of oxygen followed by acid treatment provided artemisinin **1**.

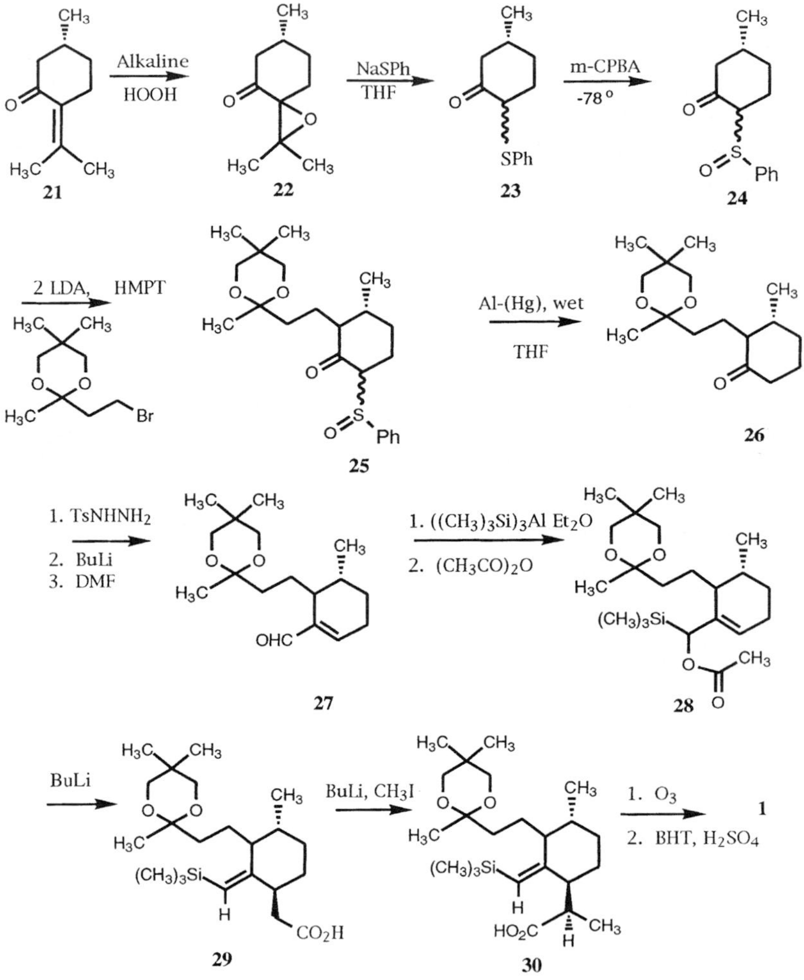

Scheme 6

4.3. From 3(R)-Methyl-6-phenylsulfinyl-cyclohexanone

In their synthesis of artemisinin AVERY et al. (43) utilized a single asymmetric center of (−)-isopulegol, oxidizing **7** to pulegone **21** (Scheme 6). Formation of the epoxide **22**, and cleavage by sodium thiophenolate allowed removal of the side chain by a concomitant reverse aldol condensation. Oxidation of the thioether **23** provided a sulfoxide **24** which was selectively alkylated with the atoms needed to form ring A. The sulfoxide of the product **25** was removed by reductive cleavage to provide a ketone, **26**. Addition of a formyl group via a hydrazone provided aldehyde, **27**, which was reduced and acetylated. Claisen rearrangement of the ester **28** produced **29**. This material was stereospecifically methylated to provide **30** with all of the requisite carbon atoms. It was this material which provided the substrate for BÜCHI's ozonization. Mild acid treatment effected the hydrolysis of the ketal and trimethylsilylether, followed by cyclization to **1**.

4.4. From (+)-Isolimonene

RAVINDRANATHAN et al. employed an intramolecular Diels-Alder reaction to prepare a tricyclic intermediate with the required stereochemistry (44) (Scheme 7). Hydroboration of the readily available (+)-isolimonene **31** provided alcohol **32**, as a mixture of epimers. Trans-etherification with 1-ethoxy-2-methylbutadiene provided **33**, which cyclized at 210° to the tricyclic **34**. After epoxidation, the oxirane was reduced with lithium aluminum hydride to yield a tertiary alcohol **35**. The ether was oxidized to a mixture of lactones separable by chromatography to provide **36**. Hydrolysis, periodate oxidation and esterification by diazomethane provided aldehyde **37b**, a derivative of which XU et al. had earlier converted to artemisinin (4).

4.5. From Artemisinic Acid

The preceeding syntheses of **1** are technically difficult and impractical for the preparation of large quantities of artemisinin or its derivatives. XU et al. (45) converted artemisinic acid **2**, which is available in large quantities from A. annua, into **1** and deoxyartemisinin, **3**, (Scheme 8) by steps similar to those reported (42).

ROTH and ACTON (46, 47) achieved a notable improvement in the process by photooxidation of **18b** to the hydroperoxide **38** (Scheme 9).

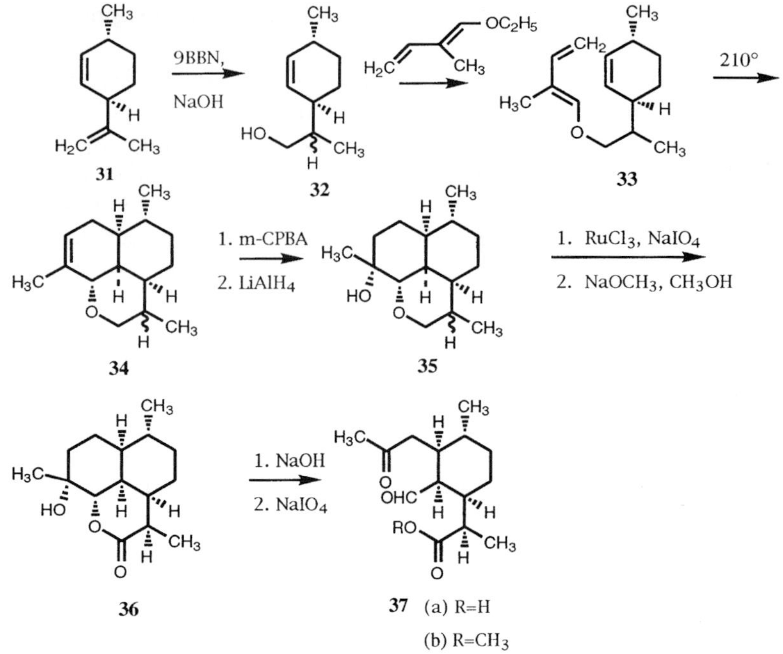

Scheme 7

When the crude reaction mixture is allowed to stand for 4 days at room temperature a remarkable transformation occurs, involving an oxidative ring opening and recyclization. The product may be extracted into an organic solvent and purified by crystallization to give **1** in 30% yield.

A valuable approach to intermediates has been provided by HAYNES and VONVILLER (*48*) who found that alicyclic allyl hydroperoxides, such as that from pinene, are readily cleaved with ferric chloride to dicarbonyl compounds (Scheme 10). Thus, the hydroperoxide of artemisinic acid, **38**, is converted to the dicarbonyl compound, **39**, by treatment with $Cu(OSO_2CF_3)_2$ and $Fe(phenanthroline)_3(PF_6)_3$; **39**, in turn, was converted by acid to artemisitene (dehydroartemisinin), **5** (Scheme 11) (*49*).

YE and WU (*50*) (Scheme 12), by a series of reductive steps, followed by ozonization and acid cyclization, converted **40** into a mixture including **41**, which ruthenium periodate treatment converted into **1**.

References, pp. 202–214

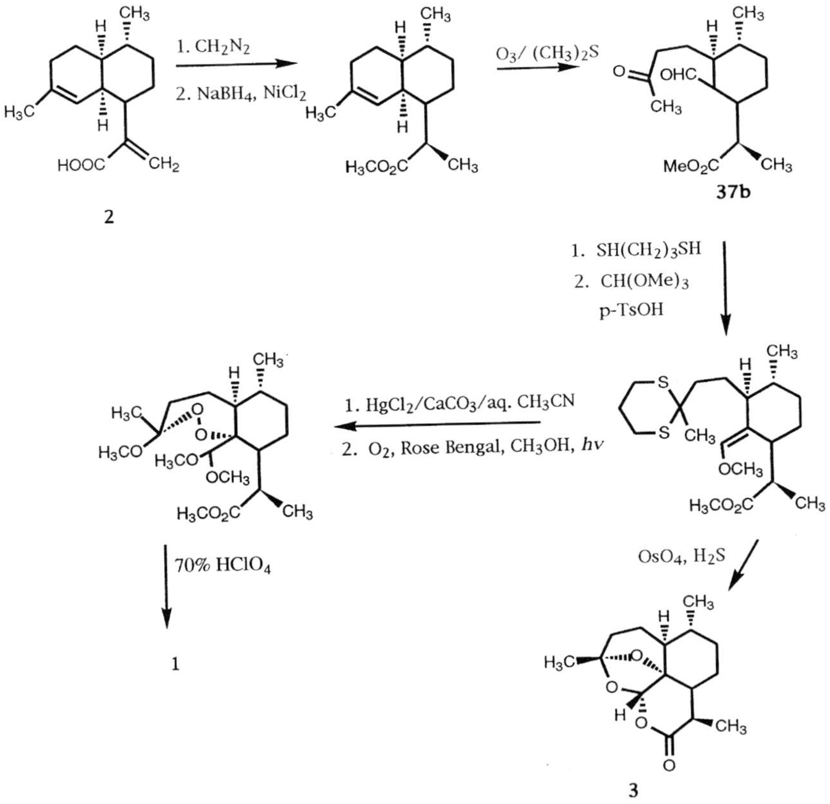

Scheme 8

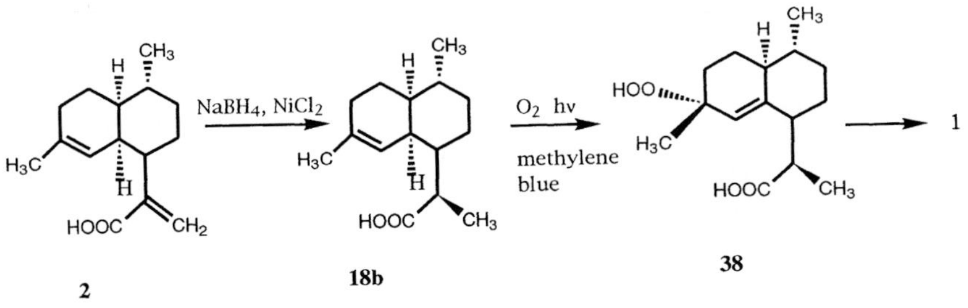

Scheme 9

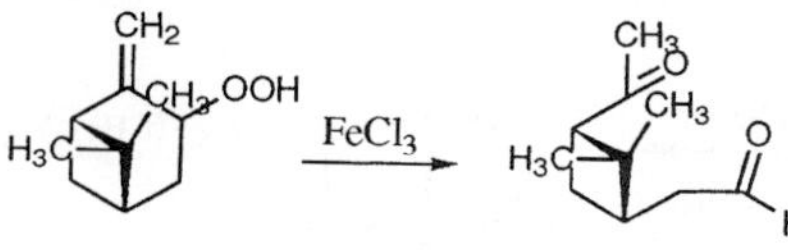

Scheme 10

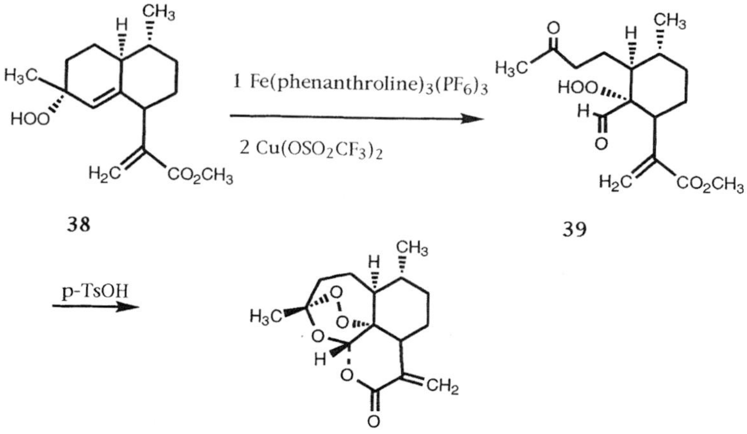

Scheme 11

4.6. From Arteannuin B

LANSBURY and NOWAK (51) showed that dihydroarteannuin B, **4**, was converted by butyllithium and tungsten hexachloride to an unsaturated lactone (Scheme 13). The stereochemistry of the lactone was established by an X-ray crystal structure determination. Ozonolysis provided a ketoaldehyde, **42**, which, after selective ketalization, was converted to the enol ether, **43**. Thus, the stereochemistry of the lactone in **42** was unimportant in the reaction sequence. The peroxide grouping was introduced by a photochemically sensitized (Rose Bengal) oxidation followed by a camphorsulfonic acid catalyzed condensation.

References, pp. 202–214

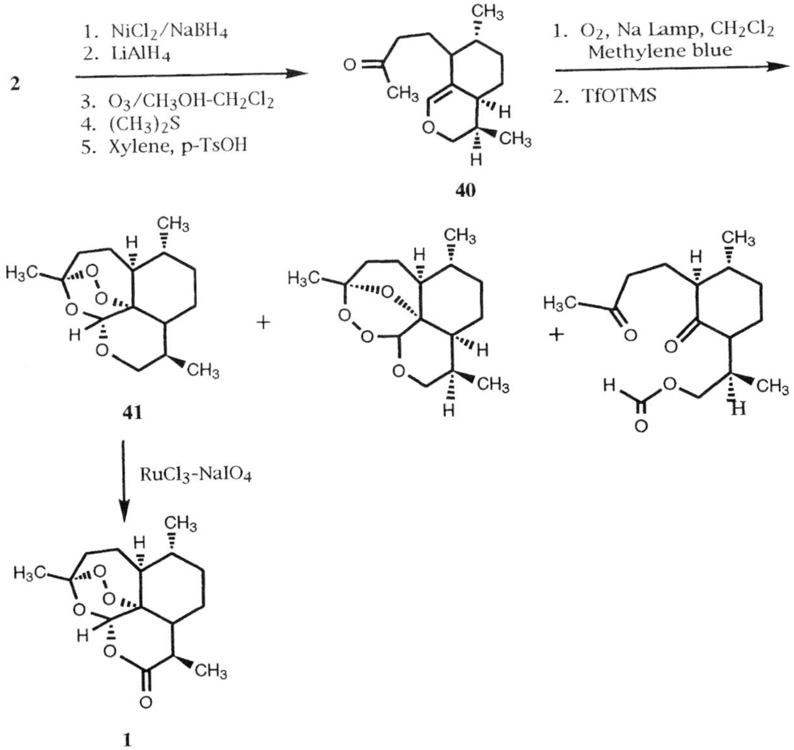

Scheme 12

5. Physical Measurements and Analyses

5.1. NMR

5.1.1. ^1H

Although the structures of artemisinin and of several transformation products were determined by x-ray crystallography, those of most transformation products were deduced from spectroscopic data, primarily 1D and 2D ^1H and ^{13}C NMR data. Because early assignments of both nuclei in artemisinin differed consideraby *(34, 52, 53, 54)* BLASKO *et al.* *(37)* redetermined them, using the superior techniques which had become available. The use of different numbering systems by various groups requires care in comparing assignments.

Scheme 13

Sodium borohydride reduction of the lactone group of **1** produces an epimeric mixture of dihydroartemisinins, **44a** and **45a**. The epimeric acetal methyl ethers have been assigned configurations consistent with the ^{1}H characteristics anticipated from stereochemistry (55). With the movement of the methoxyl group (R = CH$_3$) at C-10 of **44b** from the axial to the equatorial position of **45b**, the axial proton at C-9 is shifted upfield (δ 2.59 to 2.36 ppm), while $J_{9,10} = 3$ Hz (an axial-equatorial coupling) increases to 9.3 Hz (characteristic of an axial-axial relation). Similar values were observed for β-arteether, **44c**, supported by a NOESY experiment (56).

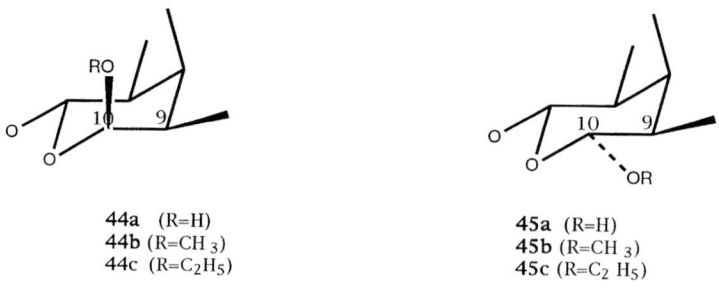

44a (R=H)
44b (R=CH$_3$)
44c (R=C$_2$H$_5$)

45a (R=H)
45b (R=CH$_3$)
45c (R=C$_2$H$_5$)

These assignments allowed study of the equilibration of dihydroartemisinin epimers. On dissolution in chloroform or methanol, the β epimer of the crystal (**44a**, R = H) forms a mixture of C-10 epimers (**44a** and **45a**) with the composition dependent upon the solvent: the α:β ratio is 1:1 chloroform and 2:1 in methanol.

5.1.2. ^{13}C

Because the peroxide group is not readily detected by spectral means, the ^{13}C chemical shifts of the terminal carbon atoms of the peroxide group in artemisinin (C-3, δ = 105; C-12a, δ = 80) have been particularly valuable in demonstrating the presence of the peroxide bridge within the series. The chemical shifts of the corresponding carbons in the oxide, deoxyartemisinin, **3**, (δ = 109; δ = 82) are somewhat downfield from those in **1** (*57*).

5.2. Circular Dichroism

The absence of absorption beyond the vacuum ultraviolet renders these compounds potentially valuable for the study of the weak absorption bands of cyclic peroxides. LIU and DUAN (*58*) reported that artemether exhibited a positive CD band at approximately 250 nm, whereas a second compound, **46**, containing a 5-membered ring peroxide had a CD-band at 232 nm; they proposed a theoretically based relationship between the geometry of the peroxide and the sign of the CD curve associated with the long wavelength transition. LIANG (*59*) reported CD curves for artemisinin, **1**, and deoxyartemisinin, **3** but none for a mixture of dihydroartemisinin epimers. One of us measured the CD spectrum of arteether, **44c**, but was unable to detect a band at 250 nm (*60*).

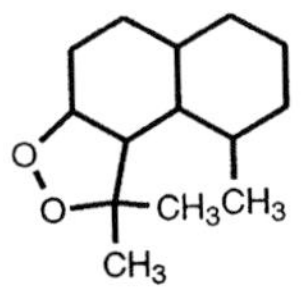

46

5.3. Infrared

Infrared measurements of artemisinin and its derivatives proved useful in establishing the presence of the δ-lactone (1745 cm^{-1}). The claim that the peak at 722 cm^{-1} is characteristic of the peroxide group (61) led BILL, JEFFORD and their associates (62) to investigate the vibrational spectra of ^{16}O and ^{18}O isotopomers of bicyclic 1,2,4-trioxanes such as **47**. Since the bands at 780 ± 20 and 880 ± 10 cm^{-1} are shifted by the isotopes, they must be characteristic of the trioxane, arising from a combination of C-O and O-O stretching vibrations. However, none of the bands are characteristic group frequencies for the peroxide group.

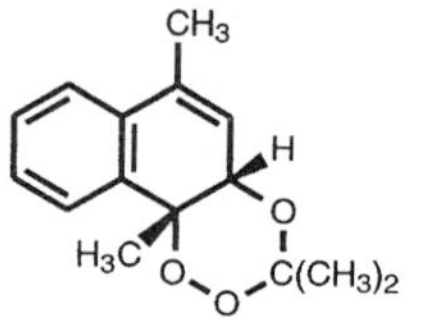

47

5.4. Mass Spectroscopy

In the absence of other spectral means of detecting peroxides, determination of the molecular weight of transformation products or synthetic materials is quite valuable, but peroxides are known to be unstable under the conditions of mass spectrometery (63). Using electron impact ionization, some workers have been able to detect the molecular ion of artemisinin (64), but others have not (65). Chemical ionization provides molecular ions more reliably and has proven a valuable support for the study of synthetic and transformation products. Ionization by methane or ammonia is widely used, but isobutane is more satisfactory. However, hemiacetals such as dihydroartemisinin give no molecular ion. Somewhat surprisingly, thermospray techniques provide spectra of dihydroartemisinin in which the molecular ion is the strongest peak above m/z >139 (66). Under conditions in which the gas chromatographic inlet to the mass spectrometer is between 320–350° the pyrolysis product **48** is observed ($C_{14}H_{22}O_3$, m/z 238), which provides a useful means of detection (67).

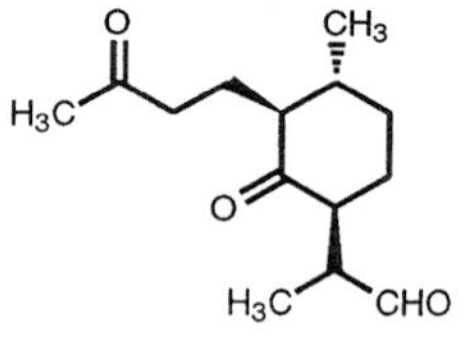

48

References, pp. 202–214

5.5. X-Ray Crystallography

In addition to the original crystal structure noted earlier (Section 3), structures have been reported for dihydroartemisinin, **44a**, artemisinic acid **2**, (*55*) arteether, **44c**, (*68*), and the two thermal decomposition products described in Section 6.1. below (*69*).

5.6. Quantitative TLC

Thin layer chromatography provides a convenient method for the assay of artemisinin and its derivatives. (*70, 71*). A chloroform extract of homogenized tissue or plasma provides a sample for application to the plate. After development by petroleum ether-ethyl ether (1:2), the plate is sprayed with *p*-dimethyl-aminobenzaldehyde and heated at 80°C to produce a color which is quantitated by densitometry at 600 nm. Substitution of a 2% solution of vanillin in sulfuric acid for *p*-dimethylaminobenzaldehyde produces a color that is measured at 560 nm (*72*).

5.7. Titrimetric

Since artemisinin is a lactone it can be dissolved in dilute alkali, and the remaining alkali titrated (*73*).

5.8. HPLC

5.8.1. Electrochemical Detection

Although there are no spectroscopic methods which directly detect the peroxide group, ACTON *et al.* have shown that electrochemical methods allow the detection of trace quantities of this group (*22*).

MELENDEZ *et al.* have employed reductive electrochemical detection as the detector of an HPLC column which permits the determination of as little as 5 µg of the terpene, suitable for both plant analyses and pharmacokinetics (*74, 75, 76*). Substitution of a glassy carbon electrode for the electrochemical detector allows one to detect arteether **44c** or dihydroartemisinin **44a** in the low nanogram range (*75*). In their pharmacokinetic studies of artemether, artemisinin, dihydroartemisinin and sodium artesunate in rats and humans, ZHOU *et al.* (*76*) employed an

HPLC based separation with a polarographic detector. Deoxygenating the mobile phase by boiling the solvent mixture for 1–2 hours while purging with nitrogen allowed the detection of 10–1600 ng of artemisinin and its derivatives. ZHANG et al. (77) employed pulse polarography to detect artemisinin in *A. annua*.

5.8.2. UV Detection Methods

The absence of a strong UV absorption band above 210 nm in **1** has hindered analysis of the compound in plant extracts, various tissues, and in purification. Artelinic acid, **49**, itself an effective agent against *P. berghei*, does have a UV absorbing aromatic chromophore which IDOWU et al. (78) used to detect the material in plasma after separation from impurities by HPLC.

One approach to this problem has been to treat samples containing artemisinin with alkali before chromatography, thus converting the compound into substances which absorb in the ultraviolet and can be chromatographed, of course, with different chromatographic characteristics than the parent. The alternative, treating column eluates with alkali, allows characterization of the compound itself. EDLUND et al. (79) mixed a 1M KOH solution (methanol:water, 9:1) with the eluent, either **49** or artesunate, **50**, and heated the mixture at 70°, to provide material detectable at 289 nm.

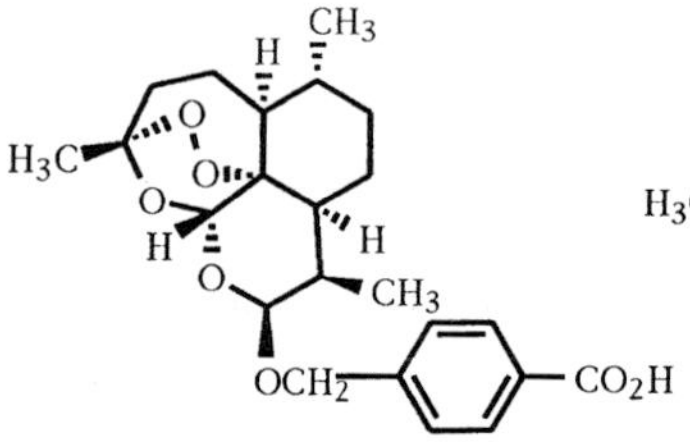

Esterification by diacetyldihydrofluorescein, previously used in TLC studies (80), provides a derivative for HPLC which allows detection in subnanogram quantities (81). Substitution of fluorometric methods would surely have reduced these levels considerably.

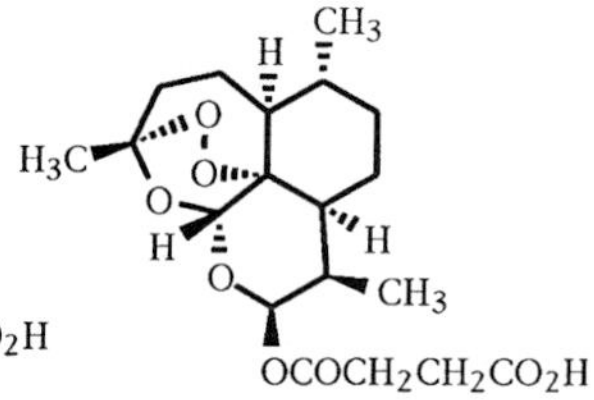

Reverse-phase HPLC is suitable for the separation of artemisitene from artemisinin (82).

References, pp. 202–214

5.8.3. Capillary Gas Chromatography

To analyze for artemisinin by gas chromatography, SIPAHIMALANI et al. (85) preceded chromatography by pyrolysis. Unfortunately, arteannuin-B could not be distinguished from thermal products derived from artemisinin (86). In their study of the metabolism of artemisinin, THEOHARIDES et al. (67) coupled gas chromatography and mass spectrometry to analyze for a pyrolysis product, (2S,3R,6R)-2-(3-oxobutyl)-3-methyl-6-[(R)-2-propanal]cyclohexanone, **48**, formed from **2**. The method allowed them to quantitate between 10 and 1000 ng of artemisinin in blood.

5.8.4. Diverse Analytical Methods

Assays have been reported by iodometry (83), polarography (84) and pulse polarography (77). Gas chromatography of the thermal degradation products of artemisinin can measure the quantity of the terpene, but fails to distinguish arteannuin-B (85). Direct GC analysis seems more appropriate (86).

5.8.5. Radiolabelling

Descriptions of preparation of radioactive artemisinin by WILZBACH'S method have not appeared; however AVERY et al. recently described (87) the synthesis of [^{14}C]-artemisinin employing an intermediate, **29**, in their synthesis of artemisinin. However, the preparation of labelled dihydroartemisinin, **44**, by reduction of **1** with NaB^{3}H$_4$ is readily achieved (88); the latter can be converted to artemether or arteether. A ^{3}H labelled sample of arteether was prepared by PU and ZIFFER (89) employing an acid catalysed addition of [^{3}H]-ethanol to anhydrodihydroartemisinin. A ^{14}C labelled sample of artemether was prepared by using ^{14}CH$_3$OH to etherify **2** (90). The use of these materials in pharmacokinetic studies is described in section 13. A sample of ^{3}H-artesunate, **50**, was prepared from **44a** using ^{3}H succinic anhydride (91).

5.8.6. Radioimmuno Assay

Radioimmuno assay allows the detection of 2–3 ng of artemisinin derivatives (e.g. **1**, **44**, and others) that contain a peroxide group (91, 92). Sheep antibodies were obtained using a conjugate of dihydroartemisinin-12-O-acetic acid with bovine serum albumin (BSA) plus complete Freund's adjuvant. Derivatives lacking the peroxide group are readily distinguished, for they are only weakly bound to the antibodies.

6. Reactions of Artemisinin and its Derivatives

6.1. Thermolysis

Artemisinin is a remarkably stable peroxide, withstanding heating in neutral solvents to 150°, but controlled pyrolysis does provide interesting products (Scheme 14) (*69, 93*). Two groups showed that a crystalline

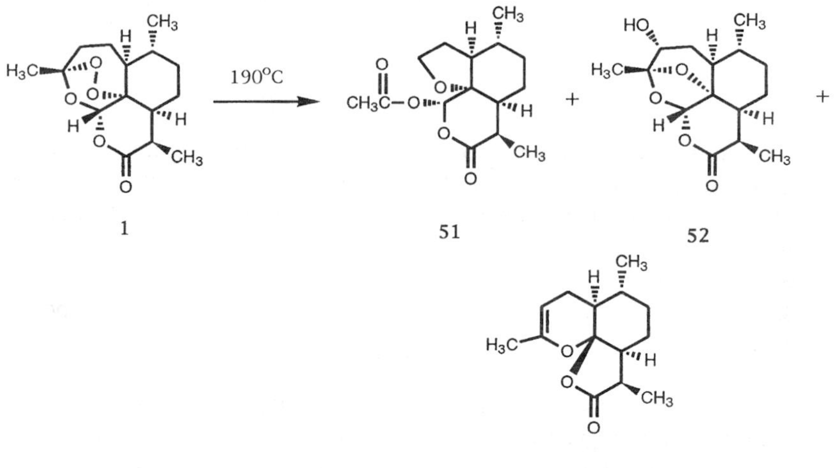

Scheme 14

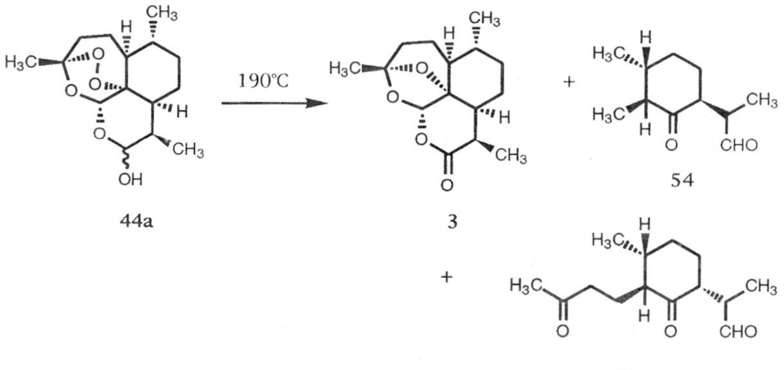

Scheme 15

References, pp. 202–214

product corresponds to **51** by crystallographic analysis, while an oil showed spectral properties corresponding to **53**. Luo et al. assigned a different structure to a second crystalline product they isolated by heating artemisinin for a longer period at 180° (*93*).

The thermal decomposition of dihydroartemisinin, **44a**, under the same conditions (Scheme 15), somewhat surprisingly produces deoxyartemisinin, **3**, and an oil, with spectral properties corresponding to a mixture of **54** and **55** (*94*).

6.2. Chemical

6.2.1. Reactions with Alkali

In aqueous alkaline solution the lactone ring of **1** opens to free the reactive groups of an aldehyde, ketone, and hydroperoxide. These groups undergo intramolecular condensations to produce a complex mixture of products. Chinese chemists have taken on the formidable task of characterizing these substances, available only in small yield. Treatment with potassium carbonate (Scheme 16) produces ester **56**, the epoxylactone **57**, and the unsaturated ketoacid **58** (*95*). However, aqueous alkali produces pyran **59**, which is isolable in only 15% yield, although the ultraviolet absorbance of the crude reaction mixture suggests that it is the predominant product (*96*). Prolonged alkaline treatment followed by

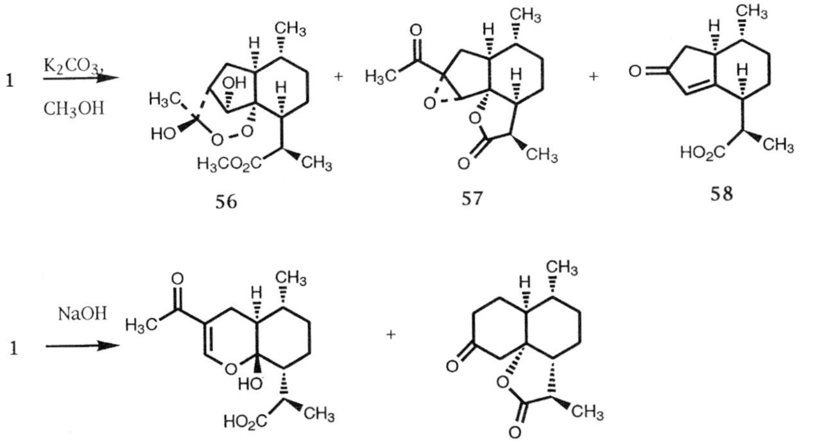

Scheme 16

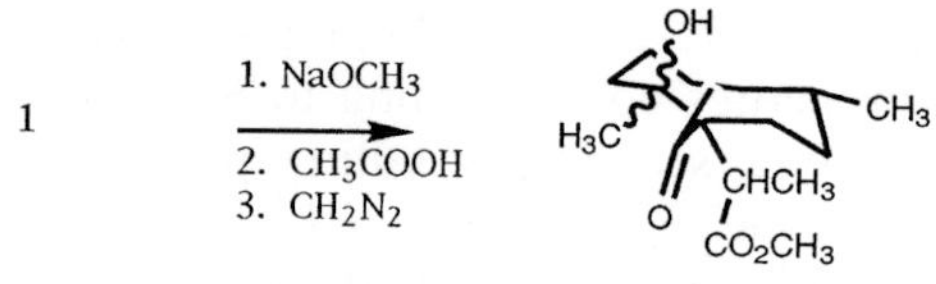

Scheme 17

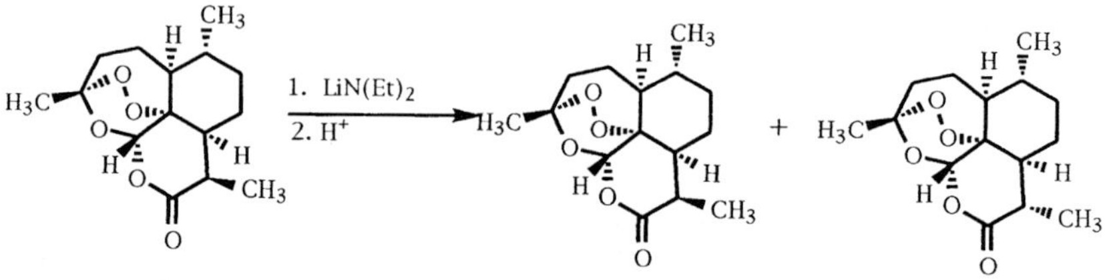

Scheme 18

acidification produces lactone **60** (*97*). Such treatment of deoxyartemisinin, **3**, produces similar mixtures.

Strenuous treatment of **1** by sodium methoxide in toluene at 105°C (Scheme 17) produced in low yield material with spectral properties consistent with the bicyclononanone structure **61**.

A dioxetane intermediate was postulated which was detected by observing its luminescence above 60–70°. Dioxetanes are one of a rare group of molecules that thermally decompose to form the excited state of the product (*98*). In returning to its ground state, light is emitted. Curiously, Shang *et al.* were unable to detect luminesence at lower temperatures.

On treatment with lithium diethylamine at −78° and reacidification **1** forms a mixture of artemisinin and 9-*epi*-artemisinin **62** (Scheme 18) (*99*).

6.2.2. Reactions with Ammonia and Amines

Torok and Ziffer (*100*), examined the reaction of artemisinin with methanolic ammonia (Scheme 19). The initial product is probably a hemiacetal-amide (R = H), **63**, in equilibrium with the hydroperoxy methyl ketones, **64** and **65**. Treatment of the crude reaction mixture with

References, pp. 202–214

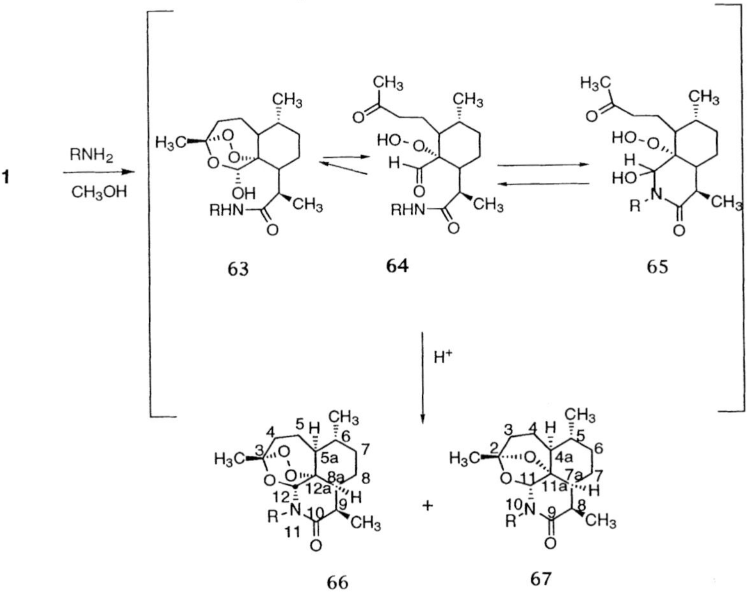

Scheme 19

	R
66	H
68	-CH$_2$CH=CH$_2$
69	-CH$_2$CH(CH$_3$)$_2$
70	-CH$_3$
71	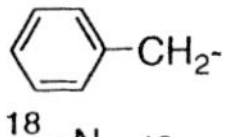 ![phenyl]-CH$_2$-
72	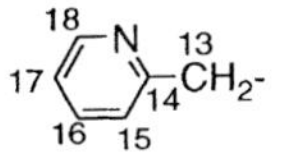 pyridyl-CH$_2$- (18-N, 17, 16, 15, 14, 13)
73	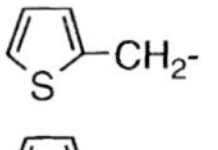 thienyl-CH$_2$-
74	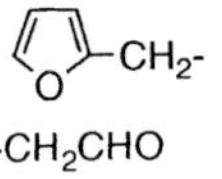 furyl-CH$_2$-
75	-CH$_2$CHO

	R
67	H
76	-CH$_2$CH=CH$_2$
77	-CH$_2$CH(CH$_3$)$_2$
78	-CH$_3$
79	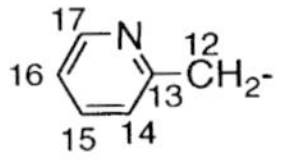 pyridyl-CH$_2$- (17-N, 16, 15, 14, 13, 12)
80	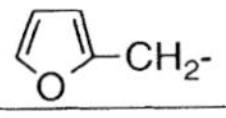 furyl-CH$_2$-

dilute sulfuric acid, BHT and silica gel, reaction conditions employed by AVERY *et al.* (*43*), produced a mixture of 11-azaartemisinin **66** and deoxyazaartemisinin, **67**. The same reaction sequence with primary amines provided a variety of N-substituted 11-azaartemisinin derivatives.

The most active derivative, N-(2'-ethanal)-11-azaartemisinin, **75**, was 26 times more active *in vitro* and 4 times more active *in vivo* than artemisinin. The *in vitro* results are given in Table 1 (Section 12.).

6.2.3. Reactions with Acid

As might well be anticipated, acid treatment of **1** produces complex mixtures of hydrolysis and rearrangement products, which have been studied with complex and conflicting reports. Simple acid-catalyzed *trans*-esterification which can be reversed to form the lactone without affecting the peroxide group has been reported without experimental detail (*101*).

Treatment **1** with acidic ethanol yielded three compounds, **81**, **82** and **83** (Scheme 20) (*102*), whose structures were established by NMR and mass spectra; *in vitro* tests of their antimalarial activities against chloroquine-resistant *P. falciparum* showed that the 1,2,4-trioxanes (**83a** and **83b**) were as active as artemisinin.

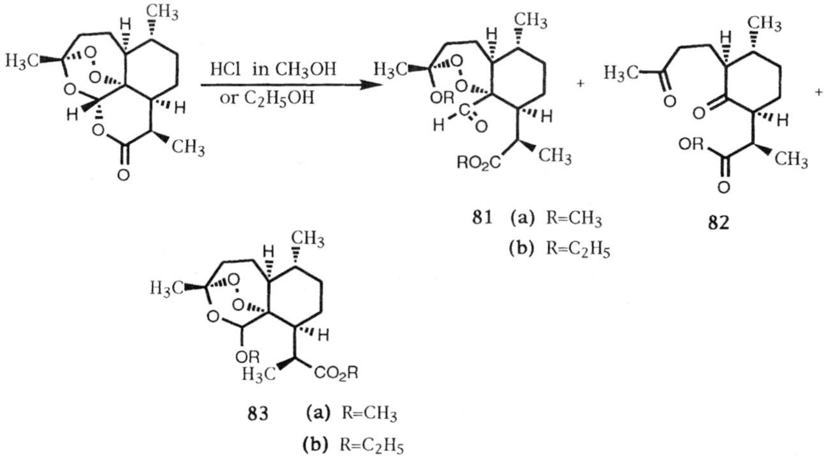

Scheme 20

6.2.3.1. Arteether, 44c

Although lipid soluble artemisinin derivatives are injected subcutaneously as oil suspensions in *in vivo* studies, oral administration is preferable. Consequently the behavior of compounds with acid has been studied. Arteether, **44c**, is an effective drug the preparation of which is

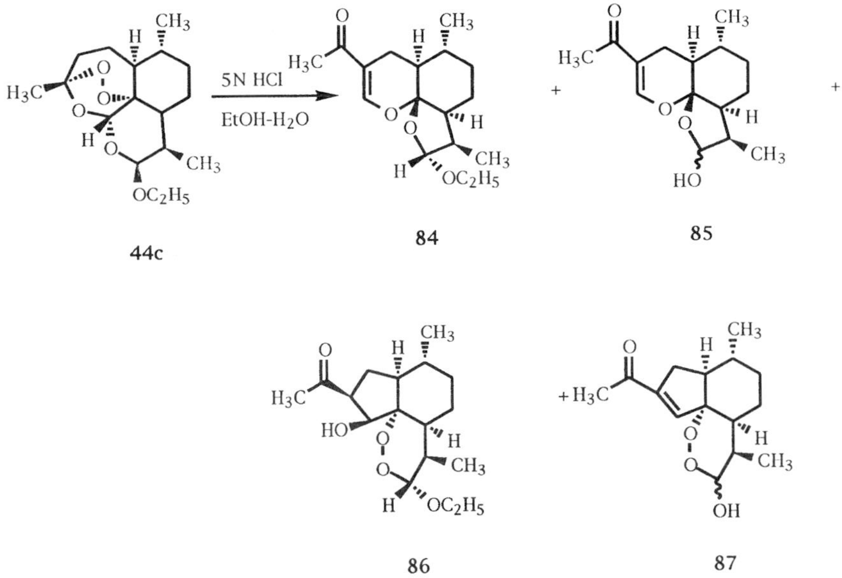

84   **85**

86   **87**

Scheme 21

described in the next section; its behavior in acid has been examined by BAKER and CHI (*103*), and by ACTON and ROTH (*104*). Treatment of **44c** with 5M HCl in aqueous ethanol at room temperature (Scheme 21) afforded **84, 85, 86** and **87**, characterized by NMR and mass spectra, the assigned stereochemical structures being supported by NOESY interactions. The peroxy groups were detected by reductive electrochemical liquid chromatography. ACTON and ROTH reported that treatment of dihydroartemisinin, **44a**, under the same reaction conditions gave rise to the same four products although there were differences in the relative amounts formed.

6.2.3.2. Dihydroartemisinin, **44a**

A host of esters, ethers, carbonates etc. of the free hydroxy group in dihydroartemisinin have been prepared by Chinese investigators and others. A listing of these derivatives and their antimalarial activities against *P. berghei* is given in Tables 14 and 15 in Section 12.2.

Dihydroartemisinin, **44a**, has been dehydrated by treatment with P_2O_5 or DCC (Scheme 22) (*105*). The resulting product, anhydrodihydroartemisinin, **88** has been employed as a starting material in the

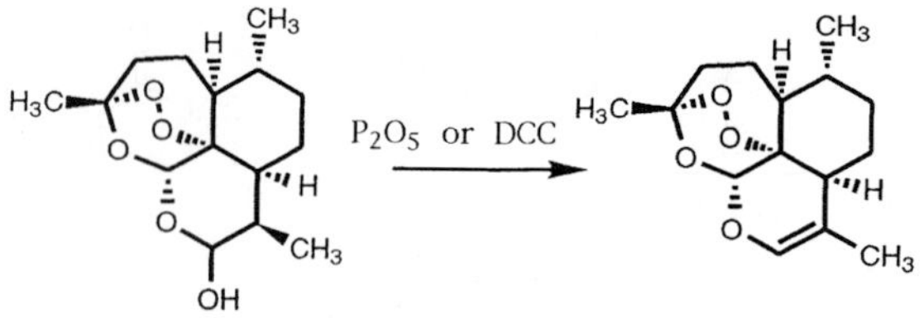

Scheme 22

syntheses of antimalarial drugs. For example two groups (*106*, *107*) have described acid catalyzed additions of alcohols to **88**.

6.2.3.3. Acid-Catalyzed Additions to Anhydrodihydroartemisinin

6.2.3.3.1. Triphenylphosphine Hydrobromide. Pu and Ziffer (*106*) catalyzed the addition of alcohols to **88** with triphenylphosphine hydrobromide (Scheme 23). Although the proton adds to C-9 predominantly from the β face, all four possible stereoisomers of arteether were formed, isolated and characterized. Several 10β-alkoxy ethers, **93**, **94** and **95**, were prepared and their *in vitro* activities determined.

6.2.3.3.2. p-Toluenesulfonic Acid. El-Feraly et al. (*107*) obtained a 3:1 mixture of arteether, **44c** and 9-*epi*-arteether, **91**, from the p-toluenesulfonic acid catalyzed addition of ethanol to **88** in absolute alcohol. The ratio of arteether to the 9-*epi*- isomer was reversed in dichloromethane.

6.2.3.4. Acid-Catalyzed Rearrangements of Artemisinin Derivatives

6.2.3.4.1. Lewis Acids. In the presence of Lewis acids, **41a** and the 10β-allyl derivative, **41b**, underwent the rearrangement shown in Scheme 24 (*108*).

6.2.3.4.2. Silica Gel-Catalyzed Rearrangements. Although dihydroartemisinin itself can be chromatographed on silica gel without decomposition or rearrangement, dihydroartemisitene, **37**, rearranges in the presence of silica gel to form **97** as shown in Scheme 25 (*109*).

References, pp. 202–214

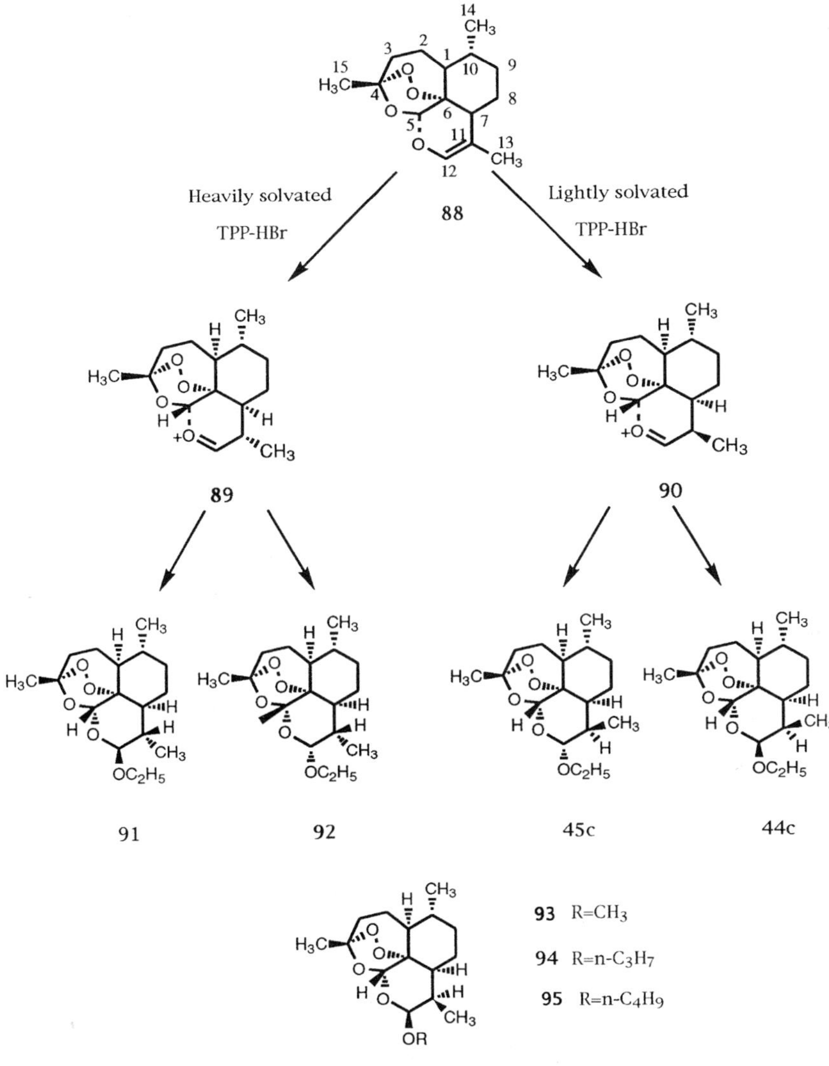

Scheme 23

YAGEN et al. (*110*) reported that heating a solution of dihydroartemisinin, **44a**, in the presence of silica gel produces **3** (Scheme 26). Under the same reaction conditions 9-β-hydroxydihydroartemisinin, **98a**, undergoes ring contraction to produce **99** (Scheme 27). However, under these reaction conditions the epimeric 9-α-hydroxydihydroartemisinin, **98b** is recovered unchanged.

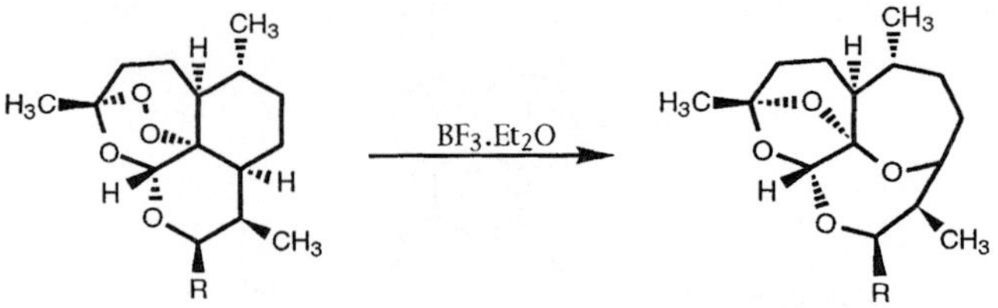

$$\underset{\textbf{41}}{\text{structure}} \xrightarrow{\text{BF}_3\cdot\text{Et}_2\text{O}} \underset{\textbf{96}}{\text{structure}}$$

41 (a) R=H
(b) R=CH$_2$CH=CH$_2$

96 (a) R=H
(b) R=CH$_2$CH=CH$_2$

Scheme 24

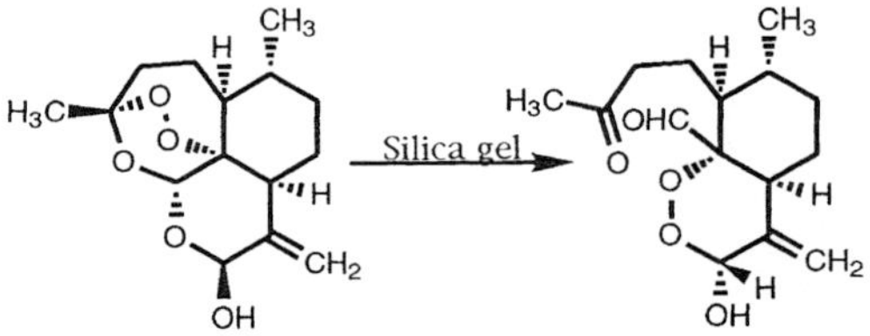

$$\underset{\textbf{37}}{\text{structure}} \xrightarrow{\text{Silica gel}} \underset{\textbf{97}}{\text{structure}}$$

Scheme 25

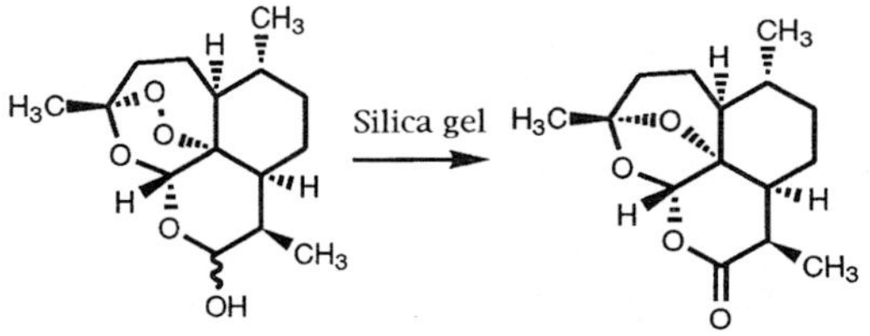

$$\underset{\textbf{44a}}{\text{structure}} \xrightarrow{\text{Silica gel}} \underset{\textbf{3}}{\text{structure}}$$

Scheme 26

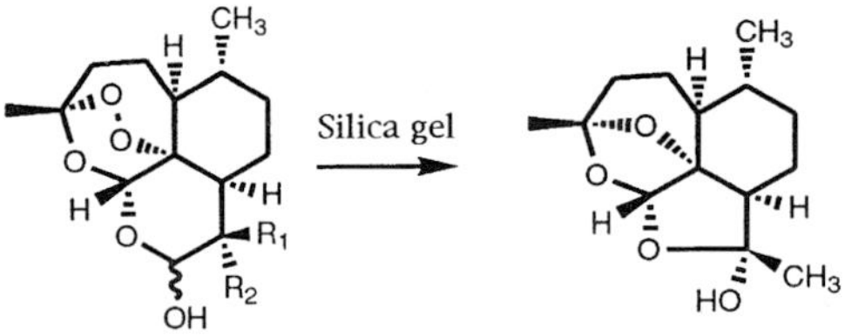

$$\underset{\textbf{98}}{\text{structure}} \xrightarrow{\text{Silica gel}} \underset{\textbf{99}}{\text{structure}}$$

(a) R$_1$=OH, R$_2$=CH$_3$
(b) R$_1$=CH$_3$, R$_2$=OH

Scheme 27

6.2.4. Reaction of Artemisinin with Reducing Agents

6.2.4.1. Lithium Aluminum Hydride

Reduction of artemisinin by LiAlH$_4$ produces at first **100** and **101**, but extended reflux leads to the further reduction of **100** to **101** (*111*) (Scheme 28).

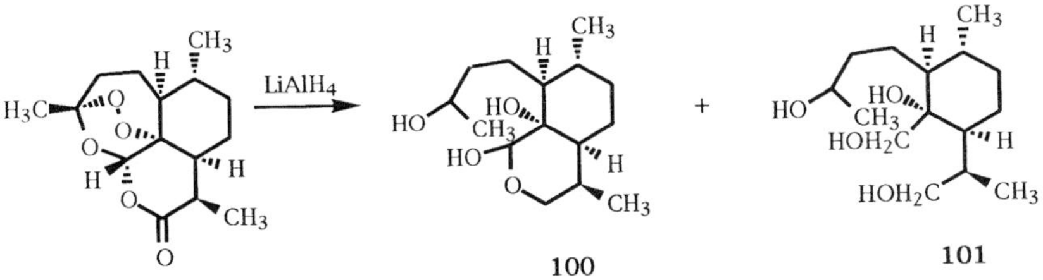

Scheme 28

6.2.4.2. Sodium Borohydride

Reduction of artemisinin by NaBH$_4$ in cold methanol produces dihydroartemisinin, **44a**, a critical intermediate in the preparation of a variety of derivatives, with very little reduction of the peroxide (*112*).

6.2.4.3. A Mixture of Sodium Borohydride and Boron Trifluoride

JUNG *et al.* employed a mixture of sodium borohydride and boron trifluoride, which PETTIT and PIATAK (*113*) had shown converts lactones into the corresponding ethers (*114*) (Scheme 29), to convert **1** into (+)-deoxoartemisinin, **38**. Reduction of the carbonyl to a methylene group occurs without loss of the peroxide grouping. The *in vitro* activity of **38** was found to be eight times greater than **1** against malaria. This

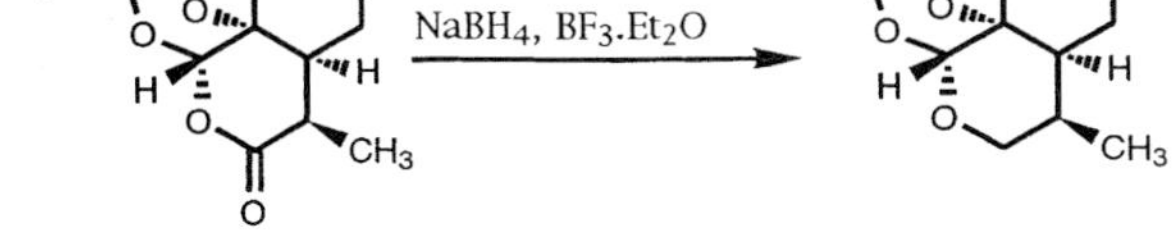

Scheme 29

discovery prompted several group to prepare 10-alkyldeoxoartemisinin derivatives as well as (+)-homodeoxoartemisinin, **97** (Section 7.).

6.2.4.4. Hydrogenation

Hydrogenation of **1** over Pd/CaCO$_3$ reduces the peroxide to an oxide, *i.e.* deoxyartemisinin, **3** (*4*). BROSSI *et al.* (*68*) reported that hydrogenation of arteether, **44c**, over Pd/CaCO$_3$ followed by treatment with p-toluenesulfonic acid yielded deoxyarteether, **102** (Scheme 30).

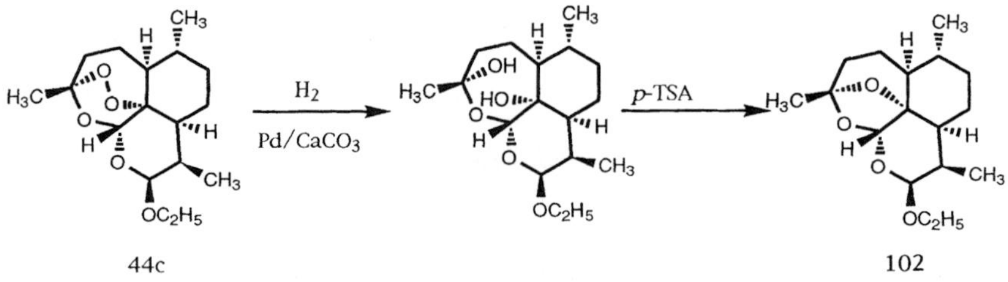

Scheme 30

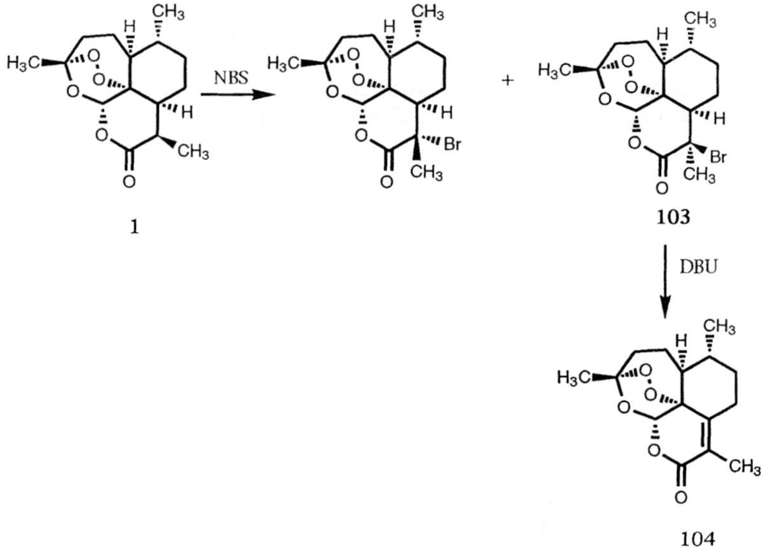

Scheme 31

References, pp. 202–214

6.2.5. Bromination

Treatment of **1** with N-bromosuccinimide produces an epimeric mixture of bromo derivatives (Scheme 31) (*99*). The 9β-bromo isomer, **103**, was converted with diazabicycloundecene into isoartemisitene, **104**.

VENUGOPALAN *et al.* (*115*) treated **88** with bromine to form a mixture of 9,10-dibromo derivatives, which led to the bromohydrins. Reaction with alcohols provided a series of ethers. That from propargyl alcohol was treated with tributyl tin hydride producing radical intermediates which cyclized to the epimeric acetals, **105** and **106** (Scheme 32).

The above dibromides were also prepared from **88** by LIN *et al.* (*116*) who reacted them with a number of heterocyclic amines to prepare water soluble artemisinin derivatives. The structures of the compounds and their antimalarial activities are given in Table 7 (Section 12.1.).

6.2.6. Fluorinated Artemisinin Derivatives

Chinese investigators prepared a number of fluorinated dihydroartemisinin derivatives which were listed in a review by LUO and SHEN (*4*); several were 2–3 times more active than the corresponding hydrogen analog. POSNER *et al.* (*117*) prepared a p-fluorobenzyl ether of a synthetic

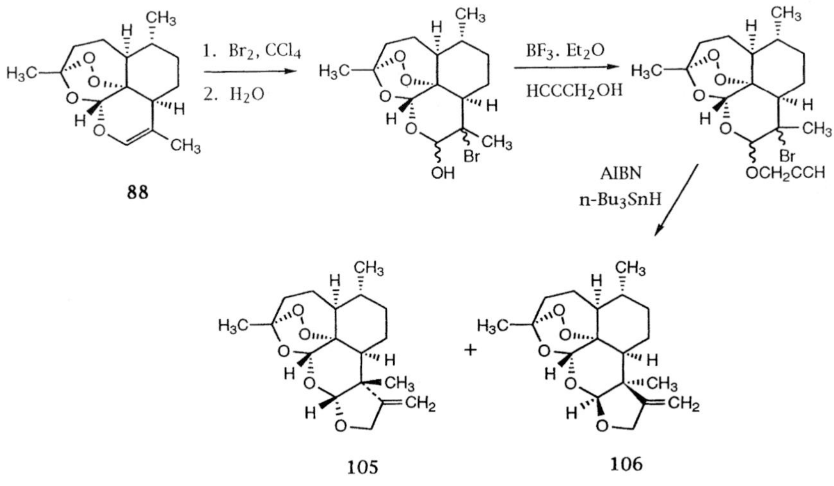

Scheme 32

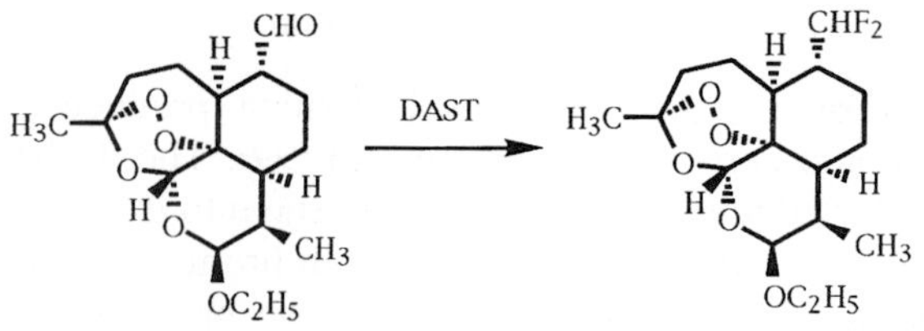

Scheme 33

1,2,4-trioxane and reported that it exhibited twice the antimalarial activity of the corresponding hydrogen analog. These results prompted Pu et al. (*118*) to prepare several geminal difluorinated artemisinin derivatives by the reaction of DAST (diethylaminosulfur trifluoride) with the corresponding carbonyl derivative (*e.g.* **107a** to **107b** in Scheme 33). Their antimalarial activities are given in Table 3 (Section 12.1.).

6.2.7. *Epoxidation of Anhydrodihydroartemisinin*

Lin et al. (*119*) were the first to epoxidize **88** with *m*-chloroperbenzoic acid; however, they isolated a mixture of 9α- and 9β-hydroxy-10β-m-chlorobenzoates instead of the expected epoxides. The *in vitro* antimalarial activity of the 10α-hydroxy isomer was comparable to that of artemisinin but that of the 10β-isomer was between one fifth to one sixth of the activity of the 10α-compound.

Petrov and Ognyanov (*120*) prepared the β-epoxide, **108**, from the reaction of **88** with the 1:2 complex of m-chloroperbenzoic acid and KF at 0°. Hufford et al. also isolated **108** from the reaction of **88** in methylene chloride with an aqueous buffered solution of *m*-chloroperbenzoic acid and aqueous sodium carbonate, determining the structure of the oxirane by x-ray crystallography (*121*). Pu et al. (*122*) showed that the α and β oxiranes were both formed in a 1:4 (α:β) mixture (Scheme 34). However, the α-oxirane, being more sensitive to moisture, was converted into the corresponding diol during the workup. Treatment of an aqueous acetone solution of **108** with dilute sulfuric acid produced 10β-hydroxydihydroartemisinin **109** which was then oxidized with chromic oxide in aqueous acetone to yield 9β-hydroxyartemisinin, **110**.

Diol **109** was difficult to purify by silica gel chromatography, undergoing an unusual silica gel-catalyzed rearrangement to form a less polar compound **99** (see Section 6.2.2.1.).

References, pp. 202–214

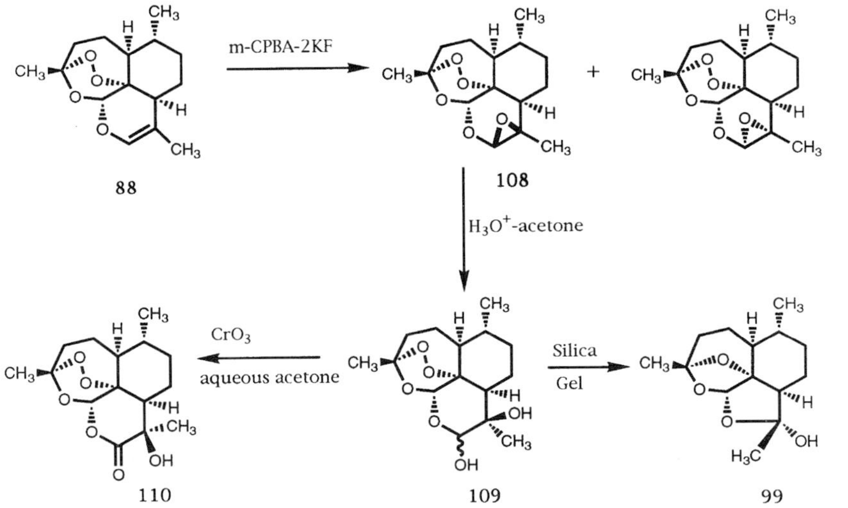

Scheme 34

6.2.8. Osmium Tetroxide Oxidation of Anhydrodihydroartemisinin

Treatment of **88** with osmium tetroxide provided entry into the corresponding 9α-hydroxyartemisinin series of compounds (Scheme 35) (*121*). HUFFORD et al. obtained a 1:1 mixture of diols, **109** and **111**, using one equivalent of osmium tetroxide. Oxidation of this mixture with Jones reagent produces 9α-hydroxy-artemisinin, **112**, while Moffat oxidation provides a mixture of 9α- and 9β-hydroxyartemisinin. PU et al. (*122*) oxidized **88** stereoselectively to **111** by employing catalytic quantities of osmium tetroxide and N-methyl morpholine as a co-oxidant.

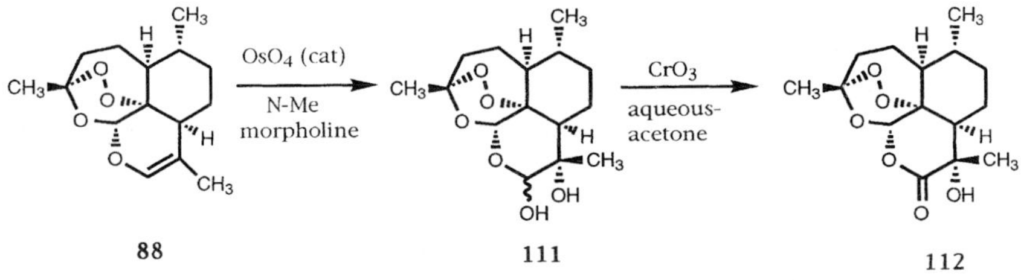

Scheme 35

7. Dihydroartemisinin Derivatives

7.1. Derivatives with Enhanced Oil Solubility

The poor solubility of artemisinin in water or oil, the two most common media for parenteral administration, prompted investigators in China and later in the U.S. to prepare some 100 semi-synthetic derivatives with improved solubilities. Virtually all of these modifications depend on reducing the lactone, the sole amenable functional group of **1**, to a lactol. In general, alkylation of a mixture of dihydroartemisinin epimers in the presence of an acidic catalyst gives products with predominantly β orientation, whereas acylation in alkaline medium preferentially yields α epimers. Many of the semi-synthetic products are more potent than the parent, with the order of activity carbonates > esters > ethers > artemisinin (*123, 124, 125*).

7.1.1. Ethers

Artemether

Artemether, **44b**, is prepared by treating a methanol solution of dihydroartemisinin with boron trifluoride-etherate (*131*). Both epimers are substantially more effective against chloroquine resistant *P. berghei* (α, SD_{50} 1.02 mg/kg; β, 1.16) than artemisinin (SD_{50} 6.2). Acute toxicity offers no problem, with subacute effects limited to minor fatty degeneration in liver cells (*126*) and, in beagles, lymphocyte reduction (*127*). The radical cure rate of 96% is to be compared with that of artemisinin, which gave 75%, and chloroquine, 11%. To reduce the high rate of recrudescence, artemether is given in combination with sulfadoxine and pyrimethamine, but a 19% rate of recrudescence persists (*128*).

Among the large number of ether derivatives of dihydroartemisinin, arteether, **44c**, shows activity comparable to artemether. The α-anomer, which is slightly more active than the β-anomer, can be prepared stereoselectively by treating dihydroartemisinin with ethyl iodide and silver oxide. Other members of the extensive series have been found to be less active than artemether against chloroquine-resistant *P. berghei* in the mice (*129*).

References, pp. 202–214

Miscellaneous Ethers

The ethers of dihydroartemisinin prepared by Chinese investigators were tabulated by LUO and SHEN (4) and their *in vivo* data are given in Table 12 in Section 12.2.

7.1.2. Esters

Esters have been prepared, largely as the α-epimer, by treating dihydroartemisinin either with acid chlorides or acid anhydrides (*125*) in pyridine, or with acids in the presence of dicyclohexylcarbodiimide or dimethylaminopyridine (*130*). Of some 40 esters examined, the most active were those of short chain aliphatic or aromatic acids. The list of esters prepared by Chinese investigators as well as the dose needed to suppress the infection of chloroquine-resistant *P. berghei* in 90% of the infected mice is given in Table 13, Section 12.2.

7.1.3. Carbonates

Carbonates of dihydroartemisinin are readily prepared by treatment with alkyl chloroformates in the presence of triethylamine in ethylene chloride or by catalysis by 4-dimethylaminopyridine, producing largely

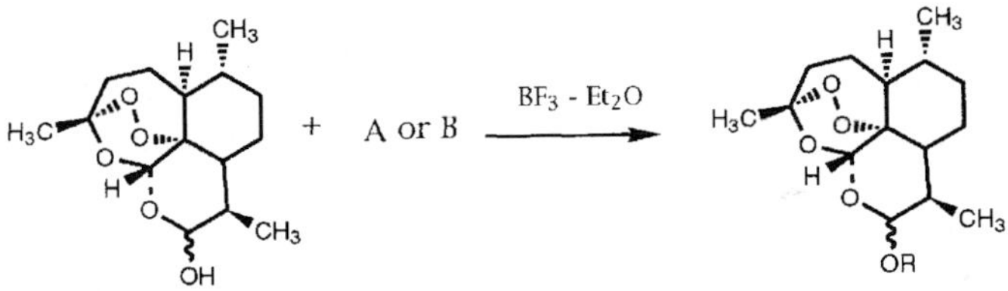

44a + A or B $\xrightarrow{BF_3 - Et_2O}$ 112

A = $HO(CH_2)_nR$

b n=2, R=CO_2CH_3
c n=3, R=CO_2CH_3
d n=1, R=4-$C_6H_4CO_2CH_3$

B = $HOCH_2CH(CH_3)CO_2R$

113
a R=CH_3
b R=H

Scheme 36

the α-epimers. The most active compounds examined were slightly more effective than artemether (*131*). Carbonates prepared by Chinese investigators and their *in vivo* activities against chloroquine-resistant *P. berghei.* are listed in Table 14, Section 12.2.

7.2. Derivatives with Enhanced Water Solubility

7.2.1. Sodium Artesunate

The half succinic acid ester of dihydroartemisinin, artesunic acid, **50**, is prepared by treating dihydroartemisinin with succinic anhydride in the presence of DMAP (*123*). The sodium salt which is readily soluble in water is effective against *P. falciparum,* ED_{50} 0.14 ng/ml (*123*) and against the asexual forms of *P. berghei* and *P. cynomolgi (132, 133)*. Although the compound is well tolerated in test animals it is nevertheless more toxic than artemisinin. However, it is far less toxic to the heart than chloroquine.

7.2.2. Sodium Artelinate and Related Derivatives

Dihydroartemisinin derivatives in which the water solubilizing function is linked by an ether, rather than an ester, are substantially more stable to hydrolysis than the esters (*134, 135*). Artelinic acid, **49**, is prepared by condensing dihydroartemisinin, **44a**, with methyl p-(hydroxymethyl)benzoate in the presence of boron trifluoride etherate followed by saponification (Scheme 36). It compares favorably with sodium artesunate, **50**, both *in vitro* against *P. falciparum* and *in vivo* against *P. berghei.*

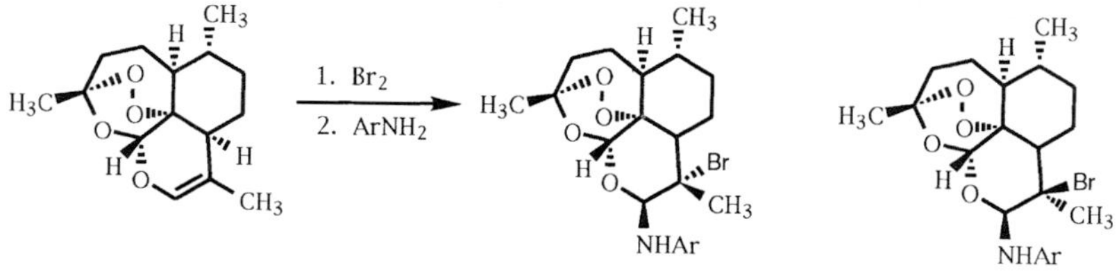

Scheme 37

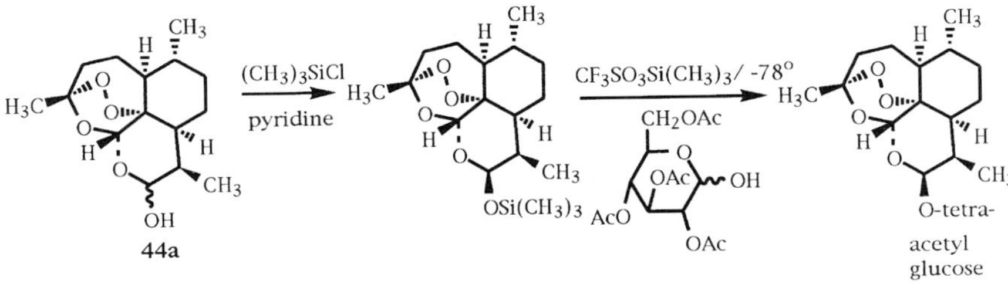

Scheme 38

The above reaction scheme was employed by LIN et al. to prepare several series of ω-hydroxy carboxylic acids containing an ether linkage between C-10 in **44a** and the carbon bearing the hydroxyl group of A or B (Scheme 36) (*136, 137*). The activities of the esters were ten times greater than those of the sodium salts of the carboxylic acids (Tables 3, 4, and 5 Section 12.2.). LIN et al. (*116*) prepared a number of water-soluble N-aryl-10-azadihydroartemisinin derivatives by treating the dibromo derivatives of **88** with a series of aryl amines (Scheme 37) (Table 7, Section 12.1.).

Attaching sugar moieties to dihydroartemisinin **44a** (Scheme 38) provided derivatives with enhanced water solubilities (*138*). The activities of the products were comparable to artemisinin; the deacetylated materials, obtained by saponification, were less active.

7.3. Artemisinin Derivatives

7.3.1. (+)-Deoxoartemisinin

Several investigators have prepared (+)-deoxoartemisinin, **3**, as in Scheme 12 or 29. LANSBURY and NOWAK employed the lactone **42** (Scheme 13) to prepare **3** as shown in Scheme 39. Selective formation of the ketal followed by treatment with sodium naphthalenide and α-chlorodimethyl ether yielded **115**. Selective reduction yielded **116**, which was oxidized with singlet oxygen and the crude reaction mixture was treated with camphor sulfonic acid to yield **3** (*51*).

JUNG et al. prepared (+)-deoxoartemisinin (**3**) from artemisinic acid, **2**, by reduction to **117** followed by reaction with singlet oxygen and treatment with acid (Scheme 40) (*114*).

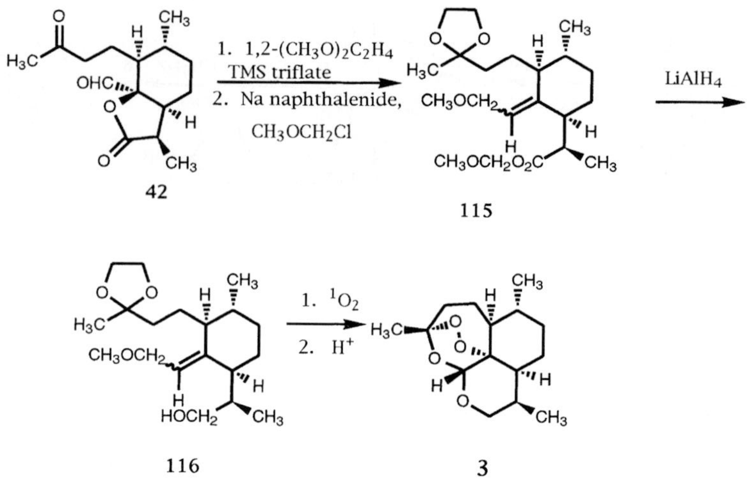

Scheme 39

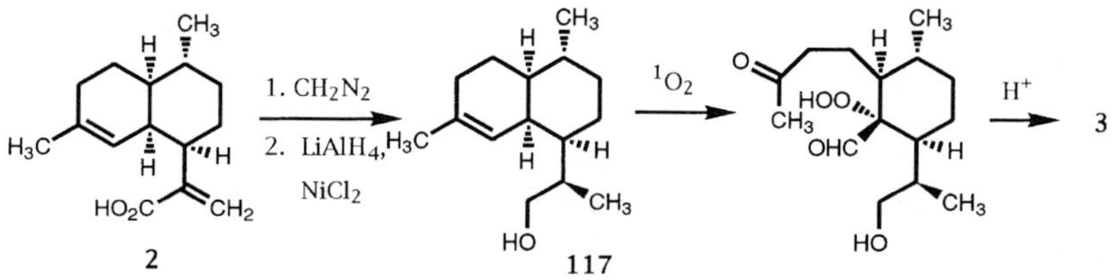

Scheme 40

7.3.2. (+)-Homodeoxoartemisinin

The enhanced antimalarial activity of **3** prompted BUSTOS *et al.* to prepare and test (+)-D-homo-deoxoartemisinin, **121a** (*139*). Artemisinic acid, **2**, by a two-step reduction, provided the aldehyde, **118**, which was converted by a Wittig reaction to **119** and by subsequent hydrolysis to provide **120**. (Scheme 41). The established treatment by singlet oxygen and acid provided the desired (+)-homodeoxoartemisinin, **121a**. Its activity was only 1/20th that of **1**.

JUNG *et al.* prepared a substituted homodeoxoartemisinin **121b** from one of the reactions products obtained in the synthesis of C-14 modified artemisinins (see Section 7.3.6.).

References, pp. 202–214

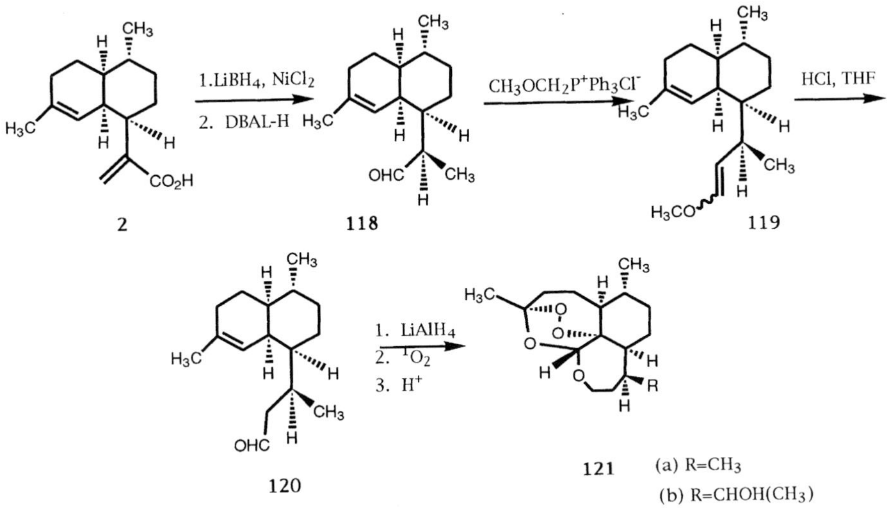

Scheme 41

7.3.3. (+)-10-Alkyldeoxoartemisinin

The enhanced activity of **3** compared to **1** also prompted JUNG (*140, 141*), and HAYNES (*142*) to prepare several 10-alkyl derivatives of deoxoartemisinin. Both groups converted artemisinic acid to **118** which was treated with Grignard reagents to yield a mixture of epimeric alcohols, **122** (Scheme 42). The allyl derivative **122b** was converted by hydroboration to the corresponding hydroxypropyl derivative. Treatment with singlet oxygen and acid provided the 10-alkyl-deoxoartemisinin as a single epimer in low yield. The 3′-hydroxy-n-propyl derivative, **123b**

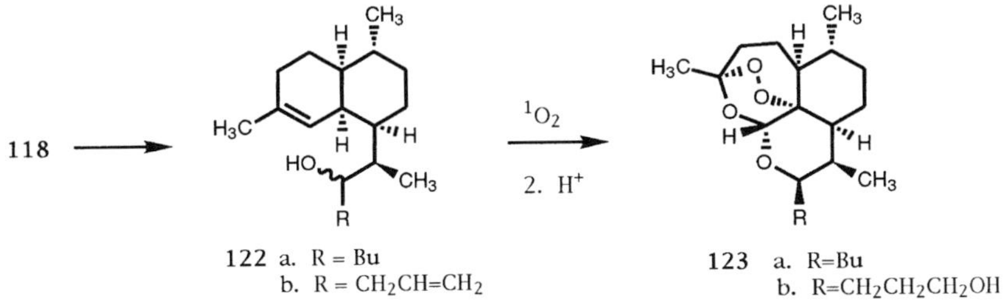

Scheme 42

was as active as deoxoartemisinin, i.e. approximately six times more active than artemisinin. Similar sequences have provided several derivatives of **123** where R = ethyl, phenyl and ω-carboxypropyl.

7.3.4. (+)-10β-Allyldeoxoartemisinin

Using a procedure developed by KISHI *et al.* (*143*) for the synthesis of C-glucosides ZIFFER *et al.* (Scheme 43) (*144*) prepared 10-allyldeoxoartemisinin, **124** from dihydroartemisinin **44a** by reaction with allyltrimethylsilane and boron trifluoride etherate, the stereochemistry of the allyl group in **124** being assigned by mechanistic considerations. Reduction of the double bond of the allyl group by diimide yielded the n-propyl compound, **125**. Test results given in Table 8, Section 12.2, show that the less polar materials were the more active antimalarials.

7.3.5. C-3 and C-9 Substituted 10-Deoxoartemisinins

AVERY *et al.* (*145*) converted many of the artemisinin derivatives they prepared with a variety of substituents at C-3 and or C-9 to the corresponding 10-deoxoartemisinin derivatives (Section 7.3.6). The reductions employed varied, since in some cases the deoxo-derivatives could only be prepared in good yield on a small scale. The *in vitro* activities of the compounds are given in Table 10, Section 12.1. A majority of the

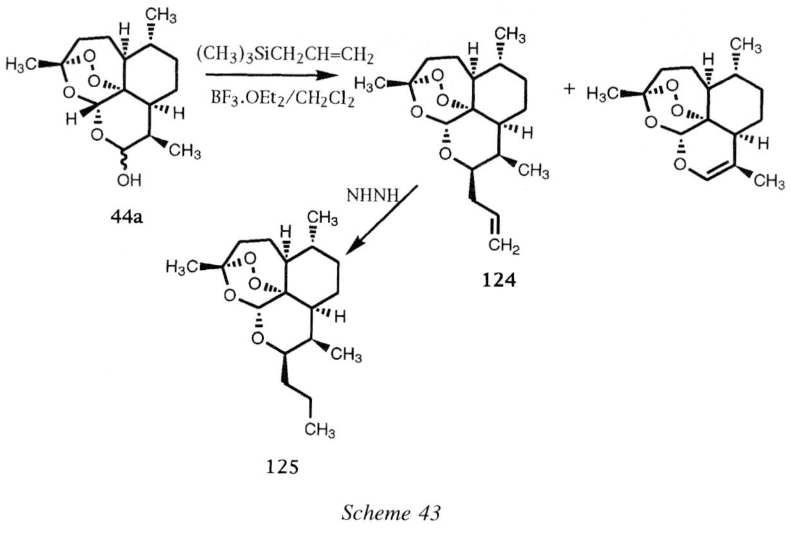

Scheme 43

compounds are several times more active than artemisinin and three were some twenty times more active than 10-deoxoartemisinin.

7.3.6. C-14 Modified Deoxoartemisinins

To obtain a derivative of deoxoartemisinin bearing a cyano substituent on C-14, JUNG *et al.* treated artemisinic acid, **2**, with methyl lithium thus producing ketone **126** which accepted a cyanide ion to provide nitrile **127** (Scheme 44) (*146*). Reduction by sodium borohydride produced alcohol **128** and the corresponding γ-lactone. The familiar treatment by singlet oxygen and acid converted the former to the substituted deoxoartemisinin **129**.

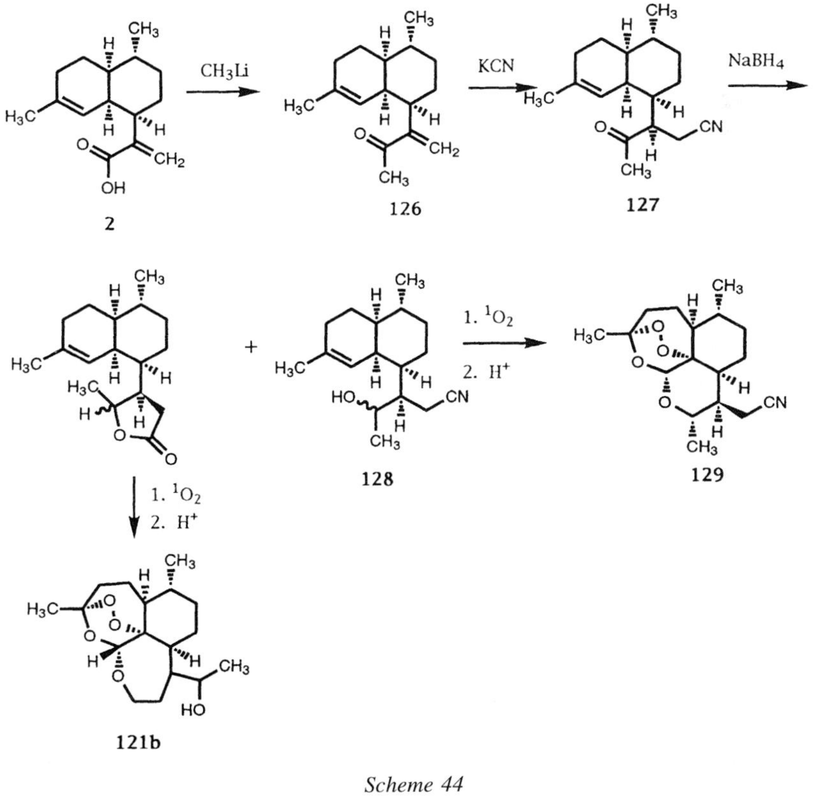

2, 126, 127, 128, 129, 121b

Scheme 44

8. Simplified Artemisinin Derivatives

In an attempt to identify the minimum structural requirements for artemisinin's activity, AVERY and coworkers embarked on a systematic program to prepare simplified analogs of **1**.

8.1. 9-Desmethylartemisinin

The intermediate **29** was ozonized and the resulting hydroperoxide treated with acid to provide 9-desmethylartemisinin, **130** (Scheme 45) (*147*).

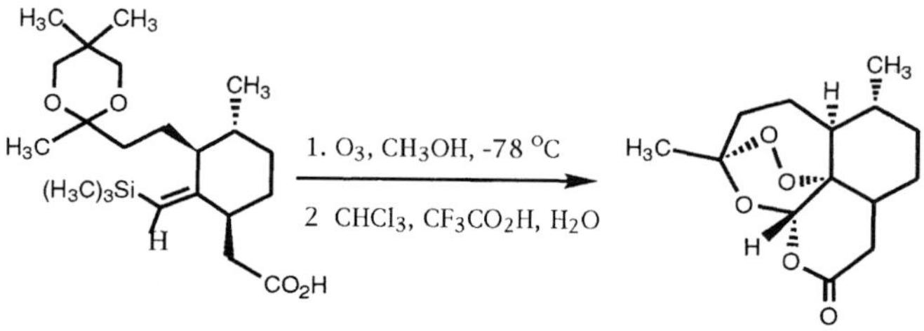

Scheme 45

8.2. 6,9-Bisnorartemisinin

In a short synthetic scheme (Scheme 46) (*148*), enamine **131** was converted to bicyclo-octenone **132** following a procedure of STILL (*149*). Selective ozonization of diene **133** provided aldehyde **134**, which was extended to **135**. Hydrolysis provided **136**, converted by ozonization and treatment with an acidic resin to the bisnor derivative **137**. Although no quantitative information on the antimalarial activity of **137** was reported, the authors stated that it displayed significant antimalarial activity against resistant strains of *P. falciparum*.

An alternative approach was employed by HAYNES *et al.* (*150*) to prepare **137**. A Lewis acid-catalyzed Diels-Alder reaction of 6-methylcyclohex-2-enone with 3,5-hexadienol provided the tricyclic **138** as a mixture of epimers (Scheme 47). Hydrogenation of the double bond of **138** proceeded in high yield as did further oxidation with Jones' reagent and methylation of the resulting acid. HAYNES *et al.* found that the

References, pp. 202–214

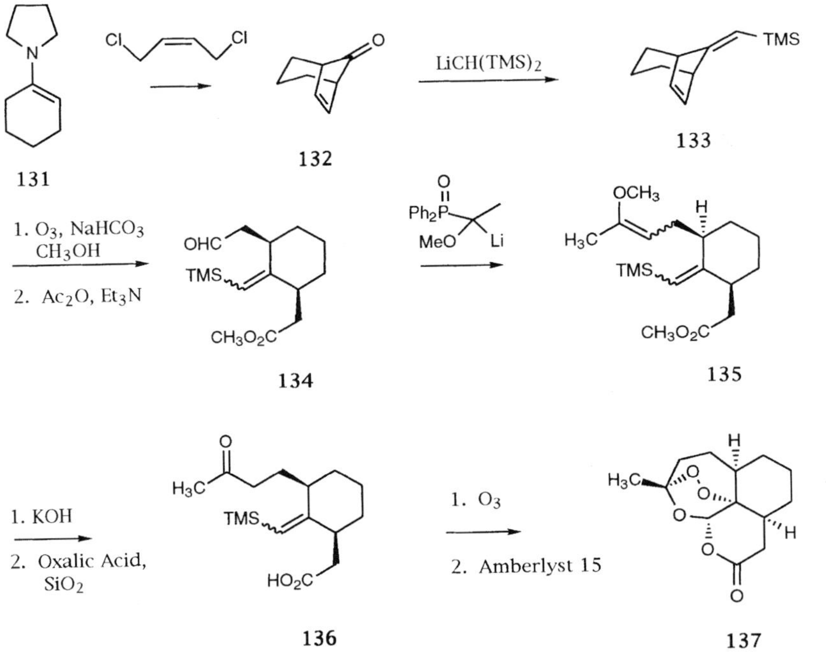

Scheme 46

photooxidation which is usually carried out in acetonitrile proceeded in much higher yield in methanol. The reaction sequence **141** to **137** proceeded in 34% yield.

8.3. (+)-8a,9-Secoartemisinin

To evaluate the importance of ring D of artemisinin which contains the lactone moiety, AVERY et al. prepared (+)-D-secoartemisinin, **146**, lacking the bond between carbons 8a and 9 (Scheme 48) (*151*). The intermediate **29** (Scheme 6) was converted to trimethylsilylether, **142**, which, after hydrolysis to ketone **143**, was converted by ozone to dioxetane **144**. Treatment of the dioxetane with boron trifluoride converted it to the cyclic peroxide **145**. In the presence of Amberlyst-15 and propionic anhydride this material rearranged to produce D-secoartemisinin **146**. Data on the antimalarial activity of **146** were not included.

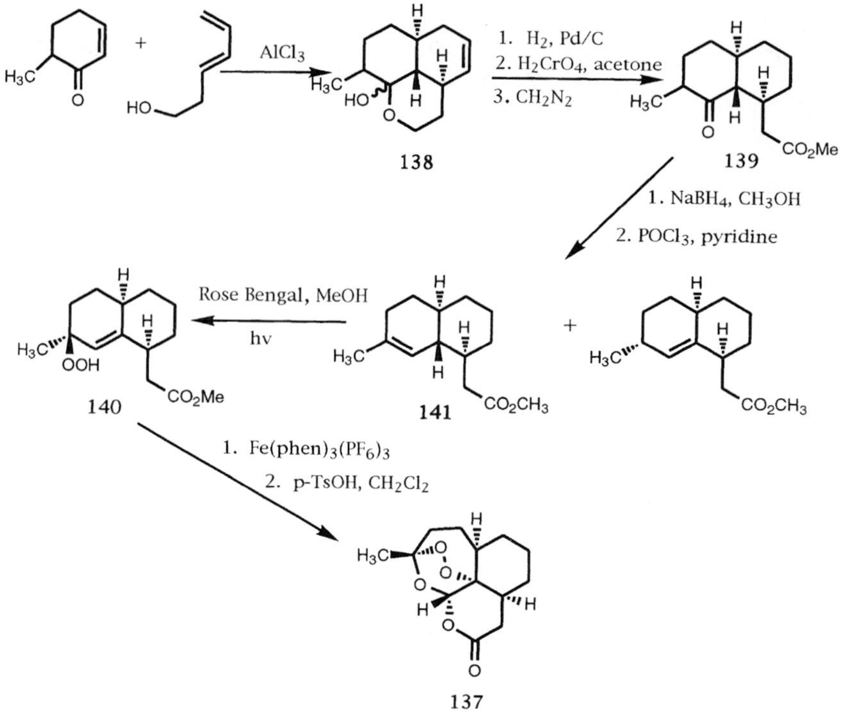

Scheme 47

8.4. (+)-4,5-Secoartemisinin

To prepare artemisinin derivatives lacking the 4,5 bond (*152*), AVERY'S group converted **8** to **147** (Scheme 49). Selective alkylation of the latter provided **148** which was converted to olefin **149**. Removal of the blocking group from the primary alcohol and oxidation provided **150**, from which the familiar ozonization and acid treatment produced the required A-seco-artemisinin **151**. The IC_{50} value for **151** was 6 ng/ml compared to values from 0.2 to 0.8 ng/ml for artemisinin *i.e.*, the compound is approximately an order of magnitude less active.

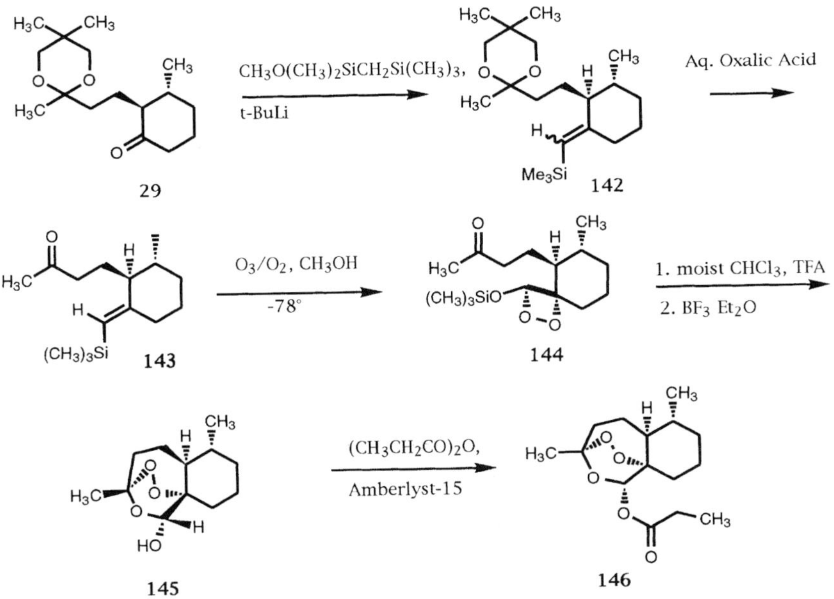

Scheme 48

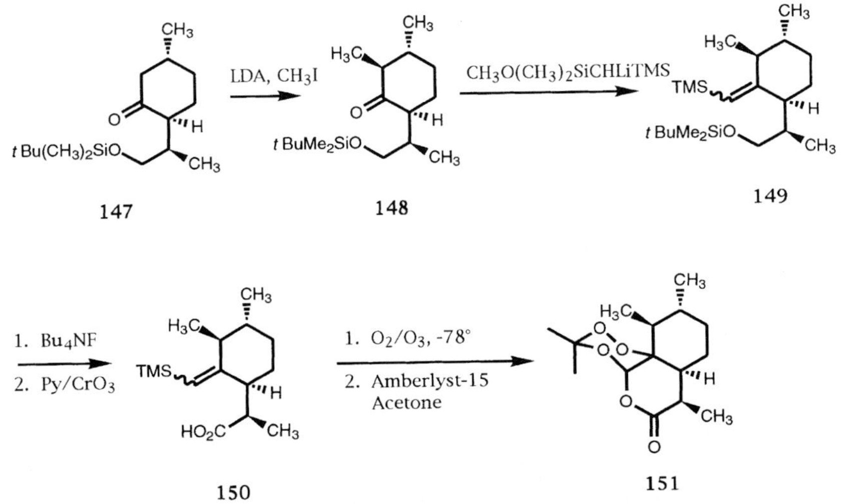

Scheme 49

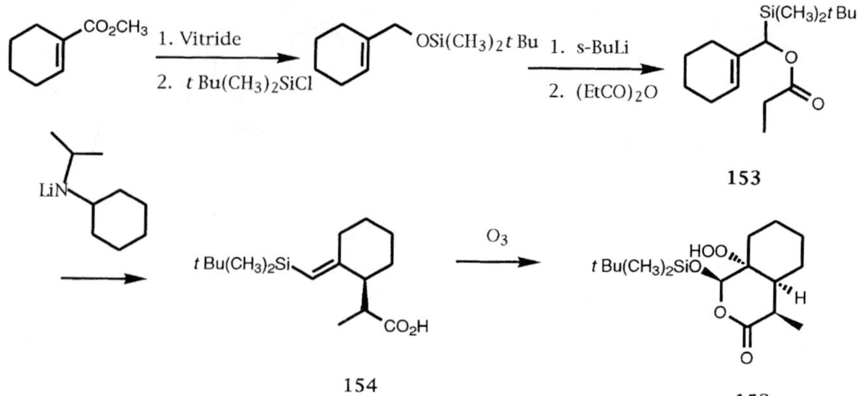

Scheme 50

8.5. (+)-Hexahydroisochroman-3-one

The synthesis of **152** (Scheme 50) (*153*) proceeded by rearrangement of the anion from **153** to yield the vinylsilane **154**. The hydroperoxide moiety was introduced during ozonolysis of **154**; ring closure occurred under the reaction conditions employed to produce **152**. The compound did not show substantial *in vitro* antimalarial activity.

8.6. 4,5-Desethanoartemisinin

IMAKURA *et al.* (*154*) examined the consequences of deleting carbons 4 and 5 in ring A on the antimalarial activity (Scheme 51). Oxidation of hydroxymenthol **8** followed by methylation with diazomethane provided a ketoester **155** suitable for a Wittig reaction to produce **156**, which, on ozonization and acid treatment, provided an analogue **157** without ring A. The biological activity of **157** was not reported.

8.7. 9-Alkyl-9-desmethylartemisinin

A successful large scale synthesis of intermediate, **29**, in their synthesis of **1** (Scheme 6) enabled AVERY *et al.* (*155*) to prepare fourteen 9-alkyl-9-desmethylartemisinin derivatives, **158**, by alkylating the dianion of acid (Scheme 52) (Biological activities given in Table 9, Section 12.1.).

References, pp. 202–214

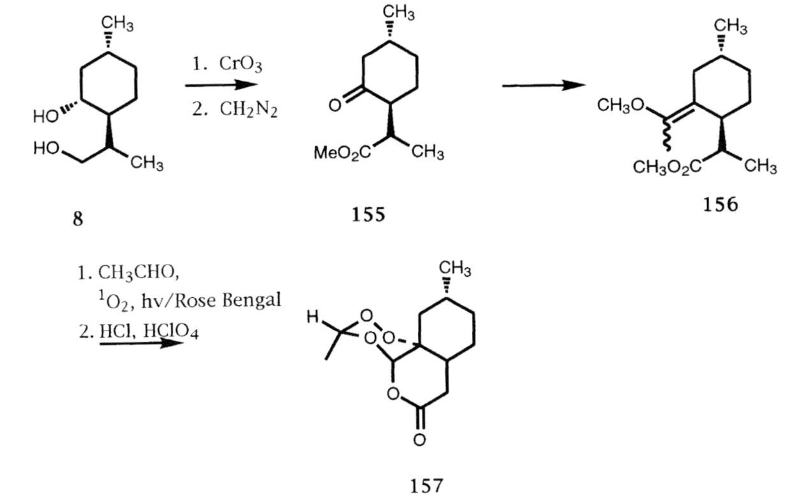

Scheme 51

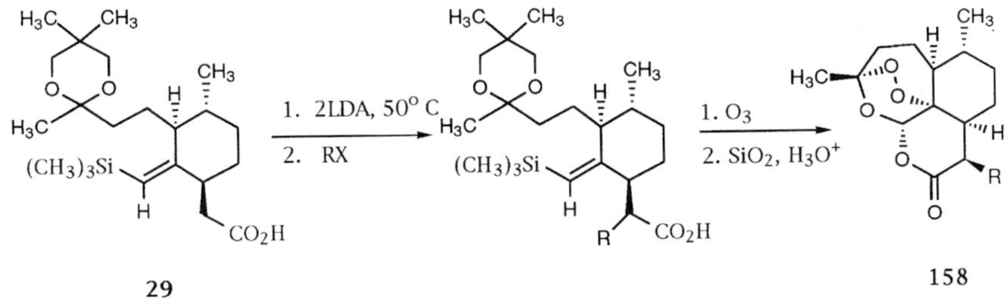

Scheme 52

8.8. C-3 and C-9 Modified Artemisinin Derivatives

AVERY et al. (*156*) prepared two series of artemisinin derivatives with different substituents at C-3, **159**, in their structure activity studies. The first contained a hydrogen in lieu of the C-13 methyl group and the second an *n*-butyl group in place of the C-13 methyl. Both syntheses were essentially identical but employed modified starting materials, i.e. **29a** and **29b** as shown in Scheme 53. The biological activities are given in Table 10 in Section 12.1.

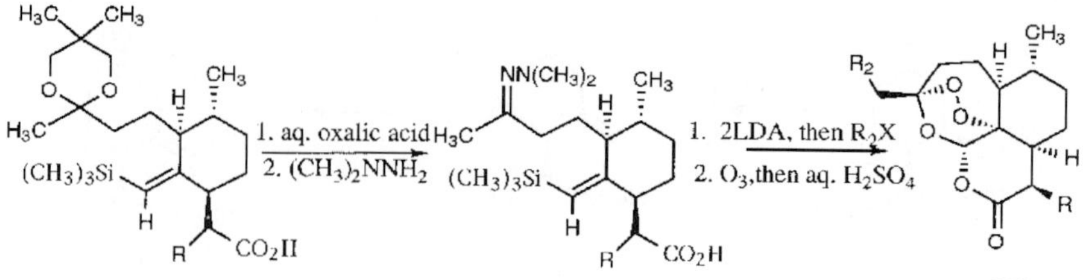

29 159

a R=H
b $R=(CH_2)_3CH_3$

Scheme 53

8.9. Carba-Analogs of Artemisinin

YE and WU (*157*) converted artemisinic acid, **2**, to **160** as shown in Scheme 54. Treatment of **160** with paraformaldehyde in the presence of boron trifluoride etherate yielded the carba-analogue of deoxoartemisinin, **161**. The latter was oxidized with ruthenium trichloride, in the presence of sodium periodate, to yield the carba-analog **162** of artemisinin. The biological activity of **162** was not reported.

AVERY *et al.* (*158*, *159*) prepared several carbaartemisinin analogs in order to evaluate the effect on the antimalarial activity of replacing the nonperoxidic trioxane ring oxygen by a methylene group. The final steps

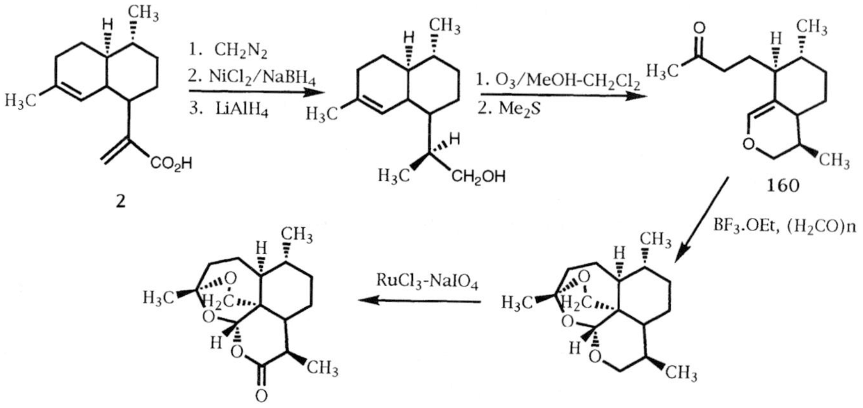

162 161

Scheme 54

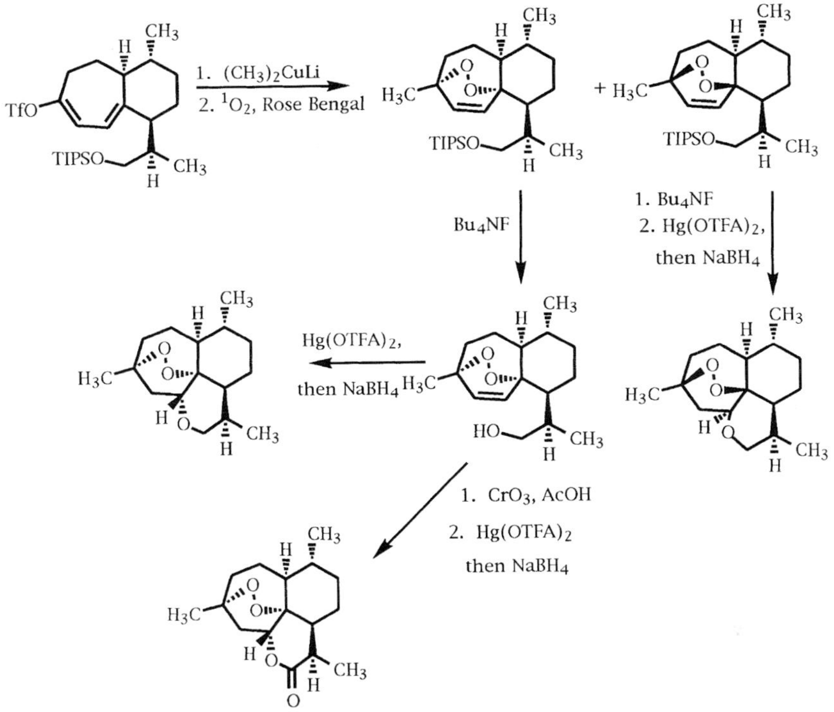

Scheme 55

in a long sequence are given in Scheme 55. All the carbaartemisinins were less active than artemisinin.

9. Quantitative Structure-Activity Analyses

The search for the minimal structural requirements for antimalarial activity in the compounds described here has also provided data for quantitative SAR. Chinese investigators first attempted such QSAR studies through HANSCH analysis and found a correlation between the lipid-solubility properties of various dihydroartemisinin derivatives and their antimalarial activity (*160*). AVERY *et al.* undertook a computer-aided QSAR study employing a Comparative Molecular Field Analysis of a series of C-9 analogs of artemisinin (*155*). The analysis suggested that steric parameters in the region of the lactone ring of artemisinin analogs are more important than electrostatic considerations. A number of 11-aza analogs of artemisinin were analysed by the same method (*161, 162*).

AVERY *et al.* suggest that other factors will have to be considered and are refining their approach.

10. Tricyclic 1,2,4-Trioxane Analogs

Triethylsilyl hydrotrioxide cleaves alkenyl esters and ethers, such as **163** in Scheme 56, to form 1,2-dioxetanes which rearrange in the presence of butyldimethylsilyl triflate to form 1,2,4-trioxanes (*163*). This reaction was incorporated into the sequence of Scheme 57 (*164*). In a

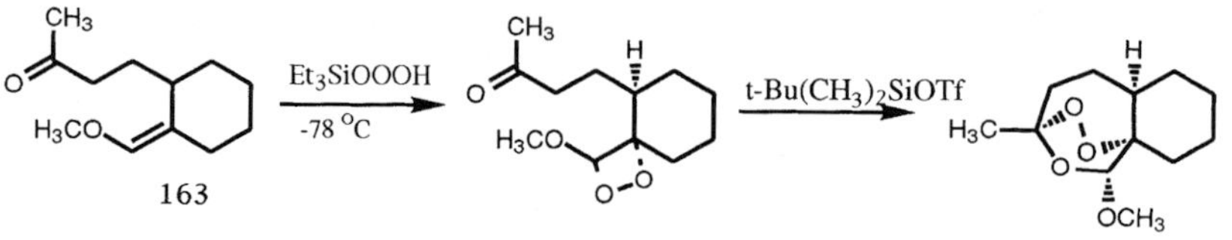

Scheme 56

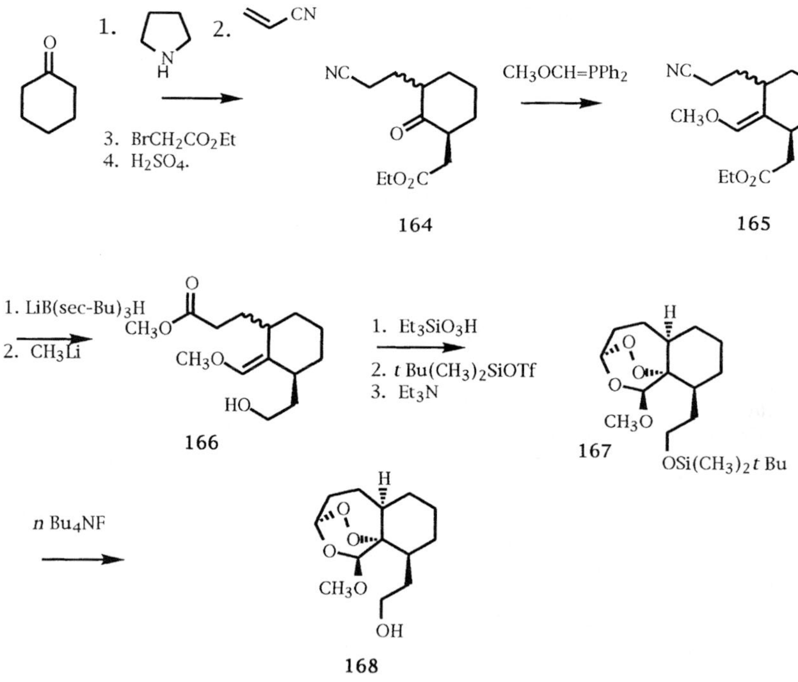

Scheme 57

one-pot reaction, sequential alkylation of cyclohexanone provided the α,α'-disubstituted cyclohexanone, **164**, which was converted by a Wittig reaction to an enol ether, **165**. The reaction of the nitrile with methyl lithium produced methyl ketone **166**, which with triethylsilicon trioxide formed a hydroperoxide. Treatment with t-butyldimethylsilyl triflate resulted in the above arrangement to a 1,2,4-trioxane, **167**. Cleavage of the silyl blocking group provided alcohol, **168**, from which a variety of esters and ethers were prepared. Of these, the *p*-fluorobenzyl ether was the most active, ten times more active *in vivo* against *P. berghei* than artemisinin.

11. Metabolism

11.1. Microbial Metabolites of Artemisinin and its Derivatives

Preliminary studies of the products of microbial metabolism are frequently employed to prepare derivatives which may be identical with those obtained from the mammalian metabolism of drugs (*165*). Such studies on artemisinin and derivatives have produced the observations in Table 1.

Table 1

Substrate	Organism	Product	Refs.
Artemisinin, **1**	*Nocardia corallina*	3, 169	*57*
Artemisinin, **1**	*Penicillium chrysogenum*	3, 169	*177*
Arteether, **82**	*Aspergillus niger*	169, 170, 171, 172, 173, 174	*166*
Arteether, **82**	*N. corallina*	170, 171	*178*
Arteether, **82**	*Cunninghamella elegans*	174, 176	*178*
Arteether, **82**	*Streptomyces lavendulae*	173, 174, 176	*167*
Anhydrodihydro-artemisinin, **87**	*S. lavendulae*	177, 180,	*168*
Anhydrodihydro-artemisinin, **87**	*Rhizopogon sp.* ATCC 36060	178, 180	*180*
Artemisinic acid, **2**	*Mucor mucedo*	3β-hydroxy artemisinic acid	*169*
Artemisinic acid, **2**	*A. flavipes*	3α-hydroxyartemisinic acid	*169*
Dihydroartemisinin, N-phenylureido	*Beauveria sulfurescens*	14-Hydroxy dihydroartemisinin, N-phenyluriedo	*170*
Arteether, **44c**	*B. sulfurescens*	172, 174, 175	*170*

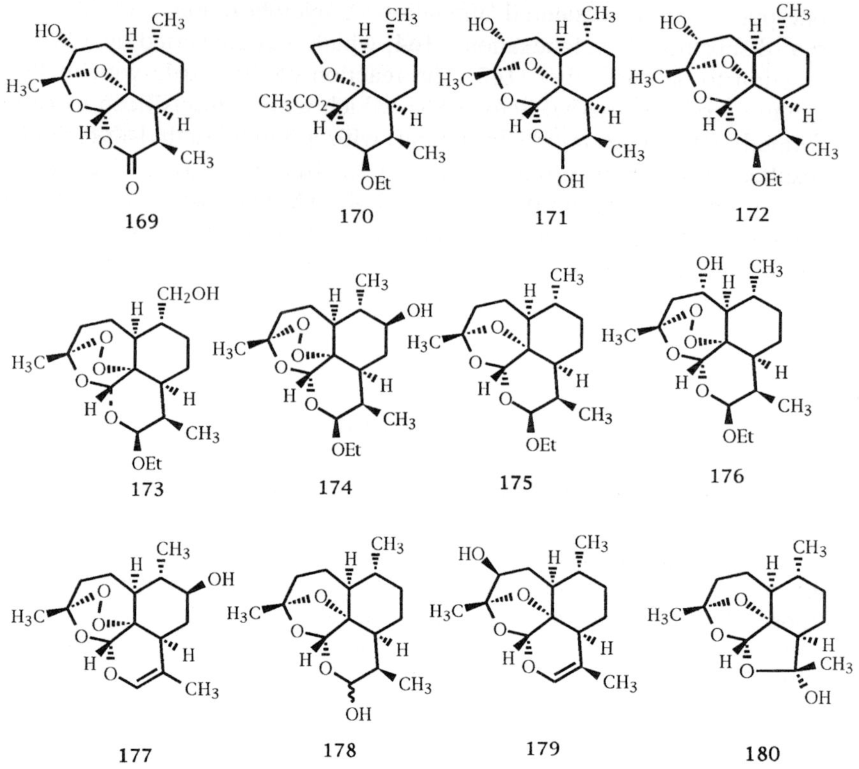

Scheme 58

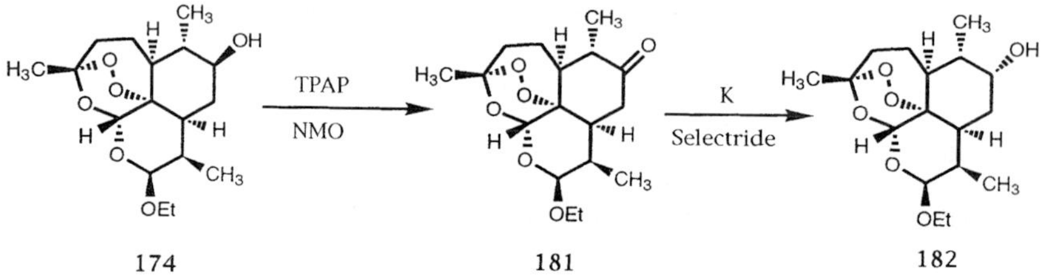

Scheme 59

The 7β-hydroxyarteether, **174**, provided by the action of *B. sulfurescens*, allowed Hu *et al.* to prepare the corresponding 7-ketoarteether, **181**, which they reduced with potassium selectride to obtain 7α-hydroxy arteether, **182** (Scheme 59) (*171*).

11.2. Mammalian Metabolites

The metabolism of arteether by rat-liver microsomes proceeds primarily first by O-dealkylation, then conversion of the peroxide into the desoxyderivative (66). Several hydroxylated derivatives are produced, as

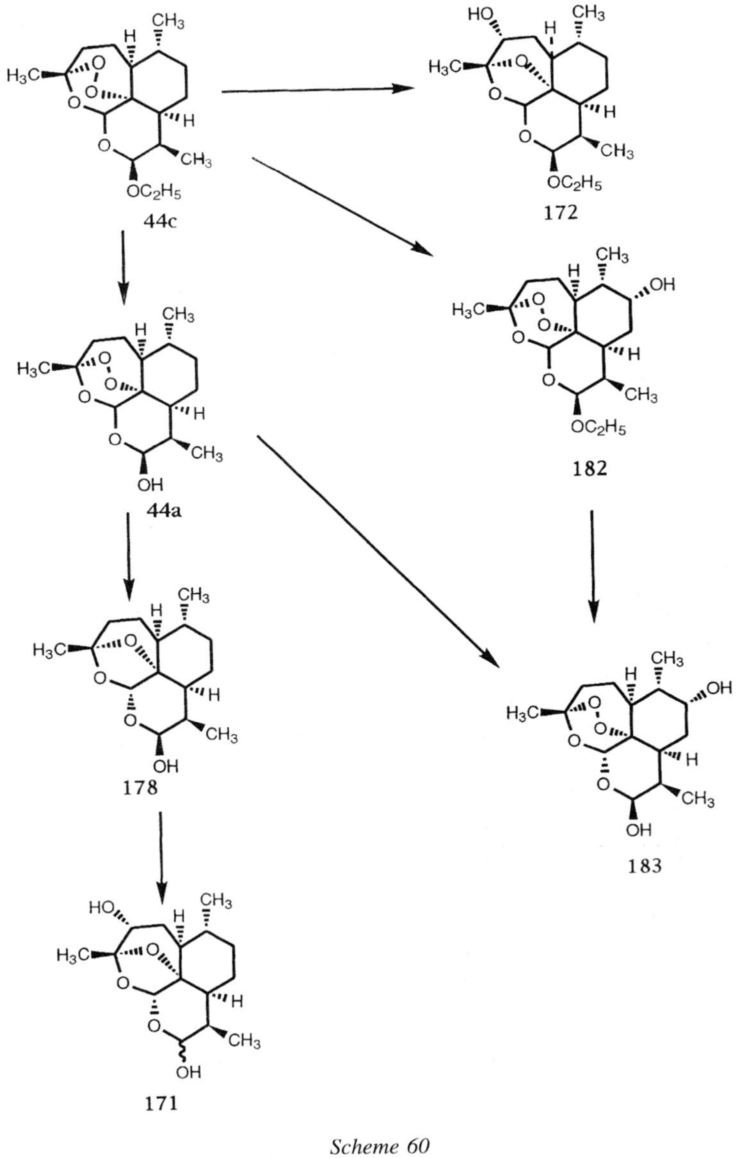

Scheme 60

shown in Scheme 60. The sequence suggests that arteether may be a "prodrug", slowly releasing the pharmacologically active entity, dihydroartemisinin.

After intravenous injection of artemisinin the highest concentration of the drug is found in the lung and kidney, moderate quantities in heart, brain and liver and only low concentrations in muscle, fat, and spleen (70). Oral administration, however, produces the highest concentration in liver, moderate concentrations in brain and plasma, and the lowest in heart, muscle and spleen. The highest rate of the metabolism of the drug occurs in the liver.

12. Test Data of Artemisinin Derivatives

A plethora of derivatives of artemisinin have been prepared for testing against resistant malarias, primarily by modifications of the lactone ring. Chinese investigators first tested their compounds *in vivo* in mice, whereas later investigators employed infected red blood cells. The most active compounds determined by *in vitro* testing were then tested *in vivo*.

12.1. *In Vitro*

To compare the many artemisinin derivatives prepared and tested over a period of years the data are reported as a ratio of the IC_{50} value for artemisinin or arteether to that of the compound in question. Thus, ratios larger than 1 indicate the compound is more active than artemisinin or arteether. An IC_{50} dose is defined as that dose which limits the growth of the parasite to 50% of that which it would attain in the absence of the

Table 2. *Relative* In Vitro *Activities of N-Substituted 11-Azaartemisinins (173)*

Compound	IC_{50} $1/IC_{50}$Compound against a chloroquine-resistant strain (FCR3) of *P. falciparum*.
64	1.0
68	0.8
69	9.0
70	2.6
72	22
73	1.1
74	1.0
75	26

References, pp. 202–214

drug. The incorporation of [^{3}H]-hypoxanthine provides a measure of the growth of the parasite in infected erythrocytes (*172*). Testing has been done primarily at the Walter Reed Army Institute of Research (in the Department of Experimental Therapeutics) against two drug resistant clones of *P. falciparum*. The W-2 clone from Indochina is resistant to chloroquine, quinine, sulfadoxine and pyrimethamine but sensitive to mefloquine, whereas the D-6 clone from African Sierra Leone is resistant to mefloquine but sensitive to chloroquine, quinine, sulfadoxine and pyrimethamine.

In vivo testing of **75** (*P. berghei* in mice) indicated it was four times more active than **1** suggesting limits to employing *in vitro* data to predict *in vivo* activity.

Comparison of the activities of the fluorinated derivatives with their hydrogen analog shows that introduction of fluorine increases the activity, but the increase is not sufficient to warrant subjecting the compounds to *in vivo* testing.

A comparison of the activities of **1** and derivatives of enhanced water-solubility is given in Tables 4a and 4b. Whereas the esters were more active than either the acids or their salts there is a need for a stable and effective water soluble artemisinin derivative to replace artesunic acid. The *in vitro* results for **195** justified *in vivo* testing, which showed it was more active than **1** or artesunic acid, **50**.

Despite the good *in vitro* activity of **205** and **207** neither they or the other compounds in this series exhibited significant *in vivo* activity against *P. berghei*.

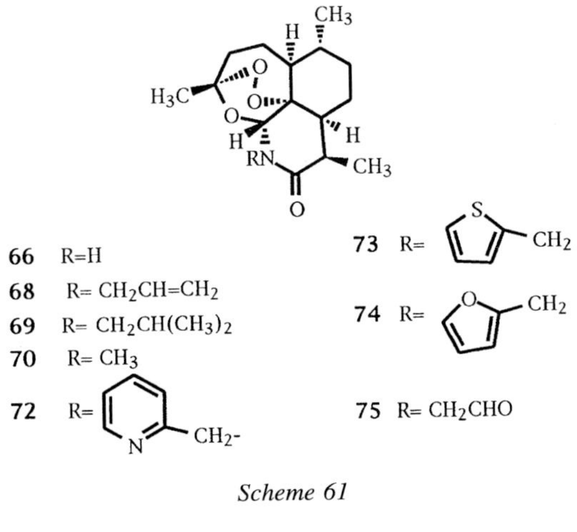

66 R=H
68 R= CH$_2$CH=CH$_2$
69 R= CH$_2$CH(CH$_3$)$_2$
70 R= CH$_3$
72 R= (pyridyl-CH$_2$-)
73 R= (thienyl-CH$_2$)
74 R= (furyl-CH$_2$)
75 R= CH$_2$CHO

Scheme 61

Table 3. *Relative* In Vitro *Activities of Fluorinated Artemisinin Derivatives and their Precursors or Corresponding H-Analogs* (*118*)

Compound	W-2 IC_{50} **1/IC_{50}** Compound	W-2 IC_{50} **1/IC_{50}** Compound	D-6 IC_{50} **1/IC_{50}** Compound	D-6 IC_{50} **1/IC_{50}** Compound
184	1.4	0.8	1.2	0.1
185	1.2	0.6	0.5	0.05
186	2.0	1.5	1.0	0.3
187	1.2	0.8	1.4	0.5
188	0.5	0.3	0.6	0.2
189	1.1	0.7	3.8	1.3
190	0.5	0.3	1.2	0.4
191	1.5		1.8	
192	0.14		0.15	
193	0.3		0.45	
194	0.06		0.08	

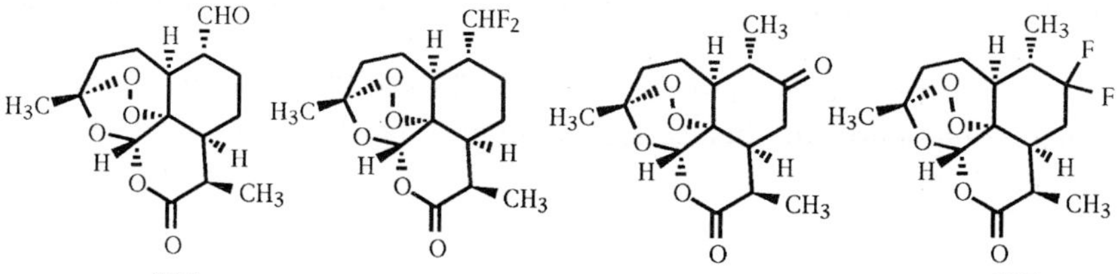

184 **185** **186** **187**

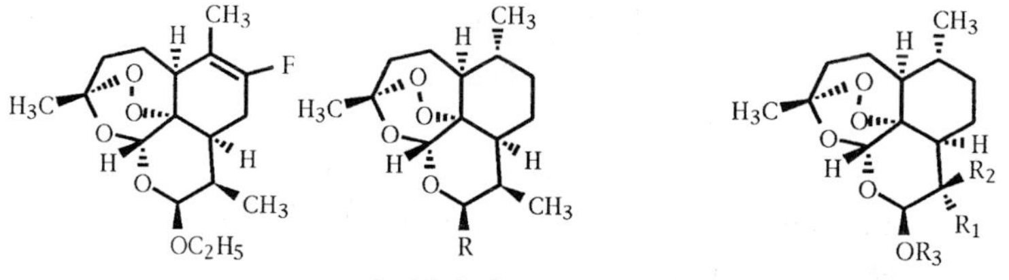

188

189 R=CH_2CHF_2

190 R=$CH_2CF_2CH_3$

191 R_1=CH_3, R_2=OH, R_3=CH_2CF_3

192 R_1=OH, R_2=CH_3, R_3=CH_2CF_3

193 R_1=OH, R_2=CH_3, R_3=C_2H_5

194 R_1=CH_3, R_2=OH, R_3=C_2H_5

Scheme 62

Table 4a. *Relative* In Vitro *Activities of Water Soluble Dihydroartemisinin Derivatives (136)*

Compound	D-6 IC$_{50}$ 1/IC$_{50}$ Compound	W-2 IC$_{50}$ 1/IC$_{50}$ Compound
195	4.9	2.5
196	0.05	0.03
197	1.6	1.0
198	0.03	0.03
199	0.96	0.7
200	0.04	0.06
201	3.8	1.8
202	1.7	0.7

Table 4b. *Relative* In Vitro *Activities of Additional Dihydroartemisinin Derivatives*

Compound	D-6 IC$_{50}$ 1/IC$_{50}$ Compound	W-2 IC$_{50}$ 1/IC$_{50}$ Compound
203	1.1	0.46
204	0.10	0.049
205	6.0	1.73
206	0.07	0.31
207	4.9	0.37
208	0.06	0.59
209	1.1	5.4
210	0.14	0.35

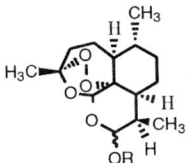

195 R = (CH$_2$)CO$_2$Et
196 R = (CH$_2$)CO$_2$K
197 R = (CH$_2$)$_2$CO$_2$CH$_3$
198 R = (CH$_2$)$_2$CO$_2$K
199 R = (CH$_2$)$_3$CO$_2$CH$_3$
200 R = (CH$_2$)$_3$CO$_2$K
201 R = CH$_2$C$_6$H$_4$CO$_2$CH$_3$
202 R = CH$_2$C$_6$H$_4$CO$_2$K

203 R = (R) CH$_2$CH(CH$_3$)CO$_2$CH$_3$
204 R = (R) CH$_2$CH(CH$_3$)CO$_2$K
205 R = (S) CH$_2$CH(CH$_3$)CO$_2$CH$_3$
206 R = (S) CH$_2$CH(CH$_3$)CO$_2$K
207 R = (R) CH(CH$_3$)CH$_2$CO$_2$CH$_3$
208 R = (R) CH(CH$_3$)CH$_2$CO$_2$K
209 R = (S) CH(CH$_3$)CH$_2$CO$_2$CH$_3$
210 R = (S) CH(CH$_3$)CH$_2$CO$_2$K

Scheme 63

Table 5. *Relative* In vitro *Activities of α-Alkylbenzylic Ethers of Dihydroartemisinin (137)*

Compound	W-2 IC_{50} $1/IC_{50}$ Compound	D-6 IC_{50} $1/IC_{50}$ Compound
211	0.53	0.75
212	10.8	8.0
213	1.13	0.63
214	4.73	4.64
215	4.40	4.09
216	3.00	2.50
217	0.40	0.40
218	0.75	0.81
219	1.45	2.37
220	4.48	4.52
221	7.0	5.0
222	20.7	9.3
223	3.0	3.3
224	0.82	0.78

The α-alkylbenzylic esters of dihydroartemisinin (Table 5) represent an effort to provide an improved analog of artelinic acid. Although several of the esters were more active than artelinic acid the authors did not report their *in vivo* activities.

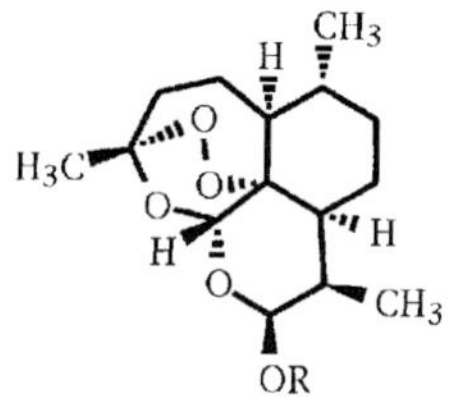

211 (R) R=CH(C_3H_7)$CH_2C_6H_5$
212 (R) R=CH(C_6H_5)$CO_2C_2H_5$
213 (R) R=CH(CH_3)p-$C_6H_4CF_3$
214 (R) R=CH($CH_2CO_2C_2H_5$)C_6H_5
215 (R) R=CH($CH_2CO_2C_2H_5$)p-$C_6H_4NO_2$
216 (R) R=CH(CH_3)(p-$C_6H_4CO_2Me$)
217 (R) R=CH(CH_3)p-$C_6H_4CO_2H$
218 (R) R=CH(CH_2CO_2H)p-$C_6H_4NO_2$

219 (S) R=CH(C_3H_7)$CH_2C_6H_5$
220 (S) R=CH(C_6H_5)$CO_2C_2H_5$
221 (S) R=CH(CH_3)p-$C_6H_4CF_3$
222 (S) R=CH($CH_2CO_2C_2H_5$)C_6H_5
223 (S) R=CH($CH_2CO_2C_2H_5$)p-$C_6H_4NO_2$
224 (S) R=CH(CH_3)(p-$C_6H_4CO_2Me$)

Scheme 64

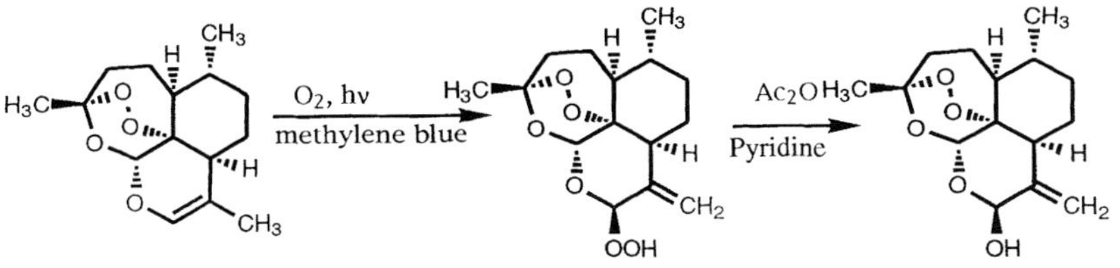

Scheme 65

Artemisitene, **5**, a minor component of the terpene mixture found in *A. annua.*, was prepared by EL-FERALY *et al.* (*105*) (Scheme 65) from **86**. Irradiation of **86** in ethanol followed by reaction with acetic anhydride in pyridine yielded **5**. ACTON *et al.* (*174*) employed **5** as the starting material for the preparation of the compounds (Scheme 66) listed in Table 6. There was a wide variation in the observed *in vitro* activities but none

Table 6. *Relative* In Vitro *Activities of 9-Substituted Artemisinin Derivatives*

Compound	D-6 IC_{50} $1/IC_{50}$ Compound	W-2 IC_{50} $1/IC_{50}$ Compound
225	0.36	0.13
226	0.48	0.12
227	0.016	0.004
228	0.004	0.002
229	0.91	0.43
230	0.19	0.12
231	0.006	0.003
232	0.26	0.13
233	0.003	0.002
234	0.038	0.022
235	0.021	0.008
236	0.77	0.21
237	0.68	0.23
238	1.1	0.28

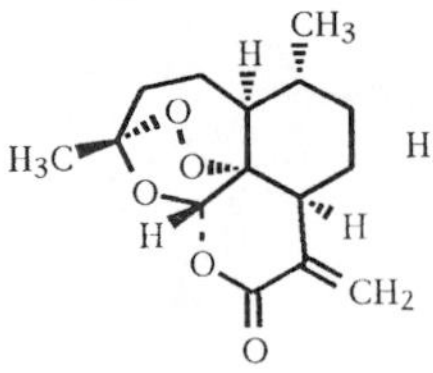
225

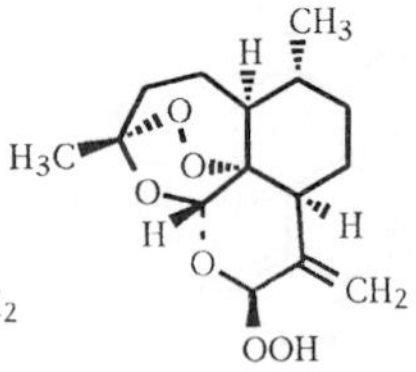
226

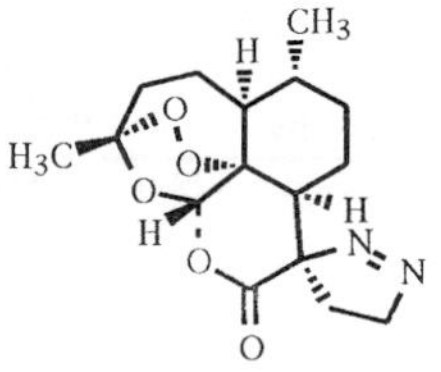
227

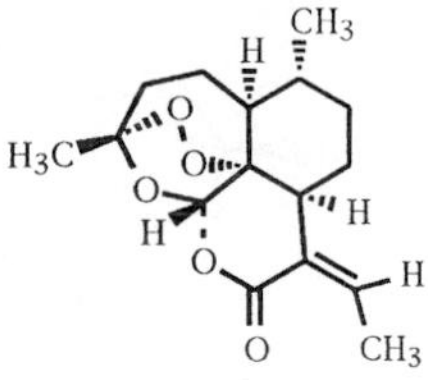
228

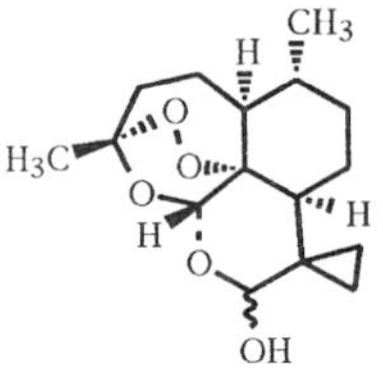
229

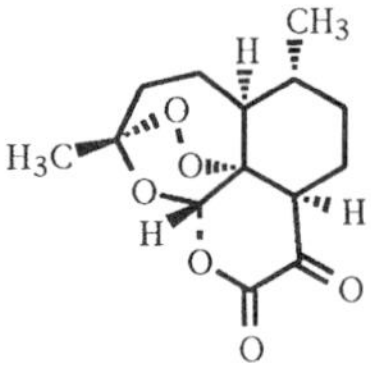
230

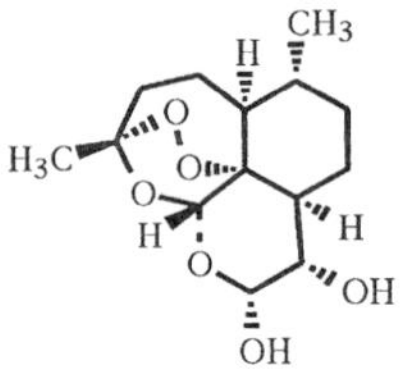
231

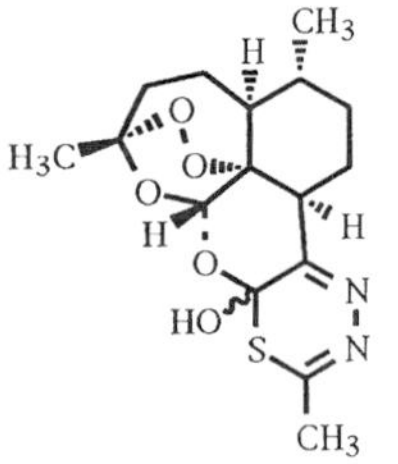
232

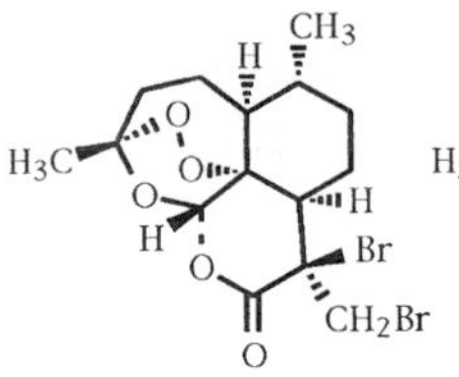
233

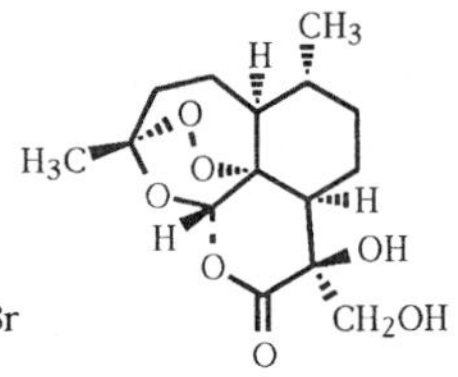
234

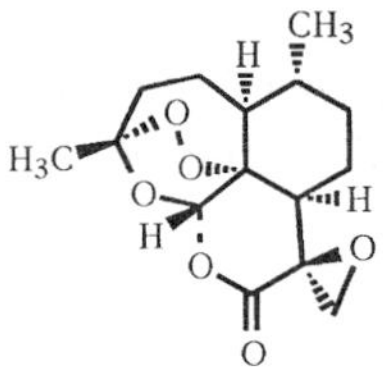
235

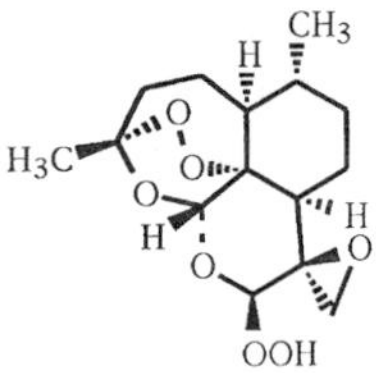
236

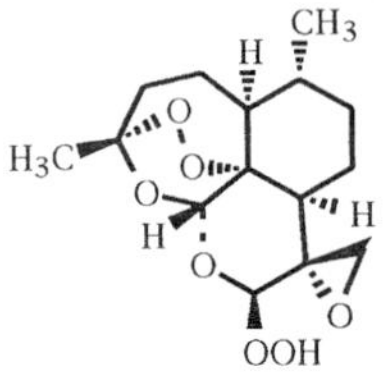
237

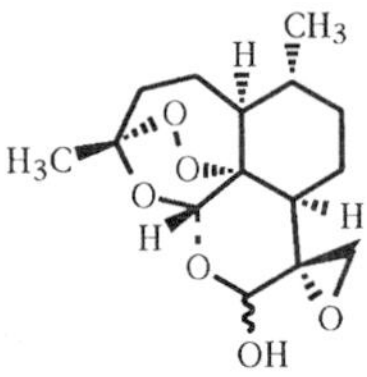
238

Scheme 66

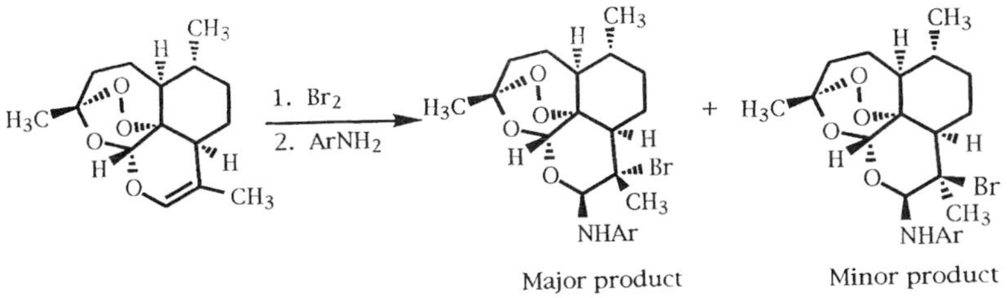

Scheme 67a

Table 7. *Relative* In Vitro *Activities of Aromatic Amine Derivatives of 9-Bromodihydroartemisinins*

Compound	D-6 IC_{50} 1/IC_{50} Compound	W-2 IC_{50} 1/IC_{50} Compound
239	0.62	2.9
240	3.5	>3.8
241	0.16	1.0
242	0.07	0.51
243	0.02	0.11

exhibited sufficient activity to warrant *in vivo* testing. HAYNES and VONWILLER (*49*) prepared **5** from artemisinic acid, **2** (see Scheme 11).

A mixture of dibromides from anhydrodihydroartemisinin was employed by LIN *et al.* (*116*) to prepare a series of heterocyclic water soluble derivatives (Table 7) listed in Scheme 67b.

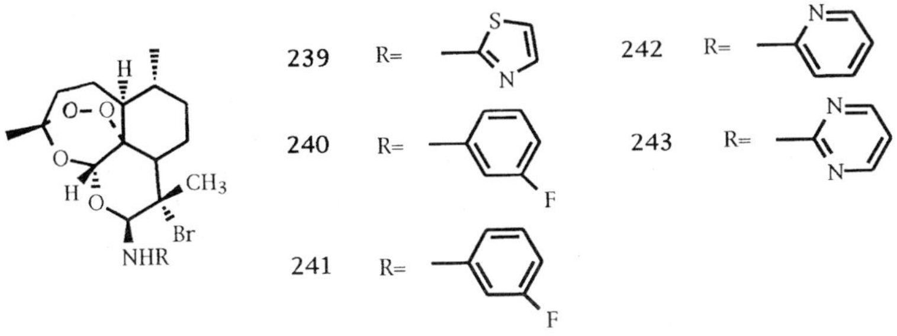

Scheme 67b

Table 8. *Relative* In Vitro *Activities of 10-Alkydeoxoartemisinins*

Compound	W-2 IC$_{50}$ 1/IC$_{50}$ Compound	D-6 IC$_{50}$ 1/IC$_{50}$ Compound
244	0.54	0.80
245	2.0	2.1
246	1.1	0.79
247	0.54	0.42
248	0.67	0.77

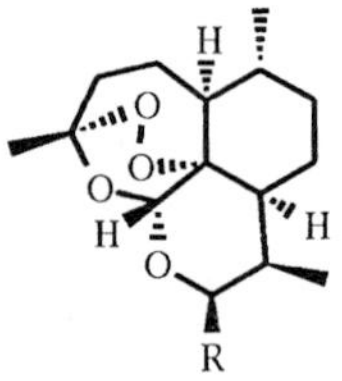

244 R= CH$_2$CH=CH$_2$ 246 R= C$_2$H$_5$ 248 R= H

245 R= C$_3$H$_7$ 247 R= CH$_2$CH$_2$OH

Scheme 68a

Table 9. *Relative* In Vitro *Activities of C-9 Analogs of Artemisinin*

Compound	W-2 IC$_{50}$ 1/IC$_{50}$ Compound	D-6 IC$_{50}$ 1/IC$_{50}$ Compound
249a	0.082	0.16
249b	0.82	0.18
249c	1.2	0.49
249d	0.78	1.0
249e	4.2	5.0
249f	0.12	0.27
249g	1.0	0.64
249h	0.18	0.21
249i	1.4	3.2
249j	1.4	3.2

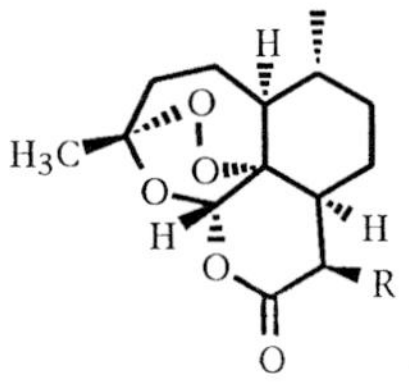

a R=C$_2$H$_5$ f R=C$_5$H$_{11}$ (n)
b R=C$_3$H$_7$ (n)   g R=C$_5$H$_{11}$ (i)
c R=C$_3$H$_7$ (i)   h R=C$_6$H$_{13}$ (n)
d R=C$_4$H$_8$ (n)   i R=C$_6$H$_{13}$ (i)
e R=C$_4$H$_8$ (i)   j R=CH$_2$CH=CH$_2$

249

Scheme 68b

References, pp. 202–214

Table 10. *Relative* In Vitro *Activities of C-3 and/or C-9 Substituted Artemisinin Derivatives*

Compound	^1R	R	D-6 IC_{50} $1/IC_{50}$ Compound	W-2 IC_{50} $1/IC_{50}$ Compound
250a	H	CH_3	1.00	1.00
250b	CH_3	H	0.88	1.12
250c	CH_2CH_3	H	21.0	6.73
250d	$(CH_2)_2CH_3$	H	0.20	0.18
250e	$CH(CH_3)_2$	H	0.53	0.45
250f	CH_2EtO_2C	H	2.32	2.32
250g	$CH_2C_6H_5$	H	0.03	0.01
250h	$(CH_2)_2p\text{-}ClC_6H_4$	H	1.14	1.27
250i	$(CH_2)_3C_6H_5$	H	2.20	2.81
250j	CH_3	$(CH_2)_3CH_3$	1.84	2.57
250k	$(CH_2)_2CH_3$	$(CH_2)_3CH_3$	0.28	0.33
250l	C_6H_5	$(CH_2)_3CH_3$	0.01	0.01
250m	$(CH_2)_2p\text{-}ClC_6H_4$	$CH_2)_3CH_3$	0.43	0.53
250n	$(CH_2)_3C_6H_5$	$(CH_2)_3CH_3$	0.39	0.48
250o	CH_2CO_2Et	$(CH_2)_3CH_3$	13.8	22.8

Relative activity = IC_{50} artemisinin/IC_{50} analog

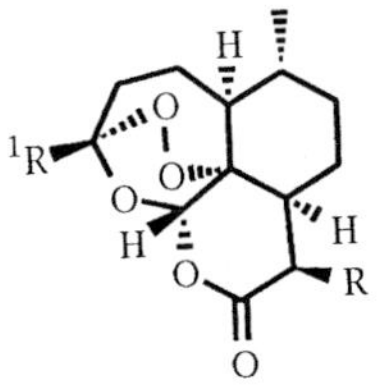

250

Scheme 69

The propyl derivative **245** proved to be approximately five times as active as artemisinin *in vivo* but produced neurological problems similar to those of arteether.

Table 11. *Relative* In Vitro *Activities of C-3 and C-9 Substituted 10-Deoxoartemisinins*

Compound	^1R	R	D-6 IC_{50} $1/IC_{50}$ Compound	W-2 IC_{50} $1/IC_{50}$ Compound
251a	CH_3	CH_3	6.59	5.67
251b	CH_3	H	2.37	1.90
251c	CH_3	C_2H_5	9.14	4.66
251d	CH_3	$(CH_2)_2CH_3$	4.73	5.50
251e	CH_3	$(CH_2)_3CH_3$	58.3	20.9
251f	CH_3	$(CH_2)_4$	1.70	1.45
251g	CH_3	$(CH_2)_3C_6H_5$	50.7	25.1
251h	CH_3	$(CH_2)_3p\text{-}ClC_6H_4$	69.9	33.2
251i	C_2H_5	H	0.10	0.10
251j	$CH_3(CH_2)_2$	H	7.22	6.85
251k	$CH_3(CH_2)_3$	H	6.53	5.56
251l	$(CH_3)_2CHCH_2$	H	1.83	2.50
251m	$C_6H_5(CH_2)_4$	H	3.36	3.80
251n	$C_6H_5(CH_2)_2$	H	0.06	0.02
251o	$p\text{-}ClC_6H_4(CH_2)_3$	H	0.13	0.28
251p	$(CH_2)_2CO_2Et$	H	4.22	5.06
251q	$(CH_2)_2CO_2H$	H	0.09	0.09

Relative activity = IC_{50} artemisinin/IC_{50} analog

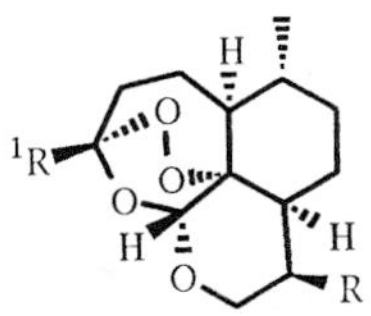

251

Scheme 70

12.2. In vivo

Chinese investigators first tested their compounds in mice infected with chloroquine-resistant *P. berghei* and reported the activities in terms of the dose necessary to suppress the infection in 90% of the infected animals, *i.e.* SD_{90}. Table 12 below summarized some of their data as reported by Luo and Shen (*4*).

Table 12. In Vivo *Activities of Dihydroartemisinin Ethers against Chloroquine-Resistant P. berghei*

Compound No.	SD$_{90}$ (mg/kg)	Compound	SD$_{90}$ (mg/kg)
44b	1.02	44m	1.39
45b	1.16	44n	1.39
44c	1.95	44o	>41
44e	1.70	44p	inactive
44f	2.24	44q	4.10
44g	1.5 mg/kg inactive	44r	2.28
44h	5.60	44s	3.42
44i	5.41	44t	>2.5
44j	>20	44w	10
44k	4.30	44x	18.7
44l	>24	44y	2.24
		44z	1.74
		44za	8.91
		44zb	10.0

44e, R = n-C$_3$H$_7$
44f, R = i-C$_3$H$_7$
44h, R = i-C$_5$H$_{11}$
44i, R = C(CH$_3$)$_2$C$_2$H$_5$
44j, R = (CH)$_7$CH$_3$
44k, R = cyclopentyl
44l, R = cyclohexyl
44n, R = CH$_2$CHCH$_2$
44o, R = CH$_2$CH$_2$OH
44q, R = CH$_2$CH$_2$OCH$_3$
44s, R = CH$_2$C$_6$H$_5$
44w, R = CH$_2$C$_6$H$_4$-p-OCH$_3$
44x, R = CH$_2$CF$_3$
44y, R = CH$_2$CH$_2$Cl
44z, R = CH$_2$CH$_2$Br
44za, R = CH$_2$C$_6$H$_4$-m-F

44g, R = n-C$_4$H$_9$
44m, R = cyclohexyl
44p, R = CH$_2$CH$_2$OH
44r, R = CH$_2$CH$_2$OCH$_3$
44t, R = CH$_2$C$_6$H$_5$
44zb, R = CH$_2$C$_6$H$_4$-m-F

Scheme 71

Table 13. In Vivo *Activities of Dihydroartemisinin Esters on Chloroquinine-Resistant P. berghei*

Compound	SD_{90} (mg/kg)	Compound No.	SD_{90} (mg/kg)
44aa	1.20	44uu	1.05
44bb	0.66	44vv	4.1
44cc	0.65	44ww	4.4
44dd	~0.50	44xx	16.7
44ee	0.51	44yy	1.37
44ff	0.48	44zz	0.65
44gg	0.67	44aaa	0.69
44hh	>20	44bbb	0.88
44ii	0.46	44ccc	0.63
44jj	0.51	44ddd	0.80
44kk	2.1	44eee	2.5
44ll	0.69	44fff	0.57
44mm	0.48	44ggg	1.17
44nn	2.43	44hhh	1.12
44oo	1.73	44iii	1.20
44pp	0.95	44jjj	4.14
44qq	0.74	44kkk	3.10
44rr	0.67	44lll	1.50
44ss	0.54	44mmm	2.97
44tt	0.65	44nnn	0.62

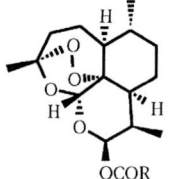

44   R= **aa** through **nnn**

Scheme 72

References, pp. 202–214

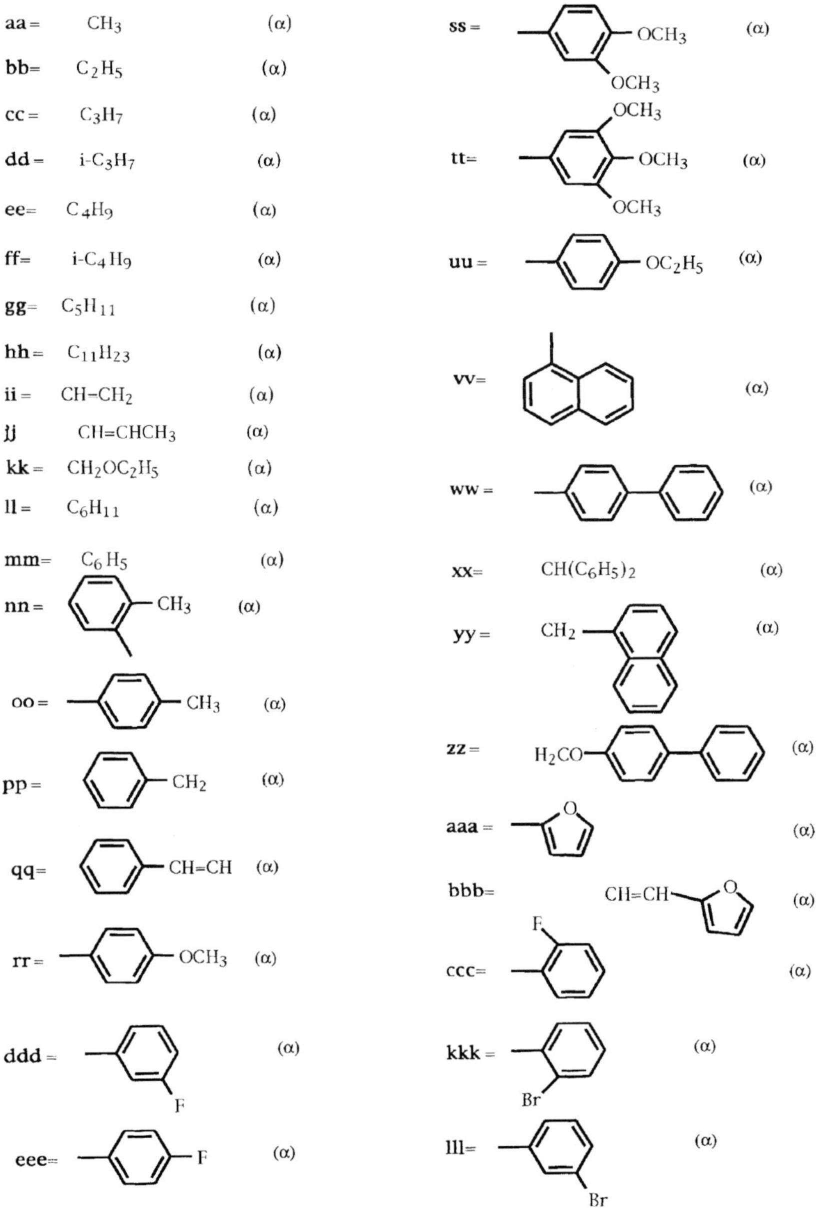

aa =	CH$_3$	(α)
bb =	C$_2$H$_5$	(α)
cc =	C$_3$H$_7$	(α)
dd =	i-C$_3$H$_7$	(α)
ee =	C$_4$H$_9$	(α)
ff =	i-C$_4$H$_9$	(α)
gg =	C$_5$H$_{11}$	(α)
hh =	C$_{11}$H$_{23}$	(α)
ii =	CH=CH$_2$	(α)
jj =	CH=CHCH$_3$	(α)
kk =	CH$_2$OC$_2$H$_5$	(α)
ll =	C$_6$H$_{11}$	(α)
mm =	C$_6$H$_5$	(α)
nn =	(o-tolyl) —CH$_3$	(α)
oo =	(p-C$_6$H$_4$)—CH$_3$	(α)
pp =	(C$_6$H$_5$)—CH$_2$	(α)
qq =	(C$_6$H$_5$)—CH=CH	(α)
rr =	(p-C$_6$H$_4$)—OCH$_3$	(α)
ddd =	(m-F-C$_6$H$_4$)	(α)
eee =	(p-F-C$_6$H$_4$)	(α)

ss =	3,4-dimethoxyphenyl	(α)
tt =	3,4,5-trimethoxyphenyl	(α)
uu =	(p-C$_6$H$_4$)—OC$_2$H$_5$	(α)
vv =	1-naphthyl	(α)
ww =	4-biphenyl	(α)
xx =	CH(C$_6$H$_5$)$_2$	(α)
yy =	CH$_2$—naphthyl	(α)
zz =	H$_2$CO—biphenyl	(α)
aaa =	2-furyl	(α)
bbb =	CH=CH—furyl	(α)
ccc =	(o-F-C$_6$H$_4$)	(α)
kkk =	(o-Br-C$_6$H$_4$)	(α)
lll =	(m-Br-C$_6$H$_4$)	(α)

Scheme 72 (continued)

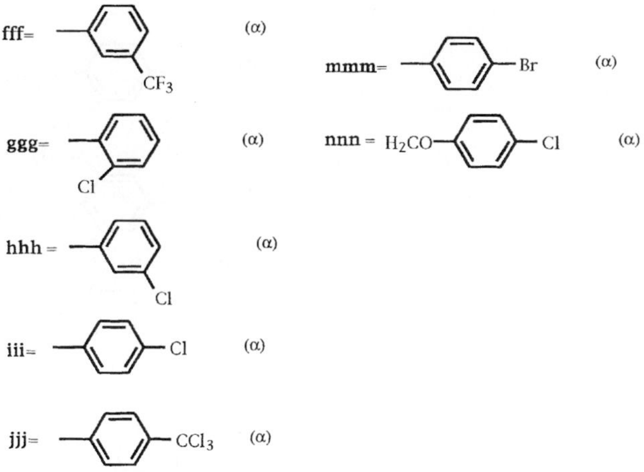

fff= (3-CF₃-C₆H₄-) (α)

mmm= (4-Br-C₆H₄-) (α)

ggg= (2-Cl-C₆H₄-) (α)

nnn = H₂CO-(4-Cl-C₆H₄-) (α)

hhh = (3-Cl-C₆H₄-) (α)

iii= (4-Cl-C₆H₄-) (α)

jjj= (4-CCl₃-C₆H₄-) (α)

Scheme 72 (continued)

Table 14. *Summary of Data on the Recrudescence, i.e. the Reappearance of Malaria, in Mice Infected with a Chloroquine-Resistant Mutant of* P. berghei

Compound	No recrudescence in 28 days (mg/kg)
44a	50
44ooo	>2.5
44ppp	2.5
44qqq	2.5

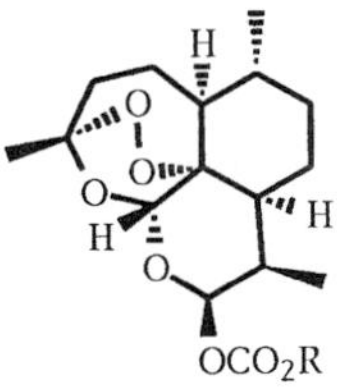

44ooo R= $(CH_2)_4CH_3$
44ppp R= $(CH_2)_5CH_3$
44qqq R= $C(CH_3)_3$

Scheme 73

13. Toxicity

Toxicity studies show artemisinin to be a well-tolerated drug (126). The LD_{50} (lethal dose for 50% of the animals) for oral administration of a water suspension in mice is 4228 mg/kg with a therapeutic index of 384. Administered as an oil suspension intramuscularly the LD_{50} is 3840 mg/kg with a therapeutic index of 4987. Acute intoxication produces restlessness, tremor, slow respiration and disappearance of the righting reflex. Smaller animals showed no apparent nervous system distress but larger animals exhibited clonic and tonic convulsions. A cession of respiration preceeded cardiac arrest. Surviving animals gradually returned to normal within 10–24 hours.

ZHAO (175) described the systemic effects of artesunate in mice, guinea pigs, rabbits, dogs and monkeys. At low dosages, between 240 and 490 mg/kg, there were no noticeable effects. Between 700 and 1000 mg/kg there was decreased activity, passive body position, tremors and convulsions. Respiration was sufficiently depressed for death to occur. High doses in guinea pigs, rabbits, dogs and monkeys result in CNS depression, convulsions and respiratory suppression. The amounts employed to produce adverse effects in the animal studies were greater than 100 times the therapeutic dose.

Artemisinin is not mutagenic in mammals either by AMES' tests (176) or in teratogenic studies of fetal rats. However artemisinin given during mid and late gestation was toxic; a dose of 1/400 of LD_{50} caused half of the fetuses to be absorbed. Artemether proved to be highly toxic to mouse and rat embryos. Dogs suffer nausea and exhaustion, with slowed cardiac rates and decreased blood pressure, but the compound is less toxic than the commonly used antimalarial drug chloroquine. Histological examination of dogs given high doses of artemisinin showed there was a slight fatty degeneration in liver cells.

At high doses arteether and artemether are neurotoxins. Dogs dosed daily for 8 days at 20 mg/kg (ten times the therapeutic dose) exhibit gait disturbances, loss of spinal reflexes and pain response reflexes as well as prominent loss of brain-stem and eye reflexes (177). Repeated high doses of arteether or artemether in dogs produce lesions of scattered neuronal degeneration and necrosis (178) characterized by swelling and rounding of nerve cell bodies, increased eosinophilia and vacuolization of the cytoplasm with a loss of Nissl substance. Only minimal changes were seen in the anterior sections, even in animals receiving the highest dose. The toxic effects of artemether and arteether were similar. The neurotoxicity of artemisinin analogs is specific for neurons but does not affect glial cells (179).

Rats suffer similar effects. MESHNICK et al. found that treating rats with high doses of arteether produces neuronal necrosis in the neuropile in specific areas (vestibular and red nuclei) of the brain (180). However, a comparison of mouse neuroblastoma cells and red cells infected with *P. falciparum* showed that the former took up much less [^{3}H]-dihydroartemisinin. This selective uptake may explain why artemisinin derivatives are selectively toxic to malaria parasites. Autoradiograms of SDS gels run from [^{3}H]-dihydroartemisinin treated neuroblastoma cell line (Neu2a) showed that neuronal proteins of molecular weight 27, 32, 40 and 81 kDa became covalently bound to dihydroartemisinin. Their results suggest that these reactions only occur at high doses or after prolonged exposures.

Despite concerns that artemether and artemisinin are neurotoxins, no neurotoxicological effects were reported in clinical studies (181, 182).

14. Pharmacology and Pharmacokinetics

For optimum treatment of patients with artemisinin and derivatives, knowledge of the effects of different modes of administration on the drug's half life, amount and frequency of administration, metabolism, *etc.*, is required. A number of different animals have been employed in early studies of the pharmacokinetics of artemisinin derivatives. The analytical techniques at the time were crude. ZHAO et al. (183) injected artemether dissolved in peanut oil intramuscularly into dogs at 10 mg or 30 mg/kg, showing that the drug was easily absorbed, with peak concentrations reached after 4 hours at the lower dose and 1.9 hrs at the higher dose; the half-lives for elimination were 4.0 hr and 6.5 hr respectively. Absorption of artemisinin at the 10 mg level was more rapid than for artemether reaching a peak after 2 hr with a half-life for elimination of 1.6 hr. Artesunate (6 mg/kg), injected as a solution in sodium bicarbonate and saline, fit a one compartment model with an elimination half-life of 0.45 hr.

Dihydroartemisinin given orally to rabbits at doses of 10, 20 and 30 mg/kg produced peak concentrations in serum in 1 to 2 hr with half-lives of 1.19, 1.00 and 1.10 hr (184). When 20 mg/kg was administered to dogs, the peak concentration in serum was reached in 2 hr with a half-life of 2.1 hr.

Artesunate was administered as a solution in sodium bicarbonate and saline through the tail vein to two groups of rats; the distribution and excretion of artesunate in rats were determined by ZHAO and SONG (185). One group of rats was sacrificed after 10 min; the highest level of the

References, pp. 202–214

drug was found in the intestine and lesser amounts in the brain. A second group was sacrificed after 60 minutes; the amounts of the drug in all tissues has decreased significantly. Since the total excretion accounted for less than one percent of that administered, the authors concluded that metabolism was the principal mode of elimination.

After intramuscular injection of arteether in dogs (beagles) BENAKIS et al. (*186*) determined that the half-life for distribution is 0.8 hr and that for elimination is 28 hr. The longer times required for elimination are an advantage in treating malaria.

TITULAER et al. (*187*) obtained pharmacokinetic data for the oral, intramuscular and rectal aministration of artemisinin to volunteers. Rapid but incomplete absorption of artemisinin given orally occurs in humans with a mean absorption time of 0.78 hr and the bioavailability relative to i.m. administration is 32%. The mean residence time for artemisinin on i. m. administration is three times that when given orally. The drug was also administered as an aqueous suspension rectally with poor results. LIN et al. reported that use of artelinic acid and other dihydroartemisinin derivatives in sustained-release transdermal medication in mice in some cases required only 10% of that given by other routes of administration (*188, 189*). The minimum curative doses is twice that for prophylactic activity. The compounds are rapidly (within 5 minutes) absorbed through the skin. The authors suggest that a sustained-release formulation or patch may provide a facile means of delivery which maintains a steady drug concentration in the blood. This delivery system may enhance the antimalarial activity of the drug and reduce the required dose.

Vietnamese scientists (*190*) also examined the pharmacokinetics of artemisinin administered orally to healthy subjects. Peak concentrations in plasma, $391 \pm 147 \,\mu g/l$, were obtained after approximately 2 hr. Elimination was rapid with a half-life of 2.6 ± 0.55 hr. Although the bioavailability of oral administration is poor they concluded that a 500 mg dose twice a day was appropriate.

15. Clinical Evaluation of Artemisinin and Derivatives

As early as 1979 the QINGHAOSU ANTIMALARIA COORDINATING RESEARCH group reported treating some 2099 cases of malaria with artemisinin (*191*) and producing clinical cures in all patients. Of the 143 patients infected with chloroquine-resistant strains of *P. falciparum*, 141 exhibited a good response. In Thailand, where multi-drug resistance is a serious problem, more than a thousand patients at the University

Hospital in Bangkok have been treated with artemisinin derivatives, with a more rapid improvement than experienced with other antimalarial drugs (*192* a, b, c, d; *193, 194*). When artemisinin derivatives were employed alone or administered for less than 5 days, the rapid clearance of fever and parasitaemia was frequently followed by relapse. Combination of high doses of mefloquine at the end of a full course of artesunate or artemether produced good cure rates (*195*).

15.1. Dihydroartemisinin

Dihydroartemisinin is a significantly more active antimalarial than artemisinin, 200 times more active against *P. falciparum in vitro* (*123*), but less tolerated when administered orally. It is evidently the main metabolite of derivatives in the body. Like artemisinin, dihydroartemisinin is concentrated 300 fold in *Plasmodium*-parasitized red blood cells (*88*).

16. Mechanisms of Action

The *Plasmodium* sporozite develops in the red blood cell by digesting hemoglobin, but the heme liberated is toxic to the parasite. It is rendered inactive by an enzyme catalyzed polymerization to form an insoluble compound, hemozoin (*196*). Quinine and quinoline derivatives may function by inhibiting this polymerization. Resistant strains of *P. falciparum* have developed ways of eliminating the drugs before an effective concentration builds up.

Several lines of evidence indicate that artemisinin and its derivatives function in a manner different from quinine. Electron microscopy of parasites treated with sodium artesunate shows changes in the mitochondria, whereas quinine causes alterations in the food vacuoles (*197, 191, 198*). The earliest and most distinctive ultrastructural changes following administration of artemether to infected *Aotus trivirgatus* (monkeys) were marked swelling of the mitochondria in the parasites (*199*). Subsequently, there are changes in electron-dense regions of chromatin in the nuclei.

When malaria-infected erythrocytes are treated with ^{14}C-artemisinin, the radioactivity is localized in the hemin/hemozoin fraction obtained from lysed cells (*200*). Indeed, hemin appears to catalyze the decomposition of artemisinin (*201*): [^{14}C]-Artemisinin has been shown to react with human albumin whereupon 20% of the drug becomes

References, pp. 202–214

covalently bound to albumin (*202*). To separate specific and non-specific reactions MESHNICK *et al.* (*203*) treated infected and uninfected cells with several [^{3}H]-labelled artemisinin derivatives and searched for radiochemically labelled proteins. In all the infected cells the same six proteins were labelled, whereas none of the proteins in uninfected cells had reacted. Thus, artemisinin appears to react with specific proteins from *P. falciparum*. Their function remains unknown.

Chemical studies have focused on the products formed from one and two electron reductions of artemisinin and analogs. Somewhat surprisingly, t-butylperoxide itself shows substantial effect upon the parasite. However, in the presence of iron chelators, both it and artemisinin are ineffective (*204, 205*), suggesting that iron catalyzed reduction of the peroxide or hydroperoxide forms oxygen radicals which are responsible for the drugs action. Consistent with this hypothesis, treatment of infected red blood cells with sodium artesunate results in complete inhibition of cytochrome oxidase activity (*197*).

POSNER *et al.* (*206, 207, 208, 209*) have carried out a series of studies in which they conclude that carbon-centered radicals are responsible for the antimalarial activity of artemisinin and other 1,2,4-trioxanes. The 1,2,4-trioxane **252** bearing a 4β-methyl group is approximately twice as active as artemisinin, whereas both the 4α-isomer and the 4,4-dimethyl derivative are less than one hundredth as active as artemisinin. The results suggest the formation of a carbon-centered radical on C-4 which could lead to a C-4-C-5 oxirane, and to 4α-hydroxydeoxyartemisinin. The putative oxirane would logically provide a mechanism by which artemisinin derivatives could react with proteins.

The chemical and biological studies allow a possible description of the action of artemisinin and its derivatives upon the parasite. Transferring an electron from Fe(II) or an enzyme involved in electron transport, e.g. cytochrome oxidase, to artemisinin or a derivative, may form a radical anion, which reacts with important proteins to block a vital biochemical reaction. These reactions are probably facilitated by the ability of infected erythrocytes to concentrate artemisinin derivatives more than 300-fold (*88*).

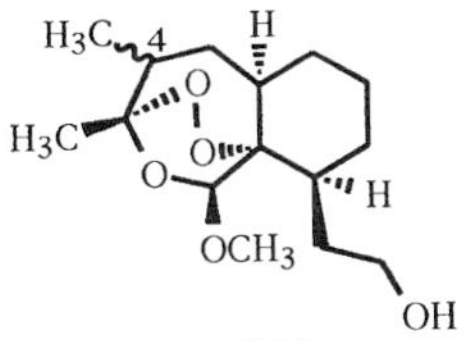

252

Scheme 74

17. Other Peroxides

17.1. Naturally Occurring Peroxides

The plant *Artabotrys unciatus* L. Meer was also known in Chinese folk medicine for treating fevers and malaria. LIANG and coworkers (*210*) isolated yingzhaosu A and B from the root of the plant, assigning their structures as **253** and **254** (Scheme 75) without determining their stereochemistry. When ZHOU and coworkers (*211*) attempted to isolate these compounds from the plant, they obtained instead another sesquiterpene peroxide, yingzhaosu C, **255**, and the sesquiterpenol yingzhaosu D, **256**. Structures were determined by analyses of their mass, infrared and ^{1}H nmr spectra, as well as by conversion of **256** to a known compound, **257**. The absolute stereochemistries of the compounds were assigned using HOREAU'S method.

XU *et al.* have reported total syntheses of Yingzhaosu A (*212*) as well as those of Yingzhaosu B (*213*), C (*214*) and D (*215*).

THEBTARANONTH *et al.* (*216*) isolated a hexane soluble diterpene, **258**, from the fruit of *Amomum krervanth* Pierre (cardamon) which is reasonably potent (EC$_{50}$ 0.8 µg/ml) against malaria. The structure and relative stereochemistry were determined by a single crystal X-ray diffraction analysis as shown in Scheme 76.

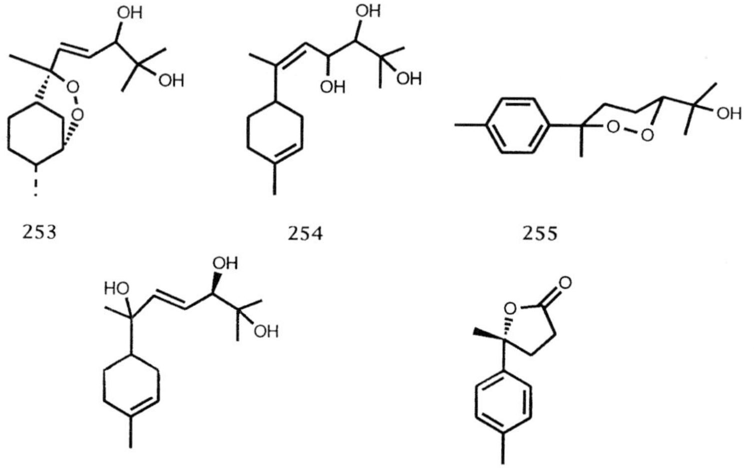

Scheme 75

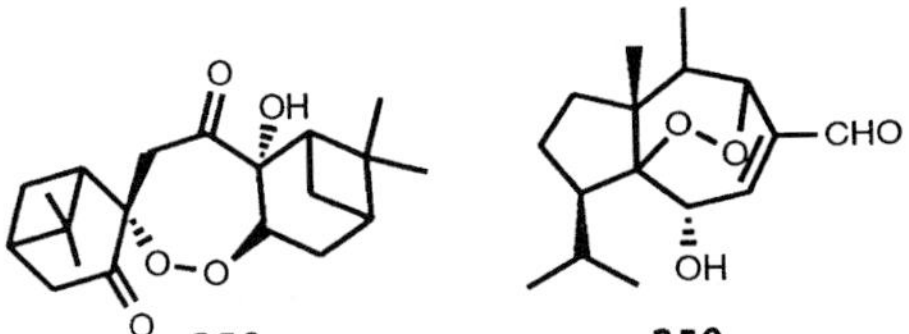

Scheme 76

HASHIDOKO et al. (217) isolated the antimicrobial sesquiterpene **259** from *Rosa rugosa* Thunb. which exhibited schizonticidal activity against *P. falciparum*.

17.2. Synthetic Peroxides

A series of synthetic peroxides with structural characteristics mimicking the yingzhaosus (218) shows activitiy against *Plasmodium spp*. The most active of these, named arteflene, **260**, synthesized as shown in Scheme 77, supported clinical studies in Nigeria and in Cameroon.

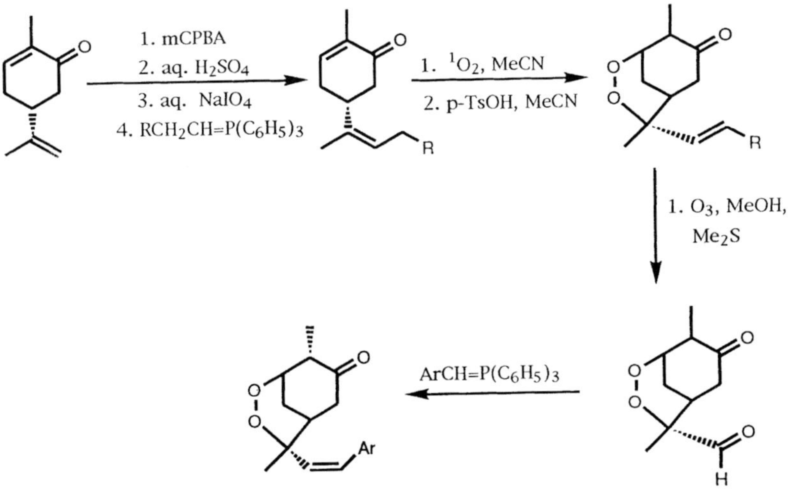

Scheme 77

Surprisingly, arteflene and its enantiomer exhibited virtually identical activities. Indeed, throughout the series of analogs, the activity was remarkably insensitive to changes in absolute and relative stereochemistry. The more lipophilic derivatives were more active than those with polar groups. Because these compounds resist degradation, they show more prolonged activity against drug resistant strains of *P. falciparum* and a lower incidence of recrudescence.

A series of bicyclo[3.2.2]nonane endoperoxides, prepared by literature methods, exhibited antimalarial activity, the most active compound having one sixth of the activity of artemisinin (*219*).

The observation (Section 10) (*220, 221*) that *t*-butyl hydroperoxide was effective against *P. vinckei* (but not *P. berghei* (*204*)) prompted a search for inexpensive synthetic antimalarials embodying this function. Several dispiro-1,2,4,4-tetraoxanes (Scheme 78), readily available by

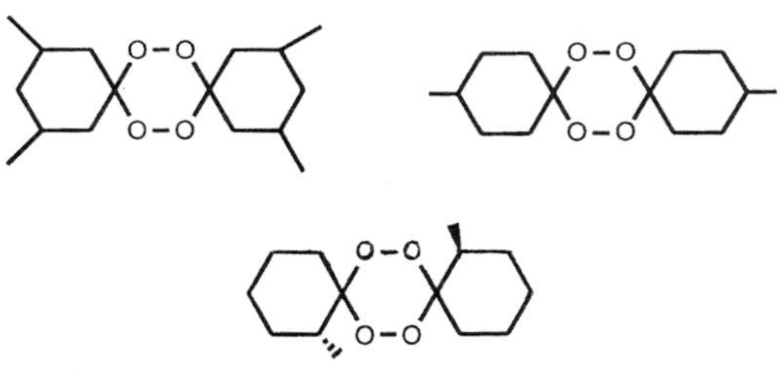

Scheme 78

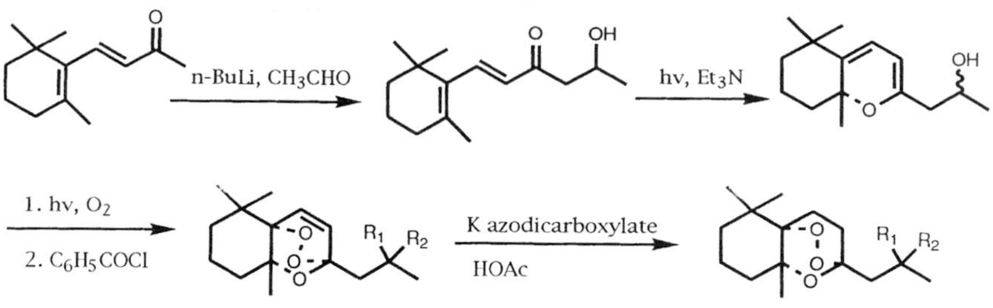

289

(a) R_1=OH; R_2=H
(b) R_1=H; R_2=OCOC$_6$H$_5$
(c) R_1=OCOC$_6$H$_5$; R_2=H

Scheme 79

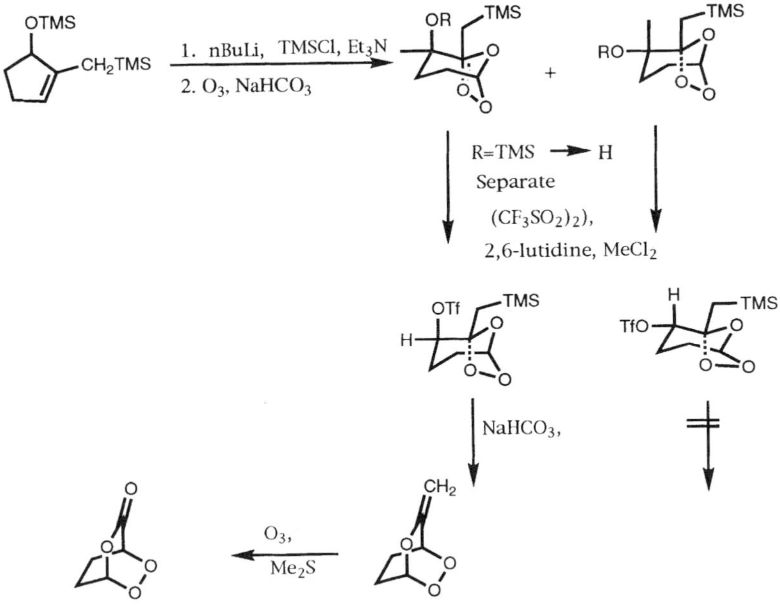

Scheme 80

acid-catalyzed peroxyketalization of substituted cyclohexanones with 30% hydrogen peroxide, proved to be 2 to 30 times less active than artemisinin (*222*) but they cured *P. falciparum* infections.

The series of synthetic tricyclic 1,2,4-trioxanes in Scheme 79 (*223*) possess IC_{50}s 25 to 100 times greater than artemisinin, i.e. the compounds were only 1/25 to 1/100th as active as artemisinin.

The rearrangement of an ozonide to a 1,2,4-trioxane (Scheme 80) yielded an inactive 1,2,4-trioxan-5-one (*224*).

18. Conclusion

The discovery that artemisinin was effective against drug-resistant strains of *P. falciparum* provided a critical lead compound in the search for drugs to treat patients with resistant strains of malaria. Since the initial reports significant progress has been made in increasing the availability of artemisinin in Vietnam, China and elsewhere in Southeast Asia. Of the host of derivatives prepared, a few are more active, with improved solubility properties. Progress in understanding artemisinin's mechanism of action is leading to the synthesis of simplified analogs in the quest for more effective and economical drugs. There is still a great need for an active long acting derivative that can be given orally.

References

1. KLAYMAN, D.L.: Qinghaosu (Artemisinin): An Antimalarial Drug from China. Science, **228**, 1049 (1985).
2. COORDINATING GROUP FOR RESEARCH ON THE STRUCTURE OF QING HAU SAU: K'o Hsueh T'ung Pao, **22**, 142 (1977); Chem. Abstr., **87**, 98788g (1977).
3. LIU, J.-M., M.Y. NI, J.-F. FAN, Y.-Y. TU, Z.H. WU, Y.-L WU, and W.-S. CHOU: Structure and Reactions of Arteannuin. Acta Chim. Sinica, **37**, 129 (1979).
4. LUO, X.-D., and C.C. SHEN: The Chemistry, Pharmacology, and Clinical Applications of Qinghaosu (Artemisinin) and its Derivatives. Med. Res. Rev., **7**, 29 (1987).
5. ZHANG, M., Y. TU, M. NI, Y. ZHONG, L. LI, S. CUI, X. WANG, and Z. JI: Studies on the Constituents of *Artemisia annua*. Planta Medica, 143 (1982).
6. XIMEN, L.: A New Drug for Malaria. China Reconstructs, **191**, 48 (1979).
7. QINGHAOSU ANTIMALARIA COORDINATING RESEARCH GROUP: Crystal Structure and Absolute Configuration of Qinghaosu. Scientia Sinica, **23**, 380 (1980); Chem. Abstr., **93**, 71991e (1980).
8. JEREMIC, D., A. JOKIC, A. BEHBUD, and M. STEFANOVIC: A New Type of Sesquiterpene-Lactone Isolated from *Artemisia annua* L. Ozonide of Dihydroarteannuin. 8th Internat. Symp. Chem. Nat. Prod., New Delhi, **2**, 222, Abstr. C-57 (1972).
9. WOERDENBAG, H.J., C.B. LUGT, and N. PRAS: *Artemisia annua* L.: A Source of Novel Antimalarial Drugs. Pharm. Weekbl. [Sci.], **12**, 169 (1990).
10. ZAMAN, S.S., and R.P. SHARMA: Some Aspects of the Chemistry and Biological Activity of Artemisinin and Related Antimalarials. Heterocycles, **32**, 1593 (1991).
11. BUTLER, A.R.: Artemisinin (Qinghaosu): A New Type of Antimalarial Drug. Chem. Soc. Rev., **21**, 85 (1992).
12. JUNG, M.: Current Developments in the Chemistry of Artemisinin and Related Compounds. Current Med. Chem., **1**, 35 (1994).
13. MESHNICK, S.R., T.E. TAYLOR, and S. KAMCHONWONGPAISAN: Artemisinin and the Antimalarial Endoperoxides: from Herbal Remedy to Targeted Chemotherapy. Microbiological Rev., 301 (1996).
14. FERNALD, M.L.: Gray's Manual of Botany. New York: American Book Company. 1950, p. 1522.
15. CHEN, T.H., H. LIN, K.C. KAO, and C.K. FAN: Cultivation of Hwa Hwo Gao *(Artemisia annua)*. Zhongcaoyao Tonyxien, **11**, 227 (1980).
16. SHUKLA, A., A.H.A. FAROOQI, Y.N. SHUKLA, and S. SHARMA: Effect of Tricontanol and Chlormequat on Growth, Plant Hormones and Artemisinin Yield in *A. annua*. Plant Growth Regul., **11**, 165 (1992); Chem. Abstr., **116**, 250407j (1992).
17. SINGH, A., V.K. KAUL, V.P. MAHA, A. SINYH, L.N. MISRA, R.S. THAKUR, and A. HUSAIN: Introduction of *Artemisia annua* in India and Isolation of Artemisinin. A Promising Antimalarial Drug. Indian J. Pharm. Sci., 137 (1986).
18. CUBUKCU, B., A.H. MERICLI, N. OZHATAY, and B. DAMADYAN: Artemisinin from Turkish *Artemisia annua*. Acta Pharm. Turc., **31**, 41 (1989); Chem. Abstr., **111**, 12390a (1989).

19. LAUGHLIN, J.C.: Agricultural Production of Artemisinin: A Review. Trans. Royal Soc. Trop. Med. and Hyg., **88**, S. 21 (1994).
20. CHEN, H.R., M. CHEN, F. ZHONG, and F. CHEN: Some Factors Affecting the Content of Artemisinin. Chinese Medicinal Reports, **7**, 393 (1986).
21. CHAN, K.L., C.K.H. TEO, S. JINADASA, and K.H. YUEN: Selection of High Artemisinin Yielding *Artemisia annua*. Lhonyyao Planta Medica, 285 (1995).
22. ACTON, N., D.L. KLAYMAN, and I. ROLLMAN: Reductive Electrochemical HPLC Assay for Artemisinin (Qinghaosu). Planta Medica, 445 (1985).
23. CHARLES, D.J., J.E. SIMON, K.V. WOOD, and P. HEINSTEIN: Mixtures of Antimalarial Agents and Artemisinin Derivatives for the Treatment of Malaria. Europ. Pat. Appl. EP290,959; Chem. Abstr., **111**, 50417x (1989).
24. LIERSCH, R., H. SOICKE, C. STEHR, and H.U. TULLNER: Formation of Artemisinin in *Artemisia annua* during One Vegetation Period. Planta Medica, 387 (1986).
25. FERREIRA, J.F.S., J.E. SIMON, and J. JANICK: Developmental Studies of *Artemisia annua*: Flowering and Artemisinin Production Under Greenhouse and Field Conditions. Planta Medica, 67 (1995).
26. NAIR, M.S.R., N. ACTON, D.L. KLAYMAN, K. KENDRICK, and D.V. BASILE: Production of Artemisinin in Tissue Cultures of *Artemisia annua*. J. Nat. Prod., **49**, 504 (1986).
27. SINGH, A., R.A. VISHWAKARMA, and A. HUSAIN: Evaluation of *Artemisia annua* Strains for Higher Artemisinin Production. Planta Medica, 475 (1988).
28. KUDAKASSERIL, G.J., L. LAM, and E.J. STABA: Effect of Sterol Inhibitors on the Incorporation of ^{14}C-Isopentenyl Pyrophosphate into Artemisinin by a Cell-Free System from *A. annua* Tissue Cultures and Plants. Planta Medica, 280 (1987).
29. FULZELE, D.P., A.T. SIPAHIMALANI, and M.R. SIPAHIMALANI: Tissue Cultures of *Artemisia annua*: Organogenesis and Artemisinin Production. Phytotherapy Res., **5**, 149 (1991).
30. WEATHERS, P.J., R.D. CHEETHAM, E. FOLLANSBEE, and K. TEOH: Artemisinin Production by Transformed Roots of *Artemisia annua*. Biotech. Lett., **16**, 1281 (1994).
31. KLAYMAN, D.L., A.J. LIN, N. ACTON, J.P. SCOVILL, J.M. HOCH, W.K. MILHOUS, A.D. THEOHARIDES, and A.S. DOBEK: Isolation of Artemisinin (Qinghaosu) from *Artemisia annua* Growing in the United States. J. Nat. Prod., **47**, 715 (1984).
32. ACTON, N., D.L. KLAYMAN, I.J. ROLLMAN, and J.F. NOVOTNY: Isolation of Artemisinin (Qinghaosu) and its Separation from Artemisitene Using the ITO Multilayer Coil Separator-Extractor; Isolation of Arteannuin B.J. Chromo., **355**, 448 (1986).
33. ELSOHLY, H.N., E.M. CROOM, F.S. ELFERALY, and M.M. ELSHEREI: A Large Scale Extraction Technique of Artemisinin from *Artemisia annua*. J. Nat. Prod., **53**, 1560 (1990).
34. ACTON, N., and D.L. KLAYMAN: Artemisitene, A New Sesquiterpene Lactone Peroxide from *Artemisia annua*. Planta Medica, 441 (1985).
35. MILHOUS, W.: Personal communication. IC_{50} = 5.56 ng/ml vs. D-6 and 7.57 ng/ml vs W-2. The IC_{50} value for artemisinin is 2.5 ng/ml for these strains of *P. falciparum*.
36. WEI, Z.X., J.U.P. PAN, and Y. LI: Artemisinin G: A Sesquiterpene from *Artemisia annua*. Planta Medica, 300 (1992).
37. BLASKO, G., G.A. CORDELL, and D.C. LANKIN: Definitive ^{1}H and ^{13}C-NMR Assignments of Artemisinin (Qinghaosu). J. Nat. Prod., **51**, 1273 (1988).
38. SCHULTE, K.H., and G. OHLOFF: Über eine außergewöhnliche Stereospezifität bei der Hydroborierung der diastereomeren (1R)-Isopulegole mit Diboran. Helv. Chim. Acta, **50**, 153 (1967).
39. BÜCHI, G., and H. WÜEST: Ozonolysis of Vinylsilanes. J. Am. Chem. Soc., **99**, 294 (1977).

40. SCHMID, G., and W. HOFHEINZ: Total Synthesis of Qinhaosu. J. Am. Chem. Soc., **105**, 624 (1983).
41. XU, X.X., J. ZHU, D.Z. HUANG, and W.S. ZHOU: Total Synthesis of Arteannuin and Deoxyarteannuin. Tetrahedron, **42**, 819 (1986).
42. ZHOU, W.S.: Total Synthesis of Arteannuin (Quinghaosu) and Related Compounds. Pure Appl. Chem., **58**, 817 (1986).
43. AVERY, M.A., W.K.M. CHONG, and C. JENNINGS-WHITE: Stereoselective Total Synthesis of (+)-Artemisinin, the Antimalarial Constituent of *Artemisia annua* L. J. Am. Chem. Soc., **114**, 974 (1992).
44. RAVINDRANATHAN, T., M.A. KUMAR, R.B. MENON, and S.V. HIREMATH: Stereoselective Synthesis of Artemisinin. Tetrahedron Lett., **31**, 755 (1990).
45. XU, X.X., J. ZHU, D.Z. HUANG, and W.S. ZHOU: The Stereocontrolled Syntheses of Arteannuin and Deoxyarteannuin from Arteannuic Acid. Acta Chim. Sinica, **41**, 574. (1983).
46. ROTH, R.J., and N. ACTON: A Simple Conversion of Artemisinic Acid into Artemisinin. J. Nat. Prod., **52**, 1183 (1989). USP 4, 992, 561 (Feb. 12, 1991).
47. ROTH, R.J., and N. ACTON: A Facile Semisynthesis of the Antimalarial Drug Qinghaosu. J. Chem. Ed., **68**, 613 (1991).
48. HAYNES, R.K., and S.C. VONWILLER: Iron (III)-Induced Cleavage of Cyclic Allylic Hydroperoxides to Dicarbonyl Compounds under Aprotic Conditions. J. C. S., Chem. Commun., 449 (1990).
49. HAYNES, R.K., and S.C. VONWILLER: Catalyzed Oxygenation of Allylic Hydroperoxides Derived from Qinghaosu (Artemisinic) Acid. Conversion of Qinghao Acid into Dehydroqinghaosu (Artemisitene) and Qinghaosu (Artemisinin). J.C.S., Chem. Commun., 451 (1990).
50. YE, B., and Y.L. WU: An Efficient Synthesis of Qinghaosu and Deoxoqinghaosu from Arteannuic Acid. J. C. S., Chem. Commun., 726 (1990).
51. LANSBURY, P.T., and D.M. NOWAK: An Efficient Partial Synthesis of (+)-Artemisinin and (+)Deoxoartemisin. Tetrahedron Lett., 1029 (1992).
52. ZHONGSHAN, W., T.T. NAKASHIMA, K.R. KOPECKY, and J. MOLINA: Qinghaosu: ^{1}H and ^{13}C Nuclear Magnetic Resonance Assignments and Luminescence. Can. J. Chem., **63**, 3070 (1985).
53. EL-FERALY, F.S., M.M. EL-SHEREI, C.D. HUFFORD, E.M. CROOM, and T.J. MAHIER: ^{13}C NMR Assignments of Artemisinin, Desoxyartemisinin and Artemether. Spectr. Lett., **18**, 843 (1985).
54. HUANG, J.J., K.M. NICHOLLS, C.H. CHENG, and Y. WANG: Two-Dimensional NMR Studies of Arteanunin; Huaxue Xuebao, **45**, 305 (1987); Chem. Abstr., **107**, 176248d (1987).
55. LOU, X., H.J.C. YEH, A. BROSSI, J.L. FLIPPEN-ANDERSON, and R. GILARDI: Configurations of Antimalarials Derived from Qinghaosu: Dihydroqinghaosu, Artemether, and Artesunic Acid. Helv. Chim. Acta, **67**, 1515 (1984).
56. BAKER, J.K., H.N. ELSOHLY, and C.D. HUFFORD: Nuclear Overhauser Effect Spectroscopy (NOESY) and $^3J_{HH}$ Coupling Measurements in the Determination of the Conformation of the Sesquiterpene Antimalarial Arteether in Solution. Spectrosc. Lett., **23**, 111 (1990).
57. LEE, I.S., H.A. ELSOHLY, E.M. CROOM, and C.D. HUFFORD: Microbial Metabolism Studies of the Antimalarial Sesquiterpene Artemisinin. J. Nat. Prod., **52**, 337 (1989)
58. LIU, J.J., and G.L. DUAN: An Ab Initio Study on the Correlation between the Absolute Configuration and the CD Spectra of Organic Peroxides. Chin. Chem. Lett., **2**, 245 (1991).

59. LIANG, X.T.: Circular Dichroism of the Peroxidic Linkage. Acta Chim. Sinica, **40**, 288 (1982).
60. ZIFFER, H.: Unpublished observation.
61. FANG, Y., Z. SHU, and D. HE: Confirmation of the Vibrational Frequency of the Peroxide Group in Arteannuin and Related Compounds. Huaxue Xuebao, **42**, 1312 (1984); Chem. Abstr., **102**, 166962q (1985).
62. MOHNHAUPT, M., H. HAGEMANN, J.P. PERLER, H. BILL, J. BOUKOUVALAS, J.C. ROSSIER, and C.W. JEFFORD: A Vibrational Study of Some 1,2,4-Trioxanes. Helv. Chim. Acta. **71**, 992 (1988).
63. SCHWARTZ, H., and H.M. SCHIEBEL: Chemistry of the Peroxides (S. PATAI, ed.), p. 105. New York: Wiley, 1983.
64. FALES, H.M., E.A. SOKOLOSKI, L.K. PANNEL, P. QUAN-LONG, D. KLAYMAN, A.J. LIN, A. BROSSI, and J.A. KELLEY: Comparison of Mass Spectral Techniques Using Organic Peroxides Related to Artemisinin. An. Chem., **62**, 2494 (1990).
65. MADHUSUDANAN, K.P., R.A. VISHWAKARMA, S. BALACHANDRAN, and S.P. POPLI: Mass Spectral Studies on Artemisinin, Dihydroartemisinin and Arteether. Indian J. Chem., **28B**, 751 (1989).
66. BAKER, J.K., R.H. YARBER, C.D. HUFFORD, I.S. LEE, H.N. ELSOHLY, and J.D. MCCHESNEY: Thermospray Mass Spectroscopy/High Performance Liquid Chromatographic Identification of the Metabolites Formed from Arteether Using a Rat Liver Microsome Preparation. Biomed. Environ. Mass Spectr., **18**, 337 (1988).
67. THEOHARIDES, A.D., M.H. SMYTH, R.W. ASHMORE, J.M. HALVERSON, Z.M. ZHOU, W.E. RIDDER, and A.J. LIN: Determination of Dihydroqinghaosu in Blood by Pyrolysis Gas Chromatography/Mass Spectrometer. An. Chem., **60**, 115 (1988).
68. BROSSI, A., B. VENUGOPALAN, L.D. GERPE, H.J.C. YEH, J.L. FLIPPEN-ANDERSON, P. BUCHS, X.D. LUO, W. NILHOUS, and W. PETERS: Arteether, a New Antimalarial Drug: Synthesis and Antimalarial Properties. J. Med. Chem., **31**, 645 (1988).
69. LIN, A.J., D.L. KLAYMAN, J.M. HOCH, J.V. SILVERTON, and C.F. GEORGE: J. Org. Chem., **50**, 4504 (1985). The structure for "4" in this paper shows an erroneous methyl group at the acetal carbon.
70. NIU, X.Y., L.Y. HO, Z.Y. REN, and Z.Y. SONG: Metabolic Fate of Qinghaosu in Rats; a New TLC Densitometric Method for its Determination in Biological Matter. Eur. J. Drug Metab. Pharmacokin., **10**, 55 (1985).
71. WANG, Y., and Y. ZHANG: Rapid Thin-Layer Densitometric Determination of Artemisinin in Injections. Yaowu Fenxci Zazhi, **3**, 353 (1983); Chem. Abstr., **100**, 74036W (1984).
72. SHU, H., G. XU, W. LI, and Y. ZENG: Colorimetric Determination of Methyldihydroartemisinin with Vanillin. Fenxi Huaxue, **10**, 678 (1982); Chem. Abstr., **99**, 43606p (1983).
73. ZHU, H.S.: Titrimetric Analysis of Artemisinin in *Artemisia annua*. Yao Hsueh Tlung Pao, **15**, 6 (1980); Chem. Abstr., **95**, 86359v (1981).
74. MELENDEZ, V., J.O. PEGGINS, T.G. BREWER, and A.D. THEOHARIDES: Determination of the Antimalarial Arteether and its Deethylated Metabolite Dihydroartemisinin in Plasma by High-Performance Liquid Chromatography with Reductive Electrochemical Detection. J. Pharm. Sci., **80**, 132 (1991).
75. ZHOU, Z.M., J.C. ANDERS, H. CHUNG, and A.D. THEOHARIDES: Analysis of Artesunic Acid and Dihydroqinghaosu in Blood by High-Performance Liquid Chromatography with Reductive Electrochemical Detection. J. Chromat., **414**, 77(1987).
76. ZHOU, Z.M., Y.X. HUANG, G.H. XIE, X.M. SUN, Y.L. WANG, L.C. FU, H.X. JIAN, X.B. GUO, and G.Q. LI: HPLC with Polarographic Detection of Artemisinin and its

Derivatives and Application of the Method to the Pharmokinetic Study of Artemether. J. Liq. Chrom., 11, 1117 (1988).
77. ZHANG, X.Q., and L.X. XU: Determination of Qinhaosu (Arteannuin) in *Artemisia annua* L. by Pulse Polarography. Yaoxue Xuebao, 20, 283 (1985); Chem. Abstr., 103, 120036h (1985).
78. IDOWU, O.R., S.A. WARD, and G. EDWARDS: Determination of Artelinic Acid in Blood Plasma by HPLC. J. Chrom. Biomed. Applicat., 495, 167 (1989).
79. EDLUND, P.O., D.J. WESTERLUND, B. WU, and Y. JIN: Determination of Artesunate and Dihydroartemisinin in Plasma by Liquid Chromatography with Post-Column Derivatization and U.V. Detection. Acta Pharm. Suec., 21, 223 (1984).
80. LUO, X.-D., M. XIE, and A.-Q. ZOU: Subnanogram Detection of Dihydroartemisinin after Chemical Derivatization with Diacetyldihydrofluorescein followed by HPLC and UV Absorption. Chromatographia, 23, 112 (1987).
81. LUO, X.-D., H.J.C. YEH, and A. BROSSI: Detection of Metabolites of Qinghaosu and its Epoxy Analog. The Chemistry of Drugs. V. Heterocycles, 22, 2559 (1984).
82. EL-DOMIATY, M.M., I.A. AL-MESHAL, and F.S. EL-FERALY: Reversed-Phase High-Performance Chromatographic Determination of Artemisitene in Artemisinin. J. Liq. Chromatog., 14, 2317 (1991).
83. YENG, M.Y.: A Modified Iodometric Method in Determination of Organic Bridged Peroxides; Iodometric Determination of Qing Hao Su. Yaowu Fenxi Zazhi, 4, 329 (1984).
84. ZHOU, Z.M., Y.X. HUANG, G.H. XIE, X.M. SUN, Y.L. WANG, L.C. FU, H.X. JIAN, X.B. GUO, and G.Q. LI: Isolation and Identification of Biotransformation Metabolites of Qin Hau Su. I Isolation and Identification of Biotransformation Metabolites in Humans. Yao Hsueh Hsueh Pao, 15, 509 (1980); Chem. Abstr., 109, 47745f (1980).
85. SIPAHIMALANI, A.T., D.P. FULZELE, and M.R. HEBLE: Rapid Method for the Detection and Determination of Artemisinin by Gas Chromatography. J. Chromatog., 538, 432 (1991).
86. WU, Z., L. DAI and G. GUO: Quantitative Analysis of the Active Constituents of Zinjiang Qinghao (*Artemisia annua*) by Capillary-Column Gas Chromatography. Zhongcaoyao, 17, 341 (1986); Chem. Abstr. 105, 232507k (1986).
87. AVERY, M.A., J.D. BONK, and J. BUPP: Radiolabelled Antimalarials: Synthesis of ^{14}C-Artemisinin. J. Labelled Compounds and Radiopharmaceuticals, 38, 263 (1996).
88. GU, H.M., D.C. WARHURST, and W. PETERS: Uptake of [^{3}H]Dihydroartemsinin by Erythrocytes Infected with *Plasmodium falciparum in vitro*. Trans. Royal Soc. Tropical Medicine and Hygiene, 78, 265 (1984). A modified version of this preparation was employed by MESHNICK, ZIFFER, et al. (unpublished data).
89. PU, Y.-M., and H. ZIFFER: Synthesis of 11-[^{3}H]-Arteether, An Experimental Antimalarial Drug. J. Labelled Compounds and Radiopharmaceuticals, 33, 1013 (1993).
90. DING, S.F., and L.X. LI: Zhongyao Tongbao, 6, 25 (1981).
91. SONG, Q.L., X.Y. LIN, K.D. ZANGE, and H.Z. ZHANY: Action of Sodium Artesunate on Tritiated Uridine Incorporation and Cell Membrane of Mouse Spleen Cells. Acta Pharm. Sinica, 8, 72 (1987).
92. SONG, Z.Y., K.C. ZHAO, X.T. LIANG, C.X. LIU and M.G. YI: Radioimmunoassay of Qinghaosu and Artesunate. Acta Pharm. Sinica, 20, 610 (1985).
93. LUO, X.D., H.J.C. YEH, and A. BROSSI: The Chemistry of Drugs. VI Thermal Decomposition of Qinghaosu. Heterocycles, 23, 881 (1985).
94. LIN, A.J., A.T. THEOHARIDES, and D.L. KLAYMAN: Thermal Decomposition Products of Dihydroartemisinin (Dihydroqinghaosu). Tetrahedron, 42, 2181 (1986).

95. ZENG, M.Y., L.N. LI, S.F. CHEN, G.Y. LI, X.T. LIANG, M. CHEN, and J. CLARDY: Chemical Transformations of Qinghaosu, a Peroxidic Antimalarial. Tetrahedron, **39**, 2941 (1983).
96. ZHOU, W.S., L. ZHANG, Z.C. FAN, and X.X. XU: Studies on Structure and Synthesis of Arteannuin and Related Compounds XX. The Structure of a New Peroxidic Arteannuin Degradation Product and the Lactone Configuration of a Related Compound. Tetrahedron, **42**, 4437 (1986).
97. SHANG, X., C.H. HE, Q.T. ZHENG, J.J. YAND, and X.T. LIAN: Chemical Transformations of Qinghaosu, a Peroxidic Antimalarial, II. Heterocycles, **28**, 421 (1989).
98. HUMMELEN, J.C., T.M. LUIDER, D. OUDMAN, J.N. KOEK, and H.W. WYNBERG: 1,2-Dioxetanes: Luminescent and Nonluminescent Decomposition, Chemistry and Potential Application. Pract. Spectrosc., 567 (1991).
99. ACTON, N., and D.L. KLAYMAN: Conversion of Artemisinin (Qinghaosu) to Iso-Artemisitene and to 9-Epi-Artemisinin. Planta Medica, 266 (1987).
100. TOROK, D.S., and H. ZIFFER: Synthesis and Reactions of 11-Azaartemisinin and its Derivatives. Tetrahedron Lett., **36**, 829 (1995).
101. LI, Y., P. YU, Y. CHEN, J. ZHANG, and Y. WU: Studies on Analogs of Qinghaosou. Some Acidic Degradations of Qinghasou. Kexue Tongbao, **31**, 1038 (1986). The paucity of experimental detail in this paper leaves uncertain the identity of the products with those of the following paper.
102. IMAKURA , Y., K. HACHIYA, T. IKEMOTO, and S. YAMASHITA: Acid Degradation Products of Qinhaosu and their structure-Activity Relationships. Heterocycles, **31**, 1011 (1990).
103. BAKER, J.K., and H.T. CHI: Novel Rearrangements of the Trioxane Ring System of the Antimalarial Arteether Upon Treatment with Acid in an Aqueous Methanol Solvent System. Heterocycles, **38**, 1497 (1994).
104. ACTON, N., and R.J. ROTH: Acid Decomposition of the Antimalarial Beta-Arteether. Heterocycles, **41**, 95 (1995).
105. EL-FERALY, F., S.A. AYALP, and M.A. AL-YAHYA: Conversion of Artemisinin to Artemisitene. J. Nat. Prod., **53**, 66 (1990).
106. PU, Y.M., and H. ZIFFER: Diastereofacial Addition to a β-Substituted Glycal, Anhydroartemisinin. Heterocycles, **39**, 649 (1994).
107. EL-FERALY, F., M.A. AL-YAHYA, K.Y. ORABI, D.R. MCPHAIL, and A.T. MCPHAIL: A New Method for the Preparation of Arteether and its C-9 Epimer. J. Nat. Prod., **55**, 878 (1992).
108. PU, Y.M., H.J.C. YEH, and H. ZIFFER: An Unusual Acid-Catalyzed Rearrangement of 1,2,4-Trioxanes. Heterocycles, **36**, 2099 (1993).
109. EL-FERALY, F.S., A. AYALP, and M.A. AL-YAHYA: Decomposition of Dihydroartemisitene on Silica Gel. J. Nat. Prod., **53**, 920 (1990).
110. YAGEN, B., Y.-M. PU, H.J.C. YEH, and H. ZIFFER, Tandem Silica Gel-Catalysed Rearrangements and Subsequent Baeyer-Villiger Reactions of Artemisinin Derivatives. J.C.S. Perkin Trans. I, 843 (1994).
111. WU, Y.L., and J.L. ZHANG: Reduction of Qinghasou (Artemisinin) with Lithium Aluminium Hydride. Youji Huaxue, 153 (1986); Chem. Abstr., **105**, 191426n (1986).
112. LIU, J.M., M.Y. NI, J.F. FAN, Y.Y. TU, Z.H. WU, Y.L. WU, and W.S. CHOU: Structure and Reactions of Arteannuin. Acta Chim. Sinica, **37**, 129 (1979); Chem. Abstr., **92**, 94594 (1979).
113. PETTIT, G.R., and D.M. PIATAK,: Steroids and Redated Natural Products. XI. Reduction of Esters to Ethers. J. Org. Chem., **27**, 2127 (1962).

114. JUNG, M., X. LI, D.A. BUSTOS, H.N. ELSOHLY, and J.D. MCCHESNEY: A Short and Stereospecific Synthesis of (+)-Deoxoartemisinin and (−)-Deoxodesoxyartemisin. Tetrahedron Lett., **30**, 5973 (1989).
115. VENUGOPALAN, B., S.L. SHINDE, and P.J. KARNIK: Role of Radical Initiated Cyclisation Reactions in the Synthesis of Artemisinin Based Novel Ring Skeletons. Tetrahedron Lett., **34**, 6305 (1993).
116. LIN, A.J., L.-Q. LI, D.L. KLAYMAN, C.F. GEORGE, and J.L. FLIPPEN-ANDERSON: Antimalarial Activity of New Water-Soluble Dihydroartemisinin Derivatives. 3. Aromatic Amine Analogues. J. Med. Chem., **33**, 2610 (1990).
117. POSNER, G.H., D.J. MCGARVEY, C.H. OH, S.R. MESHNICK, and W. ASAWAMAHASADKA: Structure-activity Relationships of Lactone Ring-Opened Analogs of the Antimalarial 1,2,4-trioxane Artemisinin. J. Med. Chem., **38**, 607 (1995).
118. PU, Y.-M., D.S. TOROK, and H. ZIFFER, X.-Q. PAN, and S.R. MESHNICK: Synthesis and Antimalarial Activities of Several Fluorinated Artemisinin Derivatives. J. Med. Chem., **38**, 4120 (1995).
119. LIN, A.J., L.Q. LI, W.K. MILHOUS, and D.L. KLAYMAN: Antimalarial Activity of Dihydroartemisinin Derivatives. 4. Stereoselectivity of 9-Hydroxy Series. Med. Chem. Res., **1**, 20 (1991).
120. PETROV, O., and I. OGNYANOV: An Approach to the Synthesis of Novel 11-Hydroxyartemisinin Derivatives. Collect. Czech. Chem. Commun., **56**, 1037 (1991).
121. HUFFORD, C.D., S.I. KHALIFA, A.T. MCPHAIL, F.S. EL-FERALY, and M.S. AHMAD: Preparation and Characterization of New C-11 Oxygenated Artemisinin Derivatives. J. Nat. Prod., **56**, 62 (1993).
122. PU, Y.-M., B. YAGEN, and H. ZIFFER: Stereoselective Oxidations of a β-Methylglycal, Anhydrodihydroartemisinin. Tetrahedron Lett., **35**, 2129 (1994).
123. CHINA COOPERATIVE RESEARCH GROUP: The Chemistry and Synthesis of Qinghaosu Derivatives. J. Traditional Chin. Med., **2**, 9 (1982).
124. LI, Y., L. YU, Y.X. CHEN, L.Q. LI, Y.Z. GAI, D.S. WANG, and Y.P. ZHENG: Studies on Analogs of Artemisinine I. Synthesis of Ethers, Carboxylate Esters, and Carbonates of Dihydroartemisinin. Yaoxue Xuebao, **16**, 429 (1981); Chem. Abstr., **97**, 92245n (1982).
125. LI, Y., P. YU, Y. CHEN, L. LI, Y.G. AI, D. WANG, and Y. ZHENG: Synthesis of Some Artemisinine Derivatives. K'o Hsueh T'ung Pao, **24**, 667 (1979); Chem. Abstr., **91**, 211376u (1979).
126. CHINESE COOPERATIVE RESEARCH GROUP: Studies on the Toxicity of Qinghaosu and its Derivatives. J. Trad. Chin. Med., **2**, 31 (1982).
127. GU, Y.-X., Y.-F. CUI, B.-A. WU, X.-C. SHI, and X. TENG: Effects of Artemether on Peripheral T, B, T-mu, and T-gama Lymphocytes in Beagle Dog. J. Trad. Chin. Med., **9**, 215 (1989).
128. NAING, U.T., U.H. WIN, D.Y.Y. NWE, U.P.T. MYINT, and U.T. SHWE: The Combined Use of Artemether, Sulfadoxine, and Pyrimethamine in the Treatment of Uncomplicated falciparum Malaria. Trans. Roy. Soc. Trop. Med. Hyg., **82**, 530 (1988).
129. CHEN, Y.-X., P.-L. YU, Y. LI, and R.-Y. JI: Studies on Anlogs of Qinghaosu. VII. The Synthesis of Ethers of Bis(dihydroqinghaosu) and Bis(dihydrodeoxyqinghaosu). Acta Pharm. Sinica, **20**, 470 (1985).
130. LI, Y., P.-L. YU, Y.-X. CHEN, and R.-Y. JI: Studies on Analogs of Arteannuin. II Synthesis of Some Carboxylic Esters and Carbonates of Dihydroarteannuin by using 4-(N,N-dimethylamino)pyridine as an Active Acylation Catalyst. Huaxue Xuebao, **40**, 557 (1982); Chem. Abstr., **98**, 4420h (1983).
131. CHINA COOPERATIVE RESEARCH GROUP: J. Trad. Chinese Med., **2**, 9 (1982).

132. LI, X.-Y., and H.-Z. LIANG: Effects of Artemether on Red Blood Cell Immunity in Malaria. Acta Pharmacol. Sinica, 7, 471 (1986).
133. YANG, Q., W. WHI, R. LI, and J. GAN: The Antimalarial and Toxic Effects of Artesunate on Animal Models. J. Trad. Chinese Med., 2, 99 (1982).
134. LIN, A.J., D.L. KLAYMAN, and W.K. MILHOUS: Antimalarial Activity of New Water-Soluble Dihydroartemisinin Derivatives. J. Med. Chem., 30, 2147 (1987).
135. LIN, A.J., D.L. KLAYMAN, and W.K. MILHOUS: Novel Antimalarial Dihydroartemisinin Derivatives. USP 4, 791, 135.
136. LIN, A.J., M. LEE, and D.L. KLAYMAN: Antimalarial Activity of New Water-Soluble Dihydroartemisinin Derivatives. 2. Stereospecificity of the Ether Side Chain. J. Med. Chem., 32, 1249 (1989).
137. LIN, A.J., and R.E. MILLER: Antimalarial Activity of New Dihydroartemisininin Derivatives. 6. α-Alkylbenzylic Ethers. J. Med. Chem., 38, 764 (1995).
138. LIN, A.J., L.-Q. LI, S.L. ANDERSEN, and D.L. KLAYMAN: Antimalarial Activity of New Dihydroartemisninn Derivatives. 5. Sugar Analogs. J. Med. Chem., 35, 1639 (1992).
139. BUSTOS, D.A., M. JUNG, H.N. ELSOHLY, and J.D. MCCHESNEY: Stereospecific Synthesis of (+)-Homodeoxoartemisinin. Heterocycles, 29, 2273 (1989).
140. JUNG, M., D.A. BUSTOS, H.N. ELSOHLY, and J.D. MCCHESNEY: A Concise and Stereoselective Synthesis of (+)-12-n-Butyldeoxoartemisinin. Syn. Lett., 743 (1990).
141. JUNG, M., D. YU, D.A. BUSTOS, H.N. ELSOHLY, and J.D. MCCHESNEY: A Concise Synthesis of 12-(3'-Hydroxy-n-propyl)-deoxoartemisinin. Bioorgan. Med. Chem. Lett., 1, 741 (1991).
142. HAYNES, R.K., and S.C. VONWILLER: Efficient Preparation of Novel Qinghaosu (Artemisinin) Derivatives: Conversion of Qinghao Acid (Artemisinic) into Deoxoqinghaosu Derivatives and 5-Carboxy-4-deoxoartesunic Acid. Syn. Lett., 481 (1992).
143. LEWIS, M.D., J.K. CHA, and Y. KISHI: Highly Stereoselective Approaches to α- and β-C-Glycopyranosides. J. Am. Chem. Soc., 104, 4976 (1982).
144. PU, Y.M., and H. ZIFFER: Synthesis and Antimalarial Activities of 12β=Allyldeoxoartemisinin. J. Med. Chem., 38, 613 (1995).
145. AVERY, M.A., S. MEHROTRA, T.L. JOHNSON, J.D. BONK, J.A. VROMAN, and R. MILLER: Structure-Activity Relationships of the Antimalarial Agent Artemisinin. 5. Analogs of 10-Deoxoartemisinin Substituted at C-3 and C-9. J. Med. Chem., 39, 4149 (1996).
146. JUNG, M., H.N. ELSOHLY, and J.D. MCCHESNEY: A Concise Synthesis of Novel C-13 Functionalized Deoxoartemisinins. Syn. Lett., 43 (1993).
147. AVERY, M.A., C. JENNINGS-WHITE, and W.K.M. CHONG: The Total Synthesis of (+)-Artemisinin and (+)-9-Desmethylartemisinin. Tetrahedron Lett., 28, 4629 (1987).
148. AVERY, M.A., C. JENNINGS-WHITE, and W.K.M. CHONG: Simplified Analogues of the Antimalarial Artemisinin: Synthesis of 6,9-Desmethylartemsinin. J. Org. Chem., 54, 1792 (1989).
149. STILL, W.C.: A Simple Synthesis of Bicyclo[4.n.1]enones by Cyclodialkylation. Synthesis, 453 (1976).
150. HAYNES, R.K., G.R. KING, and S.C. VONWILLER: Preparation of a Bicyclic Analogue of Qinghao (Artemisinic) Acid Via a Lewis Acid Catalyzed Ionic Diels-Alder Reaction Involving a Hydroxy Diene and Cyclic Enone and Facile Conversion into 6,9-Desmethylqinghaosu. J. Org. Chem., 59, 4743 (1994).
151. AVERY, M.A., W.K.M. CHONG, and G. DETRE: Synthesis of (+)-8a,9-Seco-artemisinin and Related Analogs. Tetrahedron Lett., 31, 1799 (1990).

152. AVERY, M.A., W.K.M. CHONG, and J.E. BUPP: Tricyclic Analogues of Artemisinin: Synthesis and Antimalarial Activity of (+)-4,5-Secoartemisinin and (−)-5-Nor-4,5-secoartemisinin. J. C. S., Chem. Commun., 1487 (1990).
153. AVERY, M.A., C. JENNINGS-WHITE, and W.K.M. CHONG: Synthesis of a C,D-Ring Fragment of Artemisinin. J. Org. Chem., **54**, 1789 (1989).
154. IMAKURA, Y., T. YOKOI, T. YAMAGISH, J. KOYAMA, H. HU, D.R. MCPHAIL, A.T. MCPHAIL, and K.H. LEE: Synthesis of Desethanoqinghaosu, a Novel Analogue of the Antimalarial Qinghaosu. J. C. S. Chem. Commun., 372 (1988).
155. AVERY, M.A., F. GAO, W.K.M. CHONG, S. MEHROTRA, and W.K. MILHOUS: Structure-Activity Relationships of the Antimalaria Agent Artemisinin. 1. Synthesis and Comparative Molecular Field Analysis of C-9 Analogs of Artemisinin and 10-Deoxoartemisinin. J. Med. Chem., **36**, 4264 (1993).
156. AVERY, M.A., S. MEHROTRA, J.D. BONK, J.A. VROMAN, D.K. GOINS, and R. MILLER: Structure-Activity relationships of the Antimalarial Agent Artemisinin. 4. Effect of Substitution at C-3. J. Med. Chem., **39**, 2900 (1996).
157. YE, B., and Y.L. WU: Syntheses of Carba-Analogues of Qinghaosu. Tetrahedron, **45**, 7287 (1989).
158. AVERY, M.A., P. FAN, J.M. KARLE, R. MILLER, and D.K. GOINS: Replacement of the Nonperoxidic trioxane Oxygen Atom of Artemisinin by Carbon: Total Synthesis of (+)-13-Carbaartemisinin and Related Structures. Tetrahedron Lett., **36**, 3965 (1995).
159. AVERY, M.A., P. FAN, J.M. KARLE, J.D. BONK, R. MILLER, and D.K. GOINS: Structure-Activity Relationships of the Antimalarial Agent Artemisinin. 3. Total Synthesis of (+)-13-Carbaartemisinin and Related Tetra- and Tricyclic Structures. J. Med. Chem., **39**, 1885 (1996).
160. WU, J., R.-Y. JI, and Z. Y. KYI: A Quantitative Structure-Activity Study on Artemisinine Analogs. Acta Pharm. Sinica, **3**, 55 (1982).
161. AVERY, M.A., J.D. BONK, W.K.M. CHONG, S. MEHROTRA, R. MILLER, W.K. MILHOUS, D.K. GOINS, S. VENKATESAN, C. WYANDT, I. KHAN, and B.A. AVERY: Structure-Activity Relationships of the Antimalarial Agent Artemisinin. 2. Effect of Heteroatom Substitution at O-11: Synthesis and Bioassay of N-Alkyl-11-aza-9-desmethylartemisinins. J. Med. Chem., **38**, 5038 (1995).
162. AVERY, M.A., F. GAO, S. MEHROTRA, W.K.M. CHONG, and C. JENNINGS-WHITE: The Organic and Medicinal Chemistry of Artemisinin and Analogs. Trends in Organic Chemistry; Trivandrum. India, **4**, 413 (1993).
163. POSNER, G.H., C.H. OH, and W.K. MILHOUS: Olefin Oxidative Cleavage and Dioxetane Formation Using Triethylsilyl Hydrotrioxide: Application to Preparation of Potent Antimalarial 1,2,4-Trioxanes. Tetrahedron Lett., **32**, 4235 (1991).
164. POSNER, G.H., C.H. OH, and W.K. MILHOUS: Extraordinarily Potent Antimalarial Compounds: New Structurally Simple, Easily Synthesized, Tricyclic 1,2,4-trioxanes. J. Med. Chem., **35**, 2459 (1992).
165. CLARK, A.M., and C.D. HUFFORD: Use of Microorganisms for the Study of Drug Metabolism: An Update. Med. Res. Rev., **11**, 473 (1991).
166. LEE, I.S., H.N. ELSOHLY, and C.D. HUFFORD: Microbial Metabolism Studies of the Antimalarial Drug Arteether. Pharm. Res., **7**, 199 (1990).
167. HUFFORD, C.D., I.-S. LEE, H.N. ELSOHLY, H.T. CHI, and J.K. BAKER: Structure Elucidation and Thermospray High-Performance Liquid Chromatography/Mass Spectroscopy (HPLC/MS) of the Microbial and Mammalian Metabolites of the Antimalarial Arteether. Pharm. Res., **7**, 923 (1990).

168. KHALIFA, S.I., J.K. BAKER, R.D. ROGERS, F.S. EL-FERALY, and C.D. HUFFORD: Microbial and Mammalian Metabolism Studies of the Semisynthetic Antimalarial, Anhydrodihydroartemisinin. Pharm. Res., **11**, 990 (1994).
169. ELMARAKBY, S.A., F.S. EL-FERALY, H.N. ELSOHLY, E.M. CROOM, and C.D. HUFFORD: Microbiological Transformations of Artemisinic Acid. Phytochem., **27**, 3089 (1988).
170. HU, Y., R.J. HIGHET, D. MARION, and H. ZIFFER: Microbial Hydroxylation of a Dihydroartemisinin Derivative. J. C. S. Chem. Commun., 1176 (1991).
171. HU, Y., H. ZIFFER, G. LI, and H.J.C. YEH: Microbial Oxidation of the Antimalarial Drug Arteether. Bioorganic Chem., **20** 148 (1992).
172. DESJARDINS, R.E., C.J. CANFIELD, D.E. HAYNES, and J.D. CHULAY: Quantitative Assessment of Antimalarial Activity *in vitro* by a Semi-automated Microdilution Technique. Antimicrob. Agent Chemother., **16**, 710 (1979).
173. TOROK, D.S., and H. ZIFFER: Synthesis and Antimalarial Activities of N-Substituted 11-Azaartemisinins. J. Med. Chem., **38**, 5045 (1995).
174. ACTON, N., J.M. KARLE, and R.E. MILLER: Synthesis and Antimalarial Activity of Some 9-Substituted Artemsinin Derivatives. J. Med. Chem., **36**, 2552 (1993).
175. ZHAO, Y.: Studies on Systemic Pharmacological Effects of Artesunate. J. Trop. Med. Hyg., **88**, 391 (1985).
176. AMES, B.N.: Methods for Detecting Carcinogens and Mutagens with the Salmonella/Mammalian Microsome Mutagenicity Test. Mutat. Res., **31**, 347 (1975).
177. BREWER, T.G., J.O. PEGGINS, S.J. GRATE, J.M. PETRAS, B.S. LEVINE, P.J. WEINA, J. SWEARENGEN, M.H. HEIFFER, and B.G. SCHUSTER: Neurotoxicity in Animals due to Arteether and Artemether. Trans. Royal Soc. Trop. Med., **88**, S1 33 (1994).
178. BREWER, T.G., S.J. GRATE, J.O. PEGGINS, P.J. WEINA, J.M. PETRAS, B.S. LEVINE, M.H. HEIFFER, and B.G. SCHUSTER: Fatal Neurotoxicity of Arteether and Artemether. Am. J. Trop. Med. Hyg., **51**, 251 (1994).
179. WESCHE, D.L., M.A. DECOSTER, F.C. TORTELLA, and T.G. BREWER: Neurotoxicity of Artemisinin Analogs *in vitro*. Antimicrobial Agents and Chemotherapy, **38**, 1813 (1994).
180. KAMCHONWONGPAISAN, S., P. MCKEEVER, P. HOSSLER, H. ZIFFER, and S.R. MESHNICK: Artemisinin Neurotoxicity: Neuropathology in Rats and Mechanistic Studies *in vitro*. Amer. J. Trop. Med. Hyg., **56**, 7 (1997).
181. NOSTEN, F.: Artemisinin: Large Community Studies. Royal Society Tropical Medicine and Hygiene, **88**, Supplement, **1**, 45 (1994).
182. WALKER, O., L.A. SALAKO, S.I. OMOKHODION, and A. SOWUMI: An Open Randomized Comparative Study of Intramuscular Artemether and Intravenous Quinine in Cerebral Malaria in Children. Royal Soc. Trop. Med. Hyg., **87**, 564 (1993).
183. ZHAO, K.-C., Q.-M. CHEN, and Z.-Y. SONG: Studies of the Pharmacokinetics of Qinghaosu and Two of its Active Derivatives in Dogs. Acta Pharm. Sinica, **21**, 736 (1986).
184. ZHAO, K.C., and Z.Y. SONG: The Pharmacokinetics of Dihydroqinghaosu Given Orally to Rabbits and Dogs. Acta Pharm. Sinica, **25**, 161 (1990).
185. ZHAO, K., and Z. SONG: Distribution and Excretion of Artesunate in Rats. Proc. CAMS and PUMC, **4**, 186 (1989).
186. BENAKIS, A., C. SCHOPFER, M. PARIS, CH. T. PLESSAS, P.E. KARAYANNAKOS, I. DONDAS, D. KOTSARELIS, S.T. PLESSAS, and G. SKALKEAS: Pharmacokinetics of arteether in dog. European J. Drug Metabolism and Pharmacokinetics, **16**, 325 (1991).
187. TITULAER, H.A.C., J. SUIDEMA, P.A. KAGER, J.C.F.M. WETSTEYN, Ch. B. LUGT, and F.W. H.M. MERKUS: The Pharmacokinetics of Artemisinin After Oral, Intramuscular and Rectal Administration to Volunteers. J. Pharm. Pharmacol., **42**, 810 (1990).

188. KLAYMAN, D.L., A.L. AGER JR., L. FLECKENSTEIN, and A.J. LIN: Transdermal Artelinic Acid: An Effective Treatment for *P. berghei*-Infected Mice. Am. J. Trop. Med., **45**, 602 (1991).
189. LIN, A.J., A.L. AGER JR., and D.L. KLAYMAN: Antimalarial Activity of Dihydroartemisinin Derivatives by Transdermal Application. Am. J. Trop. Med. Hyg., **50**, 777 (1994).
190. DUC, D.D., P.J. DEVRIES, N.X. KHANH, L.N. BINH, P.A. KAGER, and C.J. VAN BOXTEL: The Pharmacokinetics of a Single Dose of Artemisinin in Healthy Vietnamese Subjects. Am. J. Med. Hyg., **51**, 785 (1994).
191. QINGHAOSU ANTIMALARIA COORDINATING RESEARCH GROUP: Antimalarial Studies on Qinghaosu. Chin. Med. J., **92**, 811 (1979).
192. (a) LOOAREESUWAN, S.: Overview of Clinical Studies on Artemisinin Derivatives in Thailand. Trans. Royal Soc. Trop. Med. Hyg., **88**, Supplement 1, 9 (1994); (b) BUNNAG, D., C. VIRAVAN, S. LOOAREESUWAN, J. KARBWANG, and T. HARINASUTA: Clinical Trial of Artesunate and Artemether on Multidrug Resistant *P. falciparum* Malaria in Thailand: A Preliminary Report. Southeast Asian Journal of Tropical Medicine and Public Health, **22**, 380 (1991); (c) BUNNAG, D., C. VIRAVAN, S. LOOAREESUWAN, J. KARBWANG, and T. HARINASUTA: Double Blind Randomized Clinical Trial of Two Different Regimens of Oral Artesunate in *P. falciparum* Malaria. Southeast Asian Tropical Medicine and Public Health, **22**, 534 (1991); (d) BUNNAG, D., C. VIRAVAN, S. LOOAREESUWAN, J. KARBWANG, and T. HARINASUTA: Double Blind Randomized Clinical Trial of Oral Artesunate at Once or Twice Daily Dose in *P. falciparum* Malaria. Southeast Asian Tropical Medicine and Public Health, **22**, 762 (1991).
193. KARBWANG, J., K. SUKONTASON, W. RIMCHALA, W. NAMSIRIPONGPUN, T. TIN, P. AUPRAYOON, S. TUMSUPAPONE, D. BUNNAG., and T. HARINASUTA: Preliminary Report: a Comparative Clinical Trial of Artemether and Quinine in Severe Falciparum Malaria. Southeast Asian Tropical Medicine and Public Health, **23**, 768 (1992); (b) KARBWANG, J., K. NA BANGCHANG, A. THANAVIBUL, D. BUNNAG, T. CHONGSUPHAJAISIDDHI, and T. HARINASUTA: Comparison of Artemether and Mefloquine in Acute Uncomplicated Falciparum Malaria. Lancet, 1245 (1992).
194. LUXEMBURGER, C., F. TER KUILE, F. NOSTEN, G. DOLAN, J.H. BRADOL, L. PHAIPUN, T. CHONGSUPHAJAISIDDHI, and N.J. WHITE: Single Day Mefloquine-Artesunate Combination in the Treatment of Multi-Drug Resistant Falciparum Malaria. Trans. Royal Soc. Trop. Med. Hyg., **88**, 213 (1994).
195. ARNOLD, K.: Qinghaosu, Mefloquine, and Pyrimethamine Sulfadoxide in falciparum Malaria. Lancet, 704 (1985).
196. SLATER, A.F.G., A. CERAMI: Inhibition by Chloroquine of a Novel Haem Polymerase Enzyme Activity in Malaria Trophozoites. Nature (London), **355**, 167 (1992).
197. ZHAO,Y., W.K. HANTON, and K.-H. LEE: Antimalarial agents, 2. Artesunate, an Inhibitor of Cytochrome Oxidase Activity in *P. berghei*. J. Nat. Prod., **49**, 139 (1986).
198. ELLIS, D.S., Z.L. LI, H.M. GU, W. PETERS, B.L. ROBINSON, G. TOVEY, and D.C. WARHURST: Ultrastructural Changes Following Treatment with Artemisinin of *P. berghei* infection in Mice, with Observations of the Localization of [^{3}H]-Dihydroartemisinin in *P. falciparum in vitro*. Annals of Tropical Med. and Parasitology, **79**, 367 (1985).
199. KAWAI, S., S. KANO, and M. SUZUKI: Morphologic effects of Artemether on *P. falciparum in Aotus trivirgatus*. Am. J. Trop. Med. Hyg., **49**, 812 (1993).

200. MESHNICK, S.R., A. THOMAS, A. RANZ, C.-M. XU, and H.-Z. PAN: Artemisinin (Qinghaosu): The Role of Intracellular Hemin and its Mechanism of Antimalarial Action. Mol. and Biochem. Parasitology, **49**, 181 (1991).
201. ZHANG, F., D.K. GOSSER JR., and S.R. MESHNICK: Hemin-Catalyzed Decomposition of Artemisinin (Qinghaosu). Biochem. Pharmacol., **43**, 1805 (1992).
202. YANG, Y.-Z., W. ASAWAMAHASAKDA, and S.R. MESHNICK: Alkylation of Human Albumin by the Antimalarial Artemisinin. Biochem. Pharmacol., **46**, 336 (1993).
203. ASAWAMAHASAKDA, W., I. ITTARAT, Y.-M. PU, H. ZIFFER, and S.R. MESHNICK: Reaction of Antimalarial Endoperoxides with Specific Parasite Proteins. Antimicrobial Agents Chemotherapy, 1854 (1994).
204. CLARK, A., N.H. HUNT, W.B. COWDEN, and L.E. MAXWELL: Radical-Mediated Damage to Parasites and Erythrocytes in *P. vinckei* Infected Mice after Injection of t-Butyl Hydroperoxide. Clin. Exp. Immunol., **56**, 524 (1984).
205. MESHNICK, S.R., Y.-Z. YANG, V. LIMA, F. KUYPERS, S. KAMCHONWONGPAISAN, and Y. YUTHAVONG: Iron-Dependent Free Radical Generation from the Antimalarial Agent Artemisinin (Qinghaosu). Antimicrobial Agents and Chemotherapy, **37**, 1108 (1993).
206. POSNER, G.H., C.H. OH, D. WANG, L. GERENA, W.K. MILHOUS, S.R. MESHNICK, and W. ASAWAMAHASADKA: Mechanism-Based Design, Synthesis, and *in vitro* Antimalarial Testing of New 4-Methylated Trioxanes Structurally Related to Artemisinin: The Importance of a Carbon-Centered Radical for Antimalarial Activity. J. Med. Chem., **37**, 1256 (1994).
207. POSNER, G.H., J.N. CUMMING, P. PLOYPRADITH, and C.H. OH: Evidence for Fe(IV)=O in the Molecular Mechanism of Action of the Trioxane Antimalarial Artemisinin. J. Am Chem. Soc., **117**, 5885 (1995).
208. POSNER, G.H., D. WANG, J.N. CUMMING, C.H. OH, A.N. FRENCH, A.L. BODLEY, and T.A. SHAPIRO: Further Evidence Supporting the Importance of and the Restrictions on a Carbon-Centered Radical for High Antimalarial Activity of 1,2,4-Trioxanes Like Artemisinin. J. Med. Chem., **38**, 2273 (1995).
209. POSNER, G.H., S.B. PARK, L. GONZALEZ, D. WANG, J.N. CUMMING, D. KLINEDINST, T.A. SHAPIRO, and M.D. BACHI: Evidence for the Importance of High-Valent Fe=O and of a Diketone in the Molecular Mechanism of Action of Antimalarial Trioxane Analogs of Artemisinin. J. Am. Chem. Soc., **118**, 3537 (1996).
210. LIANG, X.T.: The Chemistry of Peroxidic Antimalarials. Adv. Chin. Med. Mater. Res. Int. Symp., 427 (1984).
211. ZHANG, L., W.S. ZHOU, and X.X. XU: A New Sesquiterpene Peroxide (Yingzhaosu C) and Sesquiterpenol (Yingzhaosu D) from *Artabotrys uniciatus* (L.) Meer. J. C. S. Chem. Commun., 523 (1988).
212. XU, X.-X., J. ZHU, D.-Z. HUANG, and W.-S. ZHOU: Total Synthesis of (+)-Yingzhaosu A. Tetrahedron Lett., **32**, 5785 (1991).
213. XU, X.-X., and X. XIE: Total Synthesis of Yingzhaosu B and its Three Diastereoisomers. Chinese J. Chem., **12**, 381 (1994).
214. XU, X.-X., and H.-Q. DONG: Enantioselective Total Synthesis of All Four Stereoisomers of Yingzhaosu C. Tetrahedron Lett., **35**, 9429 (1994).
215. XU, X.-X., and Q.-S. HU: Synthesis of the Diastereoisomeric Yingzhaosu D. Chinese J. Chem., **10**, 285 (1992).
216. KAMCHONWONGPAISAN, S., C. NILANONTA, B. TAMCHOMPOO, C. THEBTARANONTH, Y. THEBTARANONTH, Y. YUTHAVONG, P. KONGSAEREE, and J. CLARDY: An Antimalarial Peroxide from *Amomum krervanh Pierre*. Tetrahedron Lett., **36**, 1821 (1995).
217. HASHIDOKO, Y., S. TAHARA, and S.J. MIZUTANI: Antimicrobial Sesquiterpene from Damaged *Rosa rugosa* Leaves. Phytochemistry, **28**, 425 (1989).

218. HOFHEINZ, W., H. BURGIN, E. GLOCKE, C. JAQUET, R. MASCIADRI, G. SCHMID, H. STOHLER, and H. URWYLER: Ro 42-1611 (Arteflene), A New Effective Antimalarial: Chemical Structure and Biological Activity. Trop. Med. Parasitol., **45**, 261 (1994).
219. POSNER, G.H., D. WANG, L. GONZALEZ, S. TAO, J.N. CUMMING, D. KLINEDINST, and T.A. SHAPIRO: Mechanism-Based Design of Simple, Symmetrical, Easily Prepared, Potent Antimalarial Endoperoxides. Tetrahedron Lett., **37**, 815 (1996).
220. VENNERSTROM. J.L., and J.W. EATON: Oxidants, Oxidant Drugs and Malaria. J. Med. Chem., **3**, 1269 (1988).
221. CLARK, I.A., W.B. COWDEN , and G.A. BUTCHER: Lancet, 234, 1983.
222. VENNERSTROM, J.L., H.-N. FU, W.Y. ELLIS, A.L. AGER, JR., J.K. WOOD, S.L. ANDERSEN, L. GERENA, and W.K. MILHOUS: J. Med. Chem., **35**, 3023 (1992).
223. KEPLER, J.A., A. PHILIP, Y.W. LEE, H.A. MUSALLAM, and F.I. CARROLL: Endoperoxides as Potential Antimalarial Agents. J. Med. Chem., **30**, 1505 (1987).
224. BUNNELLE, W.H., T.A. ISBELL, C.L. BARNES, and S. QUALLS: Cationic Ring Expansion of an Ozonide to a 1,2,4-Trioxane. J. Am. Chem. Soc., **113**, 8168 (1991).

(Received September 27, 1996)

Marine Glycolipids

E. FATTORUSSO and A. MANGONI
Dipartimento di Chimica delle Sostanze Naturali,
Università degli Studi di Napoli Federico II, Napoli, Italy

Contents

1. Introduction 215
2. Isolation Procedures 218
3. Determination of the Structure of Glycolipids 219
   3.1. Determination of the Structure of the Sugar Portion 220
   3.2. Determination of the Structure of the Lipid Portion 222
4. Glycoglycerolipids 223
5. Glycosphingolipids 228
   5.1. Neutral Glycosphingolipids 229
   5.2. Phosphorus-Containing Glycosphingolipids 247
   5.3. Gangliosides 260
6. Other Glycolipids 281
7. Biological Activities 287
   7.1. Immunological Activity 287
   7.2. Pharmacological Activity 288
References 291

1. Introduction

Marine glycolipids, as well as those from terrestrial organisms, are amphiphylic compounds which are currently divided into two main groups: glycoglycerolipids (GGLs) and glycosphingolipids (GSLs). There is a third important group comprising glycolipids whose lipid portion is derived from mevalonate, *i.e.* steroidal and terpenic glycosides. The occurrence of polyisoprenoidic glycolipids is generally confined to

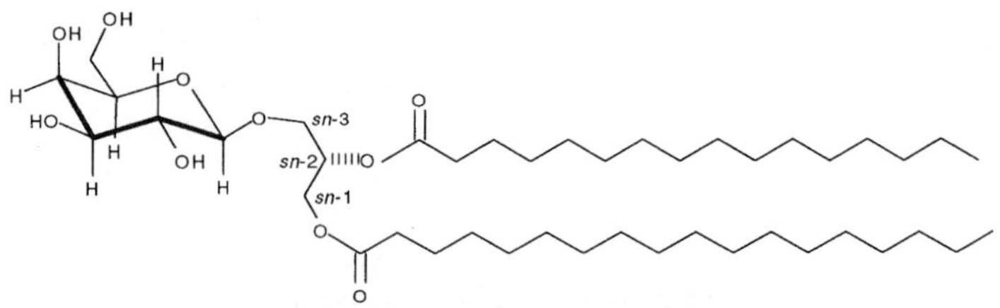

Fig. 1. Structure of a representative glycoglycerolipid

species of a few taxa, where they frequently perform peculiar biological functions; as to their occurrence in marine organisms, they are mostly present in invertebrates belonging to phylum Echinoderma.

The present review deals only with the first two classes of compounds, namely GGLs and GSLs. Glycolipids with a polyisoprenoidic aglycon are not surveyed here, since an accurate review on the steroidal glycosides from echinoderms has been published very recently (*1*).

Chemically, GGLs contain a glycerol unit glycosylated at one primary alcoholic function, with the remaining hydroxyl groups acylated by a fatty acid and/or alkylated by a long chain residue (Fig. 1). Since in all GGLs the glycerol C-2 becomes a stereogenic center, its configuration is usually indicated by numbering the glycerol molecule stereospecifically. The glycerol molecule is drawn in a Fisher projection with the hydroxyl group to the left, and the carbon atoms are numbered from top to bottom. When such a numbering is used, the prefix *sn-* is added before the word glycerol (*2*).

The structure of a representative GSL is depicted in Fig. 2. A carbohydrate chain and a fatty acyl group are linked to a long-chain aminoalcohol called the sphingoid base or long-chain base (LCB). The fatty acyl chain is amide-linked to the LCB, and together they make up what is called a ceramide; the monosaccharide or oligosaccharide group

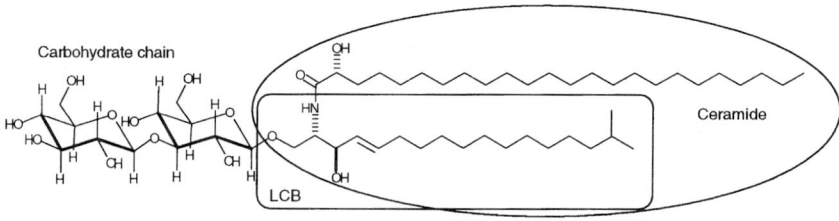

Fig. 2. Structure of a representative glycosphingolipid

References, pp. 291–301

is bound to the primary alcoholic function of the ceramide. Sphingosine (**1**) is the LCB most commonly found in higher animals, so that LCBs are often referred to as sphingosines. In plant glycolipids, the trihydroxylated LCB phytosphingosine (**2**) is frequently found; also the name phytosphingosine is often used more generally to denote any trihydroxylated LCB. Since an increasing number of different sphingoid bases are being described in the literature, a semisystematic nomenclature for LCBs has been proposed, which is based on sphinganine (**3**, 2-amino-1,3-octadecanediol) (*3*).

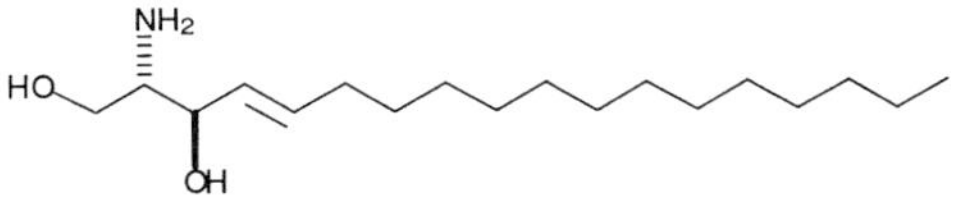

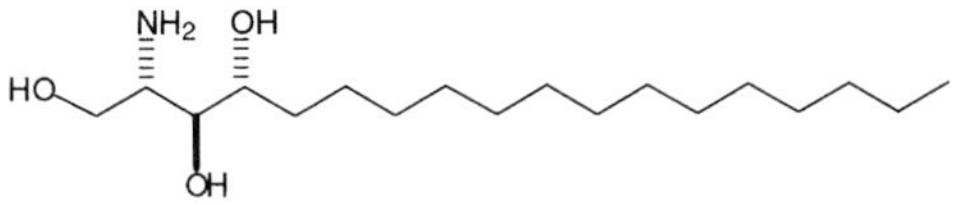

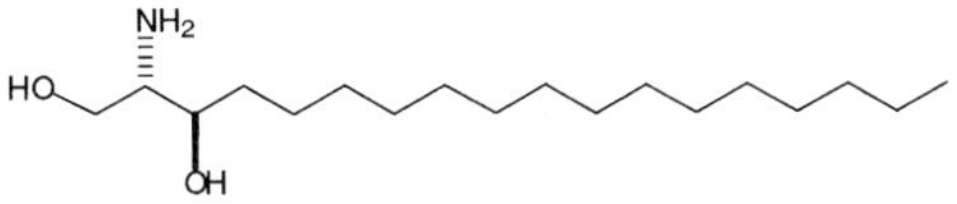

Occasionally, some marine organisms were found to contain atypical glycolipids, namely compounds which are formed of a sugar portion glycosidically connected to a non-isoprenoid lipid moiety, but which cannot be classified as GGLs or GSLs. They have been included in the Section "Other Glycolipids" of the present review.

The first reports on the chemistry of non-isoprenoidic marine glycolipids appeared in the end of the fifties (*4–22*) and did not contain exhaustive structural data on the various metabolites. Only some information on the nature of the sugars and of the lipid moiety was reported, while investigations of the homogeneity of the examined material and of the way the part structures were linked together, as well as of the stereochemistry of the whole molecules, was lacking. Unambiguous structure determinations of glycolipids from marine organisms were achieved only starting from 1973, when some glycolipids were isolated from a marine cyanobacterium and their structures clarified by chemical and spectroscopic methods (23). The present review is intended to include those marine non-isoprenoid glycolipids whose structures have been fully defined; however, if only some stereochemical detail remained to be clarified, the glycolipid has been included as well.

Readers who want to learn about historical aspects and details of marine glycolipid studies should consult previously published reviews on marine invertebrates (24, 25) and algae (26).

2. Isolation Procedures

A matter which appears to be worthy of preliminary discussion is the level of purity of a mixture of glycolipids. Virtually all natural glycolipids occur as mixtures of homologues, differing in the length and in the branching of the alkyl chains of the lipid portion of the molecule. Generally, structural studies are performed on the mixtures of homologues, while the nature of the fatty acids and/or the sphingoid bases is subsequently established by chemical degradation followed by GC/MS analysis. The difficulty (or more often the impossibility) in obtaining a chemically homogeneous glycolipid is evident, if one considers that, for example, a glycosphingolipid which appears pure by normal phase HPLC often contains 7–8 different sphingoid bases and fatty acids, and this corresponds to over 50 different compounds. Attempts at chromatographic separation of such complex mixtures are reported in the literature (27, 28), but afforded only a small number of homogeneous fractions, the other ones being still mixtures which were characterized for the major component only, or not at all.

Generally, a mixture of homologous glycolipids can be regarded as a "single" compound, the purity being referred to the biological role of the material. In fact, it is commonly acknowledged that the bioactivity depends essentially on the nature of the sugar head and the adjacent functionalized part of the molecule. In contrast, the alkyl chains serve to position the glycolipid in the membrane and affect its fluidity. In living organisms this function is satisfactorily accomplished by non-homogeneous material differing only in the skeletal structure of the alkyl moieties.

The isolation of marine glycolipids utilizes much the same methods as those used in the investigation of terrestrial organisms. Basically, the homogenate of the tissue is extracted with a mixture of methanol and chloroform. The extract contains a complex mixture consisting of almost all the low-molecular weight metabolites of the organism under investigation. It is usually subjected to partitioning between water-methanol

References, pp. 291–301

and chloroform phases according to the FOLCH method (*29*), which has been extensively employed in the last thirty years. The glycolipid material can be recovered in the MeOH-water or the chloroform phase according to its polarity. The crude material obtained after evaporation of both phases is successively subjected to appropriate chromatographic separations, strongly dependent on the polarity of the glycolipid to be isolated. The most commonly used techniques include absorption chromatography (SiO_2 or silicates) (*30–35*) molecular sieve chromatography (*33, 36, 37*), and reversed-phase chromatography (*24, 28, 30, 31, 37*); purification of ionogen glycolipids usually makes use of ion-exchange chromatography also (*38, 39*).

The most recent procedures take advantage of the amphiphilic character of glycolipids by using reversed-phase column chromatography as the first step in the purification of the crude extract: glycolipids, in spite of their remarkable polarity, are conspicuously retained by the stationary phase and are eluted together with quite apolar substances which can be easily removed through a successive adsorption chromatography on SiO_2.

An illustrative example of the isolation of neutral glycolipids is represented by the purification of glycosyl ceramides from the marine sponge *Agelas dispar* (*40*) to four compounds (**23, 24, 26,** and **28**), which were actually mixtures of homologues differing in the length and the terminus of the alkyl chains of the ceramide part of the molecules. The homogenate of the sponge was extracted with methanol and subsequently with chloroform, and the combined extracts were partitioned between water and *n*-BuOH. The organic phase was chromatographed on a RP-18 column (eluents: MeOH/H_2O and MeOH/EtOAc), and then on an SiO_2 column using as eluents EtOAc/*n*-hexane and EtOAc/MeOH. Four pure neutral GSLs were obtained by HPLC of the fraction containing the crude sphingolipids using a DIOL column (eluent: *n*-hexane/*i*PrOH/H_2O 55 : 43 : 2).

3. Determination of the Structure of Glycolipids

As mentioned in the Introduction, the first reports on complete structural assignments of glycolipids from marine sources are relatively recent and date from the seventies. From the beginning, spectroscopic data extensively assisted chemical analysis in clarification of the structures of these metabolites. Of course, the importance of the physical methods has dramatically increased during the last ten years, when the modern two-dimensional techniques improved NMR spectroscopy as a vital tool of chemical analysis which provides decisive information for the elucidation of complex structures, including the stereochemical details, on just a few milligrams of a reasonably pure product.

As a rule, the starting point of the structural analysis of a natural product is the determination of molecular formula. This is currently achieved by interpretation of mass spectral data, which are integrated when necessary with information supplied by the ^{13}C-NMR spectrum. In the field of glycolipids, which are generally mixtures of homologues of low volatility, information on the molecular composition can be obtained through the use of the quite recent FAB mass spectral technique (the negative ion mode is generally used). Subsequent steps involve the assignment of the structure to the lipid part of the molecule and to the sugar moiety, which generally are investigated in separate experiments and through quite different analytical techniques.

3.1. Determination of the Structure of the Sugar Portion

The analysis may be performed using chemical and spectroscopic methods; obviously, the latter are currently acquiring an increasing importance.

The starting point of chemical investigation is usually the identification of the monosaccharide units present in the molecule. This can be achieved by acid methanolysis of the glycolipid (*41*) followed by quantitative analysis of the thus obtained methyl monosaccharides using chromatographic methods currently employed in carbohydrate chemistry. Care must be taken in the presence of sialoglycolipids, since methanolysis induces the cleavage of the amide bond of sialic acids. Their presence can be evidenced by mild acid hydrolysis (*42*) of the starting material and successive TLC chromatography of the hydrolysate.

When the sugar portion of the molecule contains more than one saccharide unit, the subsequent steps are directed towards the identification of the positions involved in the inter-unit linkages and in the sugar-aglycon bond. This can be attained using the classical methods employed in the chemistry of glycosides. A simple and efficient analytical procedure that is frequently used is the HAKOMORI method (*43*) and a modification by SANFORD and CONRAD (*44*). It is based on the permethylation of glycolipids followed by acidic hydrolysis of the permethyl derivatives; the partially methylated monosaccharides are reduced to the corresponding alditols, acetylated and then identified and quantitized through GC or GC-MS methods. Basically, this method allows discrimination between

References, pp. 291–301

free hydroxyl groups (which are methylated in the alditol) and those involved in glycosidic bonds (acetylated in the alditol) thus providing useful data for determining the glycosylation sites and the furanose or pyranose structure of each monosaccharide of the carbohydrate chain. In a simple case, the sequence of the sugar units in the oligosaccharide chain can be also inferred. Otherwise, data supplied by the HAKOMORI method can be supplemented with those obtained by partial hydrolysis of the glycolipid, the latter producing simpler glycolipids and oligosaccharides which are chromatographically separated and analyzed by the same methods. Partial hydrolysis can be carried out through mild acidic treatment of the compound under investigation (*30, 34*) or by stereoselective enzymatic hydrolysis (*45–48*). More recently, useful information on the sugar sequence has been obtained from the interpretation of fragmentation in the mass spectrum, usually performed on the original glycolipid using the FAB technique (*30, 49, 50, 51*).

Finally, the configuration of the glycosidic bonds can be ascertained by enzymatic hydrolysis with stereoselective glycosidases or by oxidation with chromium trioxide, which for a sugar in the pyranose form is much more effective if the glycosidic bond is equatorially oriented (*52–54*). Determination of the anomeric configuration has been also accomplished by ^{1}H-NMR, long before complete structural elucidation based on NMR became feasible. In fact, anomeric protons can be readily identified even on a low field spectrometer, and their coupling constants are a clear indication of the anomeric configuration of a pyranose sugar.

From 1992 on, unambiguous structural determinations, of the whole hydrophilic portion of marine glycolipids (carbohydrate chain and functionalized part of the aglycon) performed through an extensive NMR analysis appeared in the literature (*55–58*). These NMR studies, based on the use of modern two-dimensional techniques, are often carried out using the peracetylated glycolipid because of the better signal dispersion in the ^{1}H-NMR spectrum of this derivative and the possibility of discriminating between ether and ester oxymethines proton resonance on the basis of their different chemical shift ranges (δ 3.5–4.5 and 4.7–5.7, respectively).

Even in peracetylated glycolipids, the mid-field region of the ^{1}H-NMR spectrum containing the signals of the sugar methylene and methine protons is often very crowded and is analyzed by two-dimensional techniques. A HOHAHA experiment allows one to group together protons belonging to the same monosaccharide unit giving rise to isolated spin-systems, whereas a subsequent COSY experiment is used to find out the proton sequence within each spin system. A short-range ^{1}H–^{13}C chemical shift correlation (HETCOSY or HMQC) experiment permits the identification of the relevant carbon signals in the ^{13}C NMR spectrum.

Final identification of the monosaccharide units is achieved through ^{1}H–^{1}H coupling constant analysis. Coupling constants are read directly from the multiplets or, in case of overlapping signals, are determined with the aid of the HOHAHA spectrum (57). When coupling constant analysis is not sufficient for an unambiguous structural assignment, additional information can be provided by a ROESY spectrum which evidences inter-proton spatial proximities (59). The sites of glycosylation can be identified from the chemical shifts of oxymethine protons if a peracetyl glycolipid is being used; linkages between couples of sugars and between the carbohydrate chain and the aglycon are evidenced by long-range ^{1}H–^{13}C shift correlation (COLOC or HMBC) experiment (60), which shows connectivities between proton and carbon atoms separated by two or three σ bonds, or by the ROESY spectrum (59).

Once the structure of a peracetyl glycolipid has been secured, it is convenient to perform a retrospective analysis of the NMR data of the parent compound, which usually allows assignment of the ^{1}H- and ^{13}C-NMR resonances of the original GSL, thus providing further support to the proposed structure.

It is to be noted that in most cases the absolute configuration of sugar units has not been determined. In the structural drawings, they are assumed to belong to the most commonly found D series (L series for arabinose), even though this has not been stated by the various authors explicitly.

3.2. Determination of the Structure of the Lipid Portion

The structural determination of the hydrophobic portion of a glycolipid is generally a less time-consuming work than that needed to clarify the structure of the sugar portion. It is to be noted, however, that very frequently the material to be investigated is a mixture of homologues differing in the lipid part of the molecule, and this requires the use of appropriate analytical procedures.

Analysis of glycoglycerolipids (GGLs) usually proceeds through removal of the acyl chains by hydrolysis or, better, methanolysis, followed by identification of the resulting fatty acids methyl esters using well-known GC or GC-MS procedures (30, 56).

In the case of glycosphingolipids (GSLs), the analysis is a little more complex. Sphingoid bases are normally removed from the molecule by

References, pp. 291–301

treatment with acidic methanol, then purified through a SiO_2-column chromatography and identified and quantitized as free bases or as appropriate derivatives by GC analysis (*30*). If reference compounds are not available, the identification can be accomplished by periodate/permanganate oxidation followed by methylation (*56*), or by periodate oxidation (*49*) of the sphingosine homologues, and gas chromatographic identification of the resulting fatty acid methyl esters or aldehydes, respectively.

The relative configuration of the stereocenters of the sphingosines is assigned on the basis of their ^{1}H- and ^{13}C-NMR spectra, in particular from the chemical shifts and multiplicities of the protons present in the functionalized part of the molecule (the homologues differ only in the structure of the hydrocarbon chain) compared with those reported for sphingosines of known stereochemistry (*30*). Once the relative stereochemistry has been assigned, the absolute configuration is deduced from the $[\alpha]_D$ value exhibited by the mixture of the sphingosine homologues (*56*).

The methanolysis of GSLs also provides a mixture of fatty acid methyl esters and/or α-hydroxyacid methyl esters, originally linked at position 2 of the sphingosine. They can be identified and quantitized by GC analysis (*34*). Information about the configuration at C-2 of α-hydroxy fatty acids can be obtained from the measured $[\alpha]_D$ value of the mixture of the α-hydroxy fatty acid methyl esters homologues used for GC analysis (*56*).

4. Glycoglycerolipids

The glycoglycerolipids (GGLs) most commonly found as metabolites of marine organisms are monogalactosyldiacylglycerol (MGDG), digalactosyldiacylglycerol (DGDG) and sulfoquinovosyldiacylglycerol (SQDG). They are normal constituents of membranes of photosynthetic organisms; in the marine environment they are present almost exclusively in algae and cyanobacteria (blue-green algae). In spite of their wide distribution, marine GGLs do not show a great structural variety. This is not surprising, since most unusual GGLs have been isolated from bacteria, yeasts, and fungi, and the study of marine species belonging to these classes is just beginning. Only two kinds of marine GGLs which are not closely related to MGDG, DGDG, or SQDG have been found, namely crasserides and a dixylosylalkylglycerol, and both of them have been isolated from marine sponges.

Strictly speaking, crasserides (keruffarides) (*58, 61, 62*) cannot be considered to be glycoglycerolipids or glycolipids. However, they have been included in this section on account of their peculiar structure, which mimics a 1-alkyl-2-acyl-3-glycosylglycerol with a five-membered cyclitol instead of a sugar moiety, linked to the glycerol molecule through an ether bond.

4: Monogalactosylmonoacylglycerol (MGDG)

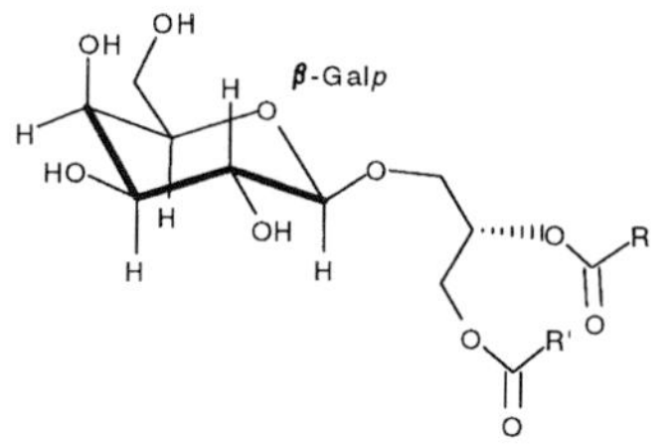

4

Occurrence: Virtually every photosynthetic organism, mainly algae (*63–69, 71*) and cyanobacteria (*63*); *Phyllospongia foliascens* (sponge) (*70*).

Physical Data: $[\alpha]_D$ (*69*), ^{1}H-NMR (*69, 71*), ^{13}C-NMR (*70, 71*).

Galactosyldiacylglycerol (also known as MGDG, **4**) is an ubiquitous metabolite of photosynthetic organisms and is therefore also present in marine plants (mainly algae). MGDGs are always present in the organism as a mixture of homologues differing in the acyl chains. Unlike MGDGs from higher plants, which are characterized by a high content of triunsaturated fatty acids (*63*), MGDGs in marine algae are often found with tetra- and pentaunsaturated fatty acids (*64, 66*). Different composition for the fatty acid linked at the *sn*-1 and *sn*-2 position in MGDGs of algae has been reported (*72*). Galactolipids isolated from the dinoflagellate *Heterosigma akashiwo* have been separated using reversed phase HPLC, and three MGDGs, homogeneous in their fatty acyl chains, have been obtained (heterosigma-glycolipids I-III, all characterized by tetra- or pentaunsaturated fatty acids) (*68, 69*). An MGDG with oxidized fatty acids (12,13-dihydroxyicosapentaenoic acid) has been isolated from the red algae *Graciliaropsis lemaneiformis* (*71*). MGDGs were also isolated from the sponge *P. foliascens* (*70*) but this appears to be the only report about isolation of MGDGs from a marine animal.

References, pp. 291–301

5: 6-O-acyl-monogalactosyldiacylglycerol

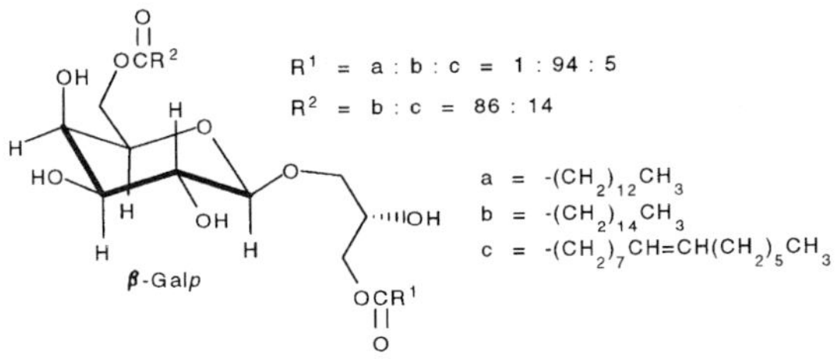
OCR²
OH
H
HO
H
OH
H
H
β-Galp
O
···OH
OCR¹
R¹ = a : b : c = 1 : 94 : 5
R² = b : c = 86 : 14
a = -(CH2)12CH3
b = -(CH2)14CH3
c = -(CH2)7CH=CH(CH2)5CH3

5

Occurrence: *Anabaena flos-aquae* f. *flos-aquae* (cyanobacterium) (*73*).

Physical Data: $[\alpha]_D = -2.1$ ($CHCl_3$); FABMS, IR, ^{1}H- and ^{13}C-NMR (*73*).

The galactosylglycerol **5** shows an unprecedented 6-O-acylation of the galactose residue, while the 2-hydroxy group of the glycerol is free. Structure **5** was demonstrated on the basis of NMR spectroscopic data. In particular, location of the two acyl groups was based on the ^{1}H–^{13}C HMBC 2D-NMR experiment, which displayed couplings between the two ester carbonyl groups and protons at C-6 and C-3′, respectively.

6: Digalactosyldiacylglycerol (DGDG)

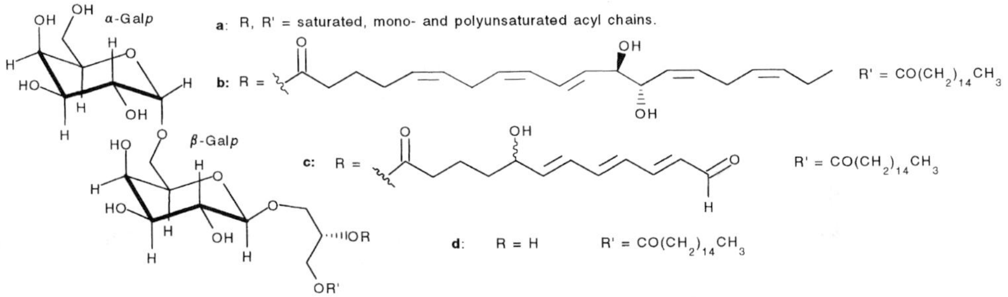
α-Galp
β-Galp
···OR
OR'
a: R, R' = saturated, mono- and polyunsaturated acyl chains.
b: R =
R' = CO(CH2)14CH3
c: R =
R' = CO(CH2)14CH3
d: R = H R' = CO(CH2)14CH3

6

Occurrence: Photosynthetic organisms (*63–69*).

Physical Data: $[\alpha]_D$, ^{1}H- and ^{13}C-NMR (*69, 71*).

Digalactosyldiacylglycerols (DGDGs, **6a**) are commonly found in photosynthetic organisms together with the corresponding MGDGs from which DGDGs are apparently derived *(71)*. Therefore, they show an acyl chain composition similar to that outlined for MGDGs, even if some desaturation of acyl chains can occur after the second galactosidation. Also the MGDG isolated from *Graciliaropsis lemaneiformis* which contains the oxidized 12,13-dihydroxyicosapentaenoic acid has its DGDG analogue (**6b**) *(71)*. In the same alga, a DGDG whose acyl chain has undergone an oxidative cleavage is also present (**6c**) *(74)*. Finally, compound **6d**, a deacylated analogue of **6c**, has been isolated for the first time from the green alga *Ulva pertusa (75)*. It is surprising that the fatty acid part of this galactolipid is homogeneous, being composed exclusively of the saturated palmitic acid.

7: Sulfoquinovosyldiacylglycerol (SQDG)

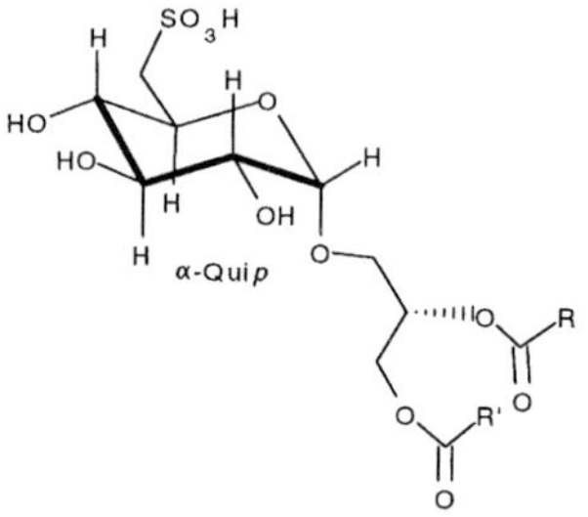

7

Occurrence: Photosynthetic organisms: algae *(64–66, 76–79)*, cyanobacteria *(80)*; *Phyllospongia foliascens* (sponge) *(70)*.

Sulfoquinovosyldiacylglycerol (SQDG, **7**) is present in large amounts in photosynthetic membranes of cyanobacteria, algae and higher plants. This compound, also known trivially as the plant sulfolipid, is the only lipid reported so far with a sulfonic acid functionality. Unlike MGDG and DGDG, in SQDG the sugar unit is linked to the diacylglycerol moiety by an α-glycosidic linkage. SQDG was also isolated from a marine animal, the sponge *Phyllospongia foliascens (70)*, which also contained MGDG. Recently, SQDG from the cyanobacteria *Lyngbya lagerheimii* and *Phormidium tenue* has been shown to possess anti-HIV activity *(80, 81)*. A synthesis of the same SQDG has been reported *(81)*. Deacyl derivatives of SQDG are reported from the algae *Ulva pertusa (75)* and *Sargassum thunbergii (82)*, and from the starfish *Anthocidaris crassispina (83)*.

References, pp. 291–301

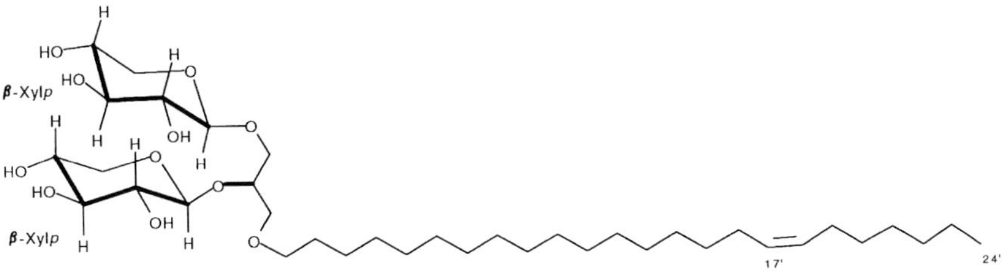

8

Occurrence: *Trikentrion loeve* (sponge) (*57*).

Physical Data: m.p. 98–100°C; [α]$_D$ = −26.9; FABMS, ^{1}H- and ^{13}C-NMR (peracetyl derivative, complete assignment) (*57*).

The ether glycolipid **8** from *T. loeve* has a structure unique among natural glycoglycerolipids, since in **8** two (instead of one as usual) glycerol hydroxy groups are glycosylated, and the sugars are two xyloses in the pyranose form. Structure **8** was elucidated by extensive use of NMR spectroscopy. Specifically, the two sugars were determined as xyloses on the basis of axial-axial coupling constants showed by all the oxymethine protons in the ring. A HOHAHA spectrum was used to find out the multiplicity of all signals, in spite of their severe overlapping. The configuration of the glycerol C-2 was deduced from the [α]$_D$ value of the *O*-alkylglycerol obtained after acid methanolysis of the glycolipid.

9: Crasserides (Keruffarides)

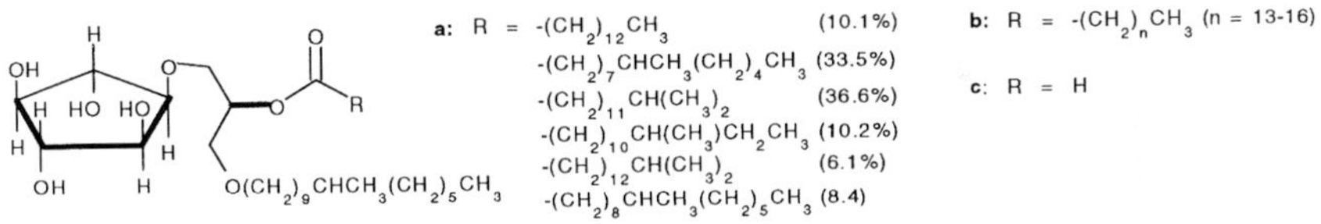

Occurrence: *Pseudoceratina crassa* (sponge) (*58*), *Verongula gigantea* (sponge) (*84*), *Aplysina fistularis fulva* (sponge) (*84*), *Aplysina cauliformis* (sponge) (*84*), *Neofibularia nolitangere* (sponge) (*84*), *Luffariella* sp. (sponge) (*62*), *Biemna* sp. (Sponge) (*62*), *Xestospongia* sp. (sponge) (*62*).

Physical Data: [α]$_D$ = +10.0 (CHCl$_3$); IR, ^{1}H- and ^{13}C-NMR (complete assignment), EIMS (peracetyl derivative).

Strictly speaking, crasserides (**9a**) are not glycoglycerolipids, since they do not contain a sugar molecule glycosidically linked to the glycerol unit. Instead, a five-membered cyclic polyalcohol (cyclitol) is linked to the glycerol moiety through an ether bond. Nevertheless, they have been included in the present review since they mimic the structure of usual glycoglycerolipids and could have a similar biological function. Crasserides are the first and at present the only natural compounds containing a five membered cyclitol. In contrast, the six-membered cyclitol, inositol, is very common and is also found in some phospholipids. Another notable feature of crasserides is the presence of an *O*-alkyl chain instead of an *O*-acyl chain at position 3 of the glycerol unit.

Crasserides (**9a**) were isolated from *P. crassa* as a mixture of homologues differing in the fatty acyl groups. Their structure was established entirely by NMR spectroscopy, except for the nature of the alkyl and acyl chains which required some degradation work. The gross structure of the molecule was deduced from a COSY spectrum, while acylation shifts measured on the peracetyl derivative **9a** were used in order to locate the ether linkages. The relative stereochemistry of the cyclitol was determined by NOE difference measurements. Configuration of the glycerol C-2 was related to that of the cyclitol using NOE difference data measured on a cyclic acetone derivative of **9a** (*58*). Finally, the absolute stereochemistry of the molecule was determined using the Mosher method (*84*).

Crasserides were independently isolated by another group from a *Luffariella* sp., and called keruffarides (**9b**). They were initially described as glycolipids containing an all-*cis* five-membered cyclitol (*61*), but the structure was subsequently revised and showed to be identical with that of crasserides, except for small differences in the alkyl chains (*62*). In addition, the same sponge contained the deacylcrasseride **9c**.

5. Glycosphingolipids

Glycosphingolipids (GSLs) are ubiquitous membrane constituents of animals and plants and are believed to possess a wide range of biological activities, including modulation of growth and regulation of differentiation. They are involved in membrane phenomena, such as cell-cell recognition, cell-cell adhesion, antigenic specificity and other kinds of transmembrane signalling (*85–87*).

GSLs are of wide occurrence as cellular constituents of marine animals, while up to now just one recent paper reports on the isolation of a GSL from a marine plant, *i.e.* the red alga *Corallina pirulifera* (*88*).

References, pp. 291–301

From the literature data, a quite simple GSL pattern appears to occur in most of the marine organisms surveyed so far allowing one to draw some chemotaxonomic generalizations.

Gangliosides, GSLs which are characterized by the presence of a residue of sialic acid in the sugar portion of the molecule and are present in all terrestrial vertebrates have been isolated only from marine invertebrates belonging to the phylum Echinoderma (sea urchins and starfishes). They are believed to play the role of reference structures in the cellular interactions related to sexual reproduction (*89*).

Phosphorus-containing glycosphingolipids, characterized by the presence in the molecule of either a phosphorylcholine or a phosphoethanolamine residue, occur in the extracts of molluscs.

A widespread distribution is characteristic of neutral GSLs, commonly referred to as cerebrosides, which are present in organisms belonging to a number of different taxa. Between them, cerebrosides from marine sponges appear to be of particular interest (*40, 56, 57, 59, 60, 90–96*) because of the α-glycosidic linkage present in most of them between the sugar chain and the aglycon. Until now, this feature has been found only in GSLs from species of the phylum Porifera. GSLs are supposed to be involved in the agamic reproductive processes of marine sponges. These primitive invertebrates possess a power of regeneration allowing a new individual to be generated from a sponge cell suspension, which can rearrange itself into a typical sponge construction. The cell recognition, which is of fundamental importance in such reproductive processes, is supposed to be mediated by the sugar heads of cell surface GSLs (*92*).

5.1. Neutral Glycosphingolipids

10: Glucosylceramide

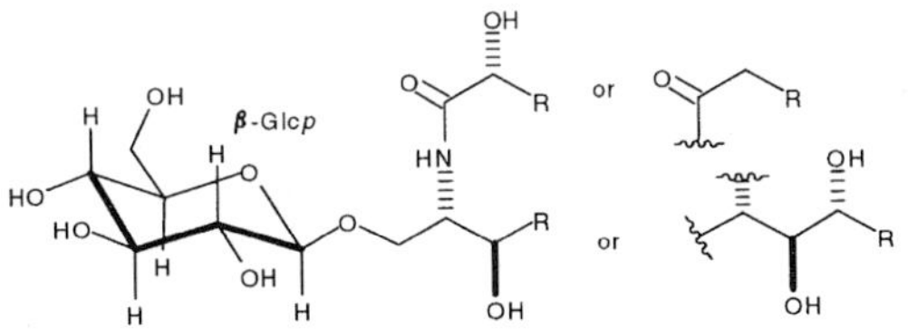

Occurrence: *Asterina pectinifera* (starfish) (*27, 97*), *Acanthaster planci* (starfish) (*98, 99*), *Asterias amurensis* (starfish) (*89, 100*), *Asterias*

amurensis versicolor (starfish) (*101*), *Astropecten latespinosus* (starfish) (*102*), *Ophidiaster ophidiamus* (starfish) (*32*), *Anthocidaris crassispina* (sea urchin) (*103*), *Cucumaria japonica* (holothurian) (*104*), *Cucumaria echinata* (holothurian) (*28*), *Pentacta australis* (holothurian) (*36*), *Chondrilla nucula* (sponge) (*90*), *Haliclona* sp. (sponge) (*91*), *Halichondria japonica* (sponge) (*92*), *Agelas mauritiana* (sponge) (*35*), *Agelas clathrodes* (sponge) (*95*), *Agelas conifera* (sponge) (*56*), *Penaeus aztecus aztecus* (arthropod) (*105*), *Metridium senile* (coelenterate) (106), *Aplysia juliana* (mollusc) (*107*).

Physical Data: ^{1}H- and ^{13}C-NMR of **10** with sphingosines and α-hydroxyacids (*28, 32*) and with phytosphingosines and α-hydroxyacids (*27*).

β-Glucosylceramide **10** is by far the most common glycosphingolipid from marine animals. It is widely distributed in echinoderms (mainly starfishes) and is present in some species of sponges; in addition, it has been occasionally found in organisms of different phyla such as arthropods, coelenterates and molluscs.

A peculiarity of glucosylceramides from starfishes is the frequent occurrence of trihydroxylated LCBs (phytosphingosines) as well as α-hydroxy fatty acids in the ceramide part; in contrast, glucosylceramides from starfish spermatozoa appear to contain mainly dihydroxylated sphingosines (*89, 100*). The stereochemistry of the trihydroxylated sphingosines from starfishes has been studied in detail by synthesizing all the possible diastereomeric phytosphingosines (*108*). A starfish cerebroside with phytosphingosine and α-hydroxyacid has been synthesized (*109*).

The glucosylceramides from the *Agelas* sponges (*35, 56, 95*) and from the starfish *O. ophidiamus* (*32*) contain exclusively a unique triunsaturated C_{18} sphingosine with a conjugated diene and a methyl branching at C-9.

All the glucosylceramides from marine animals are mixtures of homologues, and several attempts have been made to separate individual components of the mixtures by RP-18 HPLC (*27, 28, 32, 99, 101, 102*). Only in one favourable case (*32*), however, could all the components of the mixture be obtained in the pure state. More often, only a few of the fractions obtained from HPLC separation proved to be pure, and the other ones were either not examined at all or studied only with regard to the major component.

11: Galactosylceramide

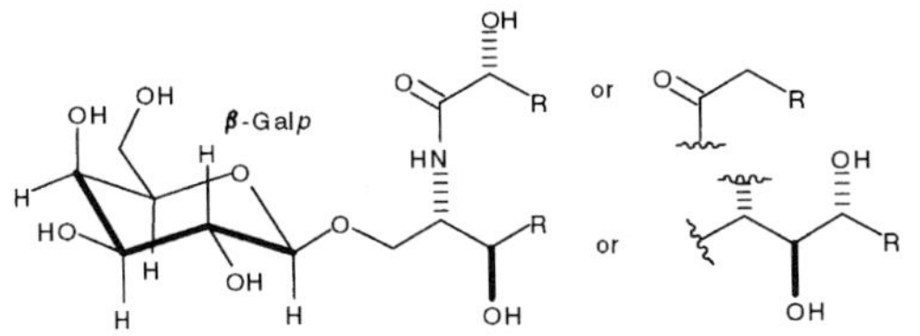

11

Occurrence: *Turbo cornutus* (mollusc) (*110*), *Chlorostoma argyrostoma turbinatum* (mollusc) (*111*), *Aplysia juliana* (mollusc) (*107*), *Chondropsis* sp. (sponge) (*93*), *Halichondria japonica* (sponge) (*92*), *Halichondria panicea* (sponge) (*94*), *Corallina pilulifera* (red alga) (*88*).

Physical Data: FABMS (*107, 111*), ^{1}H- and ^{13}C-NMR of **11** with phytosphingosine and α-hydroxyacid (*88*) and of peracetylated **11** with sphingosines and normal fatty acids (*94*).

β-Galactosylceramides (**11**) are less widespread than glucosylceramides (**10**) and among marine animals they occur only in sponges and molluscs. The sole glycosphingolipid isolated so far from a marine plant is a galactosylceramide (*88*). LCBs of galactosylceramides from molluscs and from *H. panicea* are mainly dihydroxylated. The remaining species produce galactosylceramides with trihydroxylated LCBs and α-hydroxy fatty acids. The presence in *Chondropsis* sp. and *C. pilulifera* of unusual 6-unsaturated trihydroxylated LCBs is worthy of note.

12a–12d and **13a–13f**: Halicylindrosides A_1-A_4 and B_1-B_6

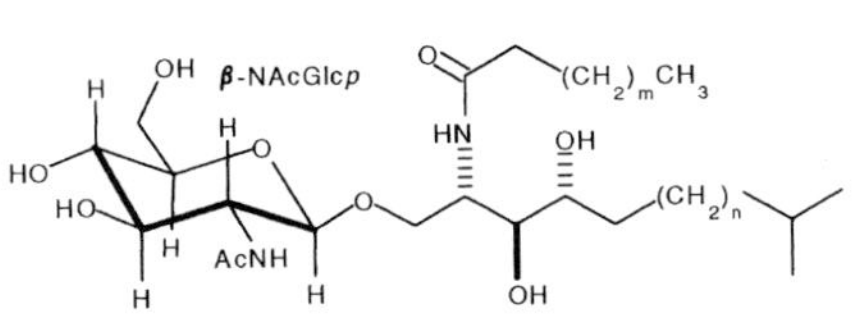

12
a m = 19, n = 10
b m = 19, n = 11
c m = 20, n = 11
d m = 21, n = 11

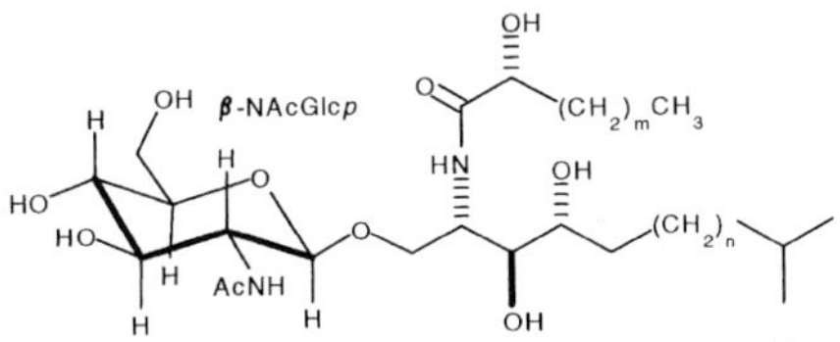

13
a m = 19, n = 8   d m = 19, n = 10
b m = 20, n = 8   e m = 20, n = 9
c m = 19, n = 9   f m = 19, n = 11

Occurrence: *Halichondria cylindrata* (sponge) (*112*).

Physical Data. Halicylindroside A_1 (**12a**): $[\alpha]_D = -20.2$ (Py); Halicylindroside A_2 (**12b**): $[\alpha]_D = -21.1$ (Py.); Halicylindroside A_3 (**12c**): $[\alpha]_D = -19.5$ (Py.); Halicylindroside A_4(**12d**): $[\alpha]_D = -22.3$ (Py); Halicylindroside B$_1$(**13a**): [α]$_D$= −9.2 (Py); Halicylindroside B$_2$(**13b**): [α]$_D$= −9.0 (Py); Halicylindroside B$_3$(**13c**):[α]$_D$= −9.7 (Py); Halicylindroside B$_4$ (**13d**): [α]$_D$= 8.5 (Py); Halicylindroside B$_5$ (**13e**): [α]$_D$= −8.6 (Py); Halicylindroside B$_6$ (**13f**): [α]$_D$= −8.3 (Py); **12a–12d** and **13a–13f**: HRFABMS, IR, ^{1}H- and ^{13}C-NMR (complete assignment) (*112*).

The unique cerebrosides halicylindrosides are the first examples of monoglycosylceramides with an *N*-acetylglucosamine. Two sets of homologous halicylindrosides (**12a–12d** and **13a–13f**) have been isolated as pure compounds by reversed-phase HPLC. The ceramide of halicylindrosides A (**12a–12d**) contains normal fatty acids, whereas that of halicylindrosides B (**13a–13f**) contains α-hydroxy fatty acids; all the LCBs are of the *iso*-type. All the structures, including the absolute configuration of sphingosines and hydroxyacids and of the sugar, were determined by spectroscopic methods. Halicylindrosides are antifungal against *Mortierella remanniana*, and weakly cytotoxic against P338 murine leukemia cells.

14a–14e: Amphicerebrosides B–F

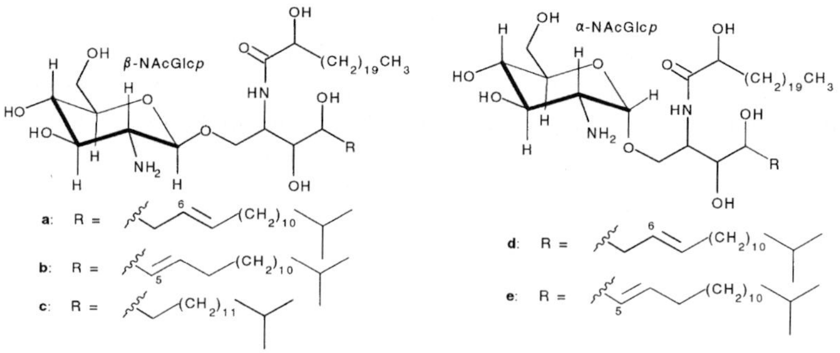

14

Occurrence: *Amphimedon viridis* (sponge) (*91*).

Physical Data: ^{1}H- and ^{13}C-NMR and IR of all six compounds as the respective peracetyl derivatives (*91*).

Isolation of the amphicerebrosides was performed on their peracetyl derivatives after acetylation of the glycolipid extract. Only amphicerebroside D (**14c**) was obtained in the pure state, while the other constituents could only be obtained in an enriched form. Amphicerebrosides B–D (**14a–14c**) are characterized by a β-glycosidic linkage of the glucosamine to the ceramide, whereas in amphicerebrosides E–F (**14d–14e**) the anomeric configuration is α. Individual compounds within each group differ in their LCBs.

Structures of the amphicerebrosides were determined mainly by NMR methods. The reported general structure, however, is not fully convincing, since the authors provide no proof that the saccharide unit of the natural (non acetylated) amphicerebrosides is a glucosamine rather than the more common *N*-acetylglucosamine. In the latter case, amphicerebrosides B–D would be very similar to halicylindrosides B (**13a–13f**) (*112*).

15a–15h: Agelasphins

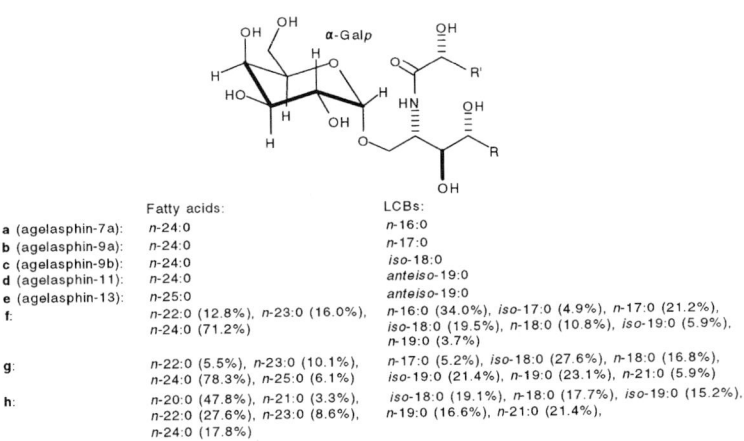

	Fatty acids:	LCBs:
a (agelasphin-7a):	*n*-24:0	*n*-16:0
b (agelasphin-9a):	*n*-24:0	*n*-17:0
c (agelasphin-9b):	*n*-24:0	*iso*-18:0
d (agelasphin-11):	*n*-24:0	*anteiso*-19:0
e (agelasphin-13):	*n*-25:0	*anteiso*-19:0
f:	*n*-22:0 (12.8%), *n*-23:0 (16.0%), *n*-24:0 (71.2%)	*n*-16:0 (34.0%), *iso*-17:0 (4.9%), *n*-17:0 (21.2%), *iso*-18:0 (19.5%), *n*-18:0 (10.8%), *iso*-19:0 (5.9%), *n*-19:0 (3.7%)
g:	*n*-22:0 (5.5%), *n*-23:0 (10.1%), *n*-24:0 (78.3%), *n*-25:0 (6.1%)	*n*-17:0 (5.2%), *iso*-18:0 (27.6%), *n*-18:0 (16.8%), *iso*-19:0 (21.4%), *n*-19:0 (23.1%), *n*-21:0 (5.9%)
h:	*n*-20:0 (47.8%), *n*-21:0 (3.3%), *n*-22:0 (27.6%), *n*-23:0 (8.6%), *n*-24:0 (17.8%)	*iso*-18:0 (19.1%), *n*-18:0 (17.7%), *iso*-19:0 (15.2%), *n*-19:0 (16.6%), *n*-21:0 (21.4%),

15

Occurrence: *Agelas mauritiana* (sponge) (*35, 113*), *Agelas clathrodes* (sponge) (*95*), *Agelas conifera* (sponge) (*56*), *Agelas longissima* (sponge) (*96*).

Physical Data: **15a**: m.p. 193.5–195.5°C; [α]$_D$ = +52.3 (Py); **15b**: m.p. 201.5–203.5°C [α]$_D$ = +49.9 (Py); **15c**: m.p. 211–212°C; [α]$_D$ = +55.0 (Py); **15d**: m.p. 189.5–190.5°C; [α]$_D$ = +51.9 (Py); **15e**: m.p. 215.5–218°C; [α]$_D$ = +48.8 (Py); **15a–15e**: HRFABMS and ^{1}H-NMR (complete assignment) (*35*).

Agelasphins are the simplest members of a class of GSLs, unique of porifera, which possess an α-galactosylceramide as the sugar directly linked to the ceramide moiety. They have been isolated for the first time from *A. mauritiana* (*113*), and later from other species of the genus

Agelas. A mixture of homologous agelasphins were present in *A. mauritiana*; HPLC separation on an RP-18 column yielded some of them in a homogeneous form (**15a–15e**) (*35*). In contrast, agelasphins from *A. clathrodes* [**15f** (*95*)], *A. conifera* [**15g**, (*56*)], and *A. longissima* [**15h**, (*96*)] were analyzed as mixtures of homologues, and their composition in fatty acids and LCBs was completely determined by degradation analysis. The structure of agelasphin was elucidated by spectroscopic (mainly NMR) methods and chemical degradation and the absolute stereochemistry was established by total synthesis (*114*).

Agelasphins possess very interesting biological properties. They show *in vivo* antitumor activity against murine B16 melanoma, but are not active against implanted leukemia P338 cells, and this suggests that they exert their antitumor activity by activating the immune system (*115*). This is confirmed by their remarkable lymphocyte proliferation (LP) stimulatory effect (*40, 115*). In order to explore the possible use of agelasphins as therapeutic agents, a series of analogues have been synthesized (*116*).

16a–16d: Lactosylceramide

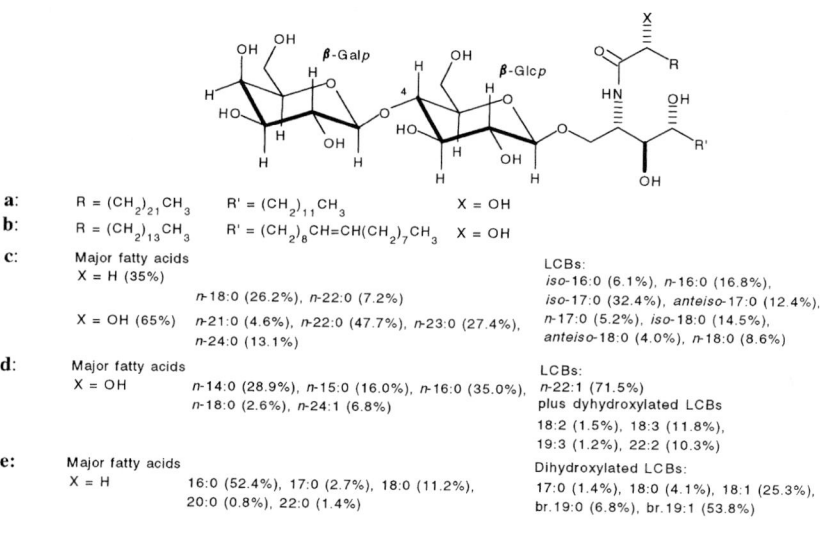

16

Occurrence: *Asterina* (= *Patiria*) *pectinifera* (starfish) (*97*), *Acanthaster planci* (starfish) (*117*), *Asterias amurensis* (starfish) (*100*), *Aplysia juliana* (mollusc) (*107*).

References, pp. 291–301

Physical Data: **16a**: m.p. 196–198°C (MeOH);[α]$_D$ = +5.3 (CHCl$_3$/MeOH 2:1); IR, -ve FABMS, ^{1}H- and ^{13}C-NMR (*117*). **16b**: m.p. 188–190°C (MeOH);[α]$_D$ = +8.5 (CHCl$_3$/MeOH 2:1); IR, -ve FABMS, ^{1}H- and ^{13}C-NMR (*117*).

Although lactosylceramides are the most common ceramide dihexosides in vertebrates, they have been found only in a few species of marine invertebrates. Lactosylceramide has been isolated from the starfish *A. pectinifera (97)* as a mixture of homologues (**16c**). Structure determination was based on sugar analysis, HAKOMORI's method and oxidation with CrO$_3$. A mixture of homologous lactosylceramides (**16d**) has been subsequently isolated from spermatozoa of *A. amurensis (100)*. The long chain bases from this mixture contained appreciable amounts of dihydroxylated LCBs in addition to a major phytosphingosine.

Two homogeneous lactosylceramides [acanthalactoside-A (**16a**) and -B (**16b**)] have been isolated from *A. planci (117)*. Final separation on RP-18 HPLC yielded several fractions, only two of which were homogeneous as shown by FAB mass spectroscopy and degradation analysis. The anomeric configurations were confirmed by ^{1}H- and ^{13}C-NMR data. In contrast with earlier work, the stereochemistry of the ceramide portions of both compounds was also established by comparison of the phytosphingosine obtained from methanolysis with reference samples. A lactosylceramide (**16e**) was also isolated from the sea hare *A. juliana*. Its ceramide portion was found to be composed exclusively of dihydroxylated LCBs and non-hydroxylated fatty acids, very different from that of lactosylceramides of starfishes.

17: Mixture of gentibiosylceramide and cellobiosylceramide

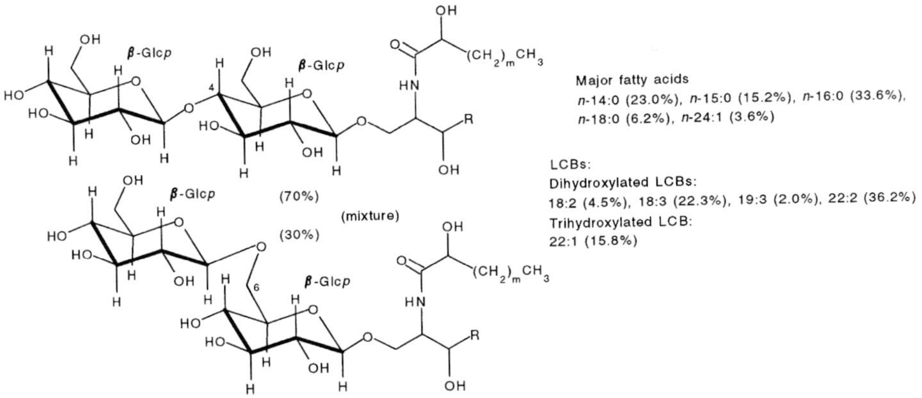

Major fatty acids
n-14:0 (23.0%), n-15:0 (15.2%), n-16:0 (33.6%), n-18:0 (6.2%), n-24:1 (3.6%)

LCBs:
Dihydroxylated LCBs:
18:2 (4.5%), 18:3 (22.3%), 19:3 (2.0%), 22:2 (36.2%)
Trihydroxylated LCB:
22:1 (15.8%)

17

Occurrence: *Asterias amurensis* (starfish) (*100*).

Physical Data: ¹H-NMR and FABMS (*100*).

Gentibiosylceramide and cellobiosylceramide were isolated as an inseparable mixture from the spermatozoa of *Asterias amurensis*. Sugar analysis, HAKOMORI'S method and ¹H-NMR spectroscopy were used for structural identification. FABMS and TLC arguments were used to rule out the possibility that the mixture was a kind of triglycosylceramide. This is the first report of gentibiosylceramide as a natural product.

18: Melibiosylceramide

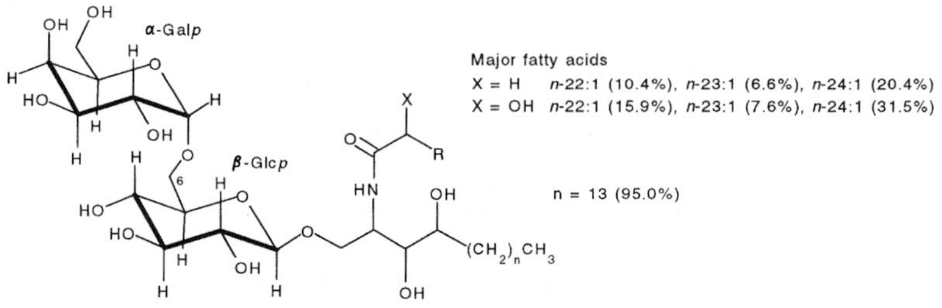

Major fatty acids
X = H n-22:1 (10.4%), n-23:1 (6.6%), n-24:1 (20.4%)
X = OH n-22:1 (15.9%), n-23:1 (7.6%), n-24:1 (31.5%)

n = 13 (95.0%)

Occurrence: *Anthocidaris crassispina* (sea urchin) (*103*)

Physical Data: ¹H-NMR (anomeric protons), FABMS. (*103*)

Compound **18** was isolated from eggs of the sea urchin *Anthocidaris crassispina*, and its structure determined by sugar analysis and methylation analysis. ¹H-NMR spectroscopy was used to confirm the structure and to assign the stereochemistry of the glycosidic linkages. In addition, the chemical shift of the anomeric proton of the sugar directly linked to the ceramide was shown to be dependent on the nature (*i.e.* dihydroxylated or trihydroxylated) of the LCB present in the molecule. This is the first report of the existence of a melibiose-containing glycosphingolipid and of the presence of the melibiose structure in the animal kingdom.

19: Galabiosylceramide GL-3

Occurrence: *Halichondria japonica* (sponge) (*92*)

Physical Data: IR, UV, FABMS, ¹H-NMR (anomeric protons) (*92*).

References, pp. 291–301

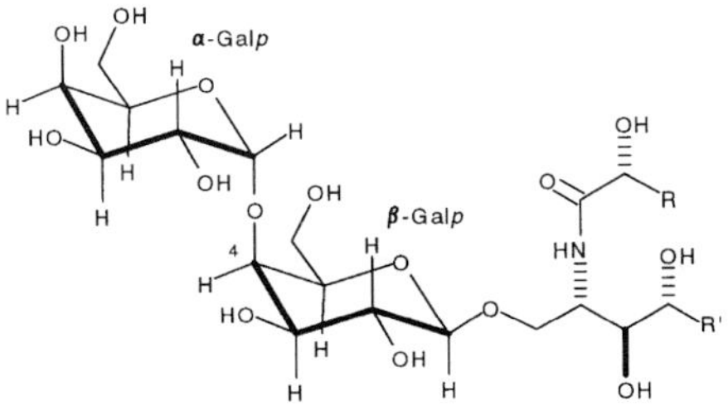

19

Fatty acids:
X = H   14:0 (0.2%), 16:0 (1.3%),
         16:1 (0.3%), 16:2 (0.2%),
         18:0 (0.8%), 18:1 (0.7%)
X = OH   21:0 (5.7%), 22:0 (58.1%),
         23:0 (11.1 %), 24:0 (12.7%),
         25:0 (7.5%)

LCB:
iso-16:0 (0.5%), n-16:0 (3.4%), iso-17:0 (7.3%),
n-17:0 (5.6%), iso-18:0 (43.9%), n-18:0 (16.3%),
anteiso-19:0 (17.5%)

Ceramide digalactoside **19** was isolated as the main glycosphingolipid of the sponge *Halichondria japonica*. A galabiosylceramide had been isolated for the first time from human kidney (*118*), but its ceramide portion was composed of dihydroxylated LCBs and non-hydroxylated fatty acids, leading to a considerably less polar GSL. Structure determination was based on sugar analysis, methylation analysis, IR and FABMS; ^{1}H-NMR was used in order to determine the anomeric configuration.

20: Digalactosylceramide and
21: Trigalactosylceramide of the Gala-6 series

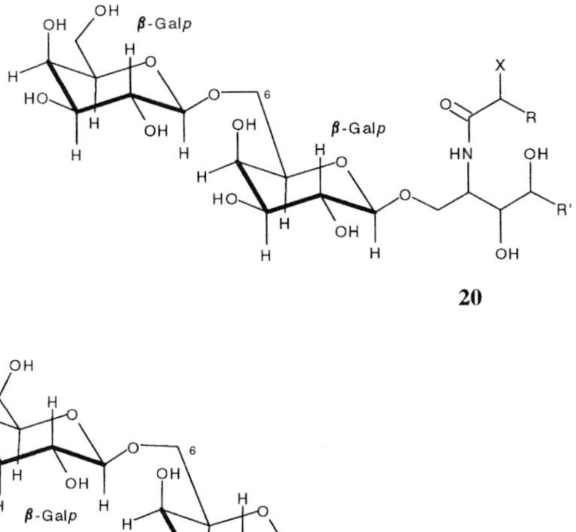

Fatty acids:
X = H   16:0 (63.9%), 16:1 (3.6%),
         17:0 (9.3%), 18:0 (14.3%)
X = OH   16:0 (7.5%)

LCBs:
16:1 (9.7%), 17:1 (9.3%), br.18:1 (13.1%)
18:2 (16.9%), 18:1 (40.3%), 22:2 (11.0%)

20

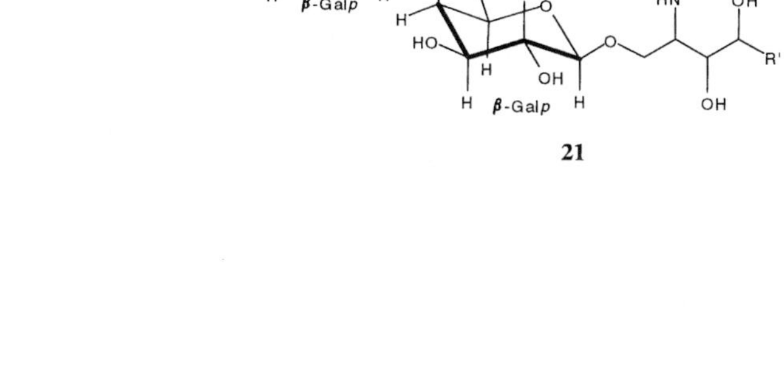

Fatty acids:
X = H   16:0 (70.0%), 16:1 (2.5%),
         17:0 (10.5%), 18:0 (7.9%)
X = OH   16:0 (8.4%)

LCBs:
16:1 (7.4%), 17:1 (9.9%), br.18:1 (15.1%)
18:2 (15.7%), 18:1 (43.5%), 22:2 (8.4%)

21

Occurrence: *Turbo cornutus* (mollusc) *(110)*, *Chlorostoma argyrostoma turbinatum* (mollusc) *(111)*.

Physical Data: ^{1}H-NMR (anomeric protons) *(110)*, FABMS *(111)*.

Compounds **20** and **21** belong to a series of neutral GSLs from *Turbo cornutus* *(110)*, which contain only galactose residues with only 1β → 6 glycosidic linkages. This class of GSLs, characteristic of molluscs, is referred to as Gala-6 series *(111)*. Methanolysis of the permethylated compound **20** was not sufficient for structure determination, since a reference sample of the resulting 2,3,4-tri-*O*-methylgalactoside was not available. Linkage between sugars was therefore established chemically by periodate oxidation, which produced only glyoxal and glyceraldehyde. Anomeric configuration was determined by enzymatic hydrolysis, ^{1}H-NMR spectroscopy, and CrO_3 oxidation. Structure **21** was determined similarly. Compounds **20** and **21** are also components of the sea snail *C. argyrostoma turbinatum* *(111)*, with similar composition in LCBs and fatty acids.

22: Tetragalactosylceramide of the Gala-6 series

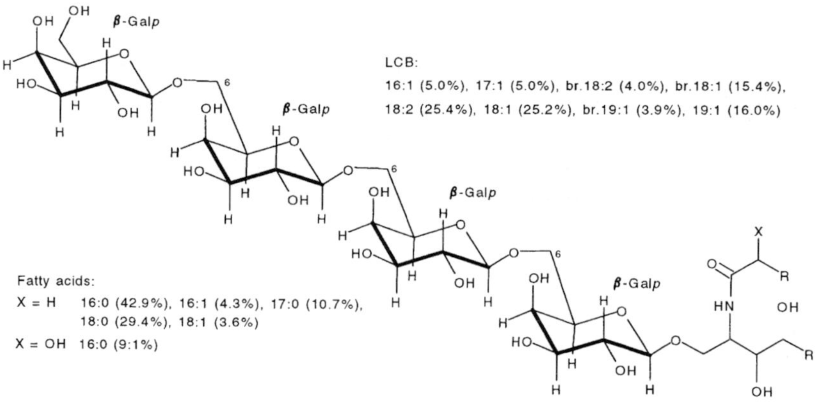

LCB:
16:1 (5.0%), 17:1 (5.0%), br.18:2 (4.0%), br.18:1 (15.4%), 18:2 (25.4%), 18:1 (25.2%), br.19:1 (3.9%), 19:1 (16.0%)

Fatty acids:
X = H   16:0 (42.9%), 16:1 (4.3%), 17:0 (10.7%), 18:0 (29.4%), 18:1 (3.6%)
X = OH 16:0 (9.1%)

22

Occurrence: *Chlorostoma argyrostoma turbinatum* (mollusc) *(111)*.

Physical Data: FABMS and ^{1}H-NMR (anomeric protons) *(111)*.

The tetragalactosyl analog of the Gala-6 series (**22**), whose presence in *T. cornutus* was only hypothesized *(110)*, was isolated and structurally characterized as a component of *C. argyrostoma turbinatum*, which also contained the other elements of the series **20** and **21** *(111)*.

References, pp. 291–301

23

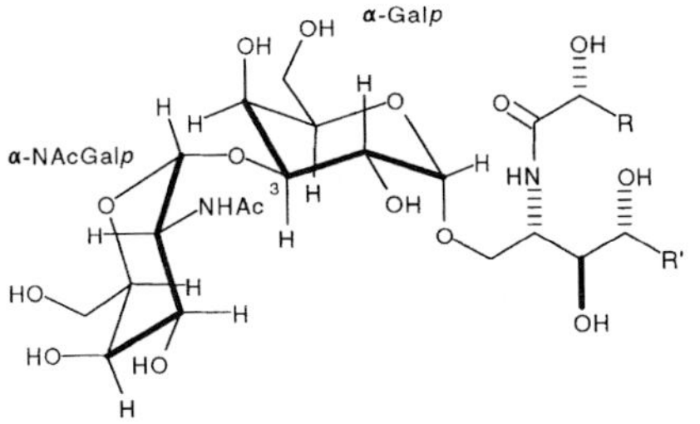

Fatty acids
n-22:0 (18.6%), n-23:0 (16.0%), n-24:0 (65.4%),

LCBs:
n-16:0 (27.9%), iso-17:0 (4.5%), n-17:0 (17.0%), iso-18:0 (21.5%)
n-18:0 (8.3%), iso-19:0 (9.4%), n-19:0 (9.0%), n-21:0 (2.4%)

23

Occurrence: *Agelas clathrodes* (sponge) (*95*), *Agelas dispar* (sponge) (*40*).

Physical Data: $[\alpha]_D = +37$(DMSO); FABMS, ^{1}H- and ^{13}C-NMR (complete assignment) (*95*).

The structure of compound **23** was elucidated (*95*) in a non-destructive way by extensive 1D and 2D NMR studies of its peracetate. ^{1}H chemical shift consideration and ^{1}H-^{1}H coupling constant analysis were used in order to establish the nature of the two saccharide units, while ROESY correlations allowed one to determine the linkage between them. The absolute stereochemistry of the ceramide portion was also established. Since the FABMS spectrum showed that **23** is a very complex mixture of homologues, no attempt was made to separate the individual compounds. Instead, a small amount of **23** was subjected to methanolysis for determination of its fatty acid and LCB composition. Diglycosylceramide **23** shares with a small number of glycosphingolipids, all isolated from sponges, the unique feature of an α-galactose as the first sugar of the saccharide chain. In addition, compound **23** shows an interesting immunostimulatory activity (*40*).

24: Longiside

Occurrence: *Agelas longissima* (sponge) (*60, 96*), *Agelas clathrodes* (sponge) (*95*), *Agelas dispar* (sponge) (*40*), *Agelas conifera* (sponge) (*56*).

Physical Data: $[\alpha]_D = +57$(peracetate, $CHCl_3$); FABMS (peracetate), ^{1}H- and ^{13}C-NMR (peracetate, complete assignment) (60).

		(60)	(96)	(95)	(56)	(40)
Fatty acids:	n-20:0	42.7%				
	n-21:0	1.5%				
	n-22:0	28.5%	18.6%	7.5%	25.2%	
	n-23:0	3.9%	15.6%	10.6%	20.1%	
	n-24:0	100%	18.3%	65.6%	76.8%	54.7%
	n-25:0				5.2%	
LCB:	n-16:0	100%		28.8%		21.5%
	n-17:0			17.1%	12.0%	13.9%
	iso-18:0		18.6%	17.1%	29.3%	13.1%
	n-18:0		23.6%	7.7%	19.5%	5.7%
	iso-19:0		11.9%	5.2%	16.3%	
	n-19:0		17.5%	14.4%	13.0%	33.2%
	n-21:0		28.4%	9.7%	10.0%	12.6%

24

The unprecedented carbohydrate moiety of digalactosylceramide **24** is characterized by a galactose unit in the unusual furanose form. Compound **24** is the most universal GSL in the *Agelas* sponges, being present in all the species examined until now, except in *A. mauritiana* (35) (however, its presence also in this species cannot be excluded). In the original paper (60), compound **24** from *A. longissima* was purified after acetylation by SiO_2 and RP-18 HPLC, and was isolated as a single homologue. In subsequent papers, compound **24** was always isolated as a complex mixture of homologues. Even when a second specimen of *A. longissima* collected in the same area was studied (96), compound **24** was a mixture of homologues, none of which was identical with the one described in (60). These results suggest that fatty acids and LCB composition of GSLs are not necessarily constant within each species, but can vary between specimens, possibly depending on local environmental conditions.

25

Fatty acids:	n-22:0	5.1%
	n-23:0	10.6%
	n-24:0	77.2%
	n-25:0	7.1%
LCBs:	n-17:0	7.8%
	iso-18:0	29.7%
	n-18:0	12.3%
	iso-19:0	20.8%
	n-19:0	22.4%
	n-21:0	7.0%

25

Occurrence: *Agelas conifera* (sponge) (56).

References, pp. 291–301

Physical Data: [α]$_D$= +35(DMSO); FABMS, ^{1}H- and ^{13}C-NMR (complete assignment) (56).

Diglycosylceramide 25, as well as most GSLs from sponges of the genus *Agelas*, is unique in that all its sugars are involved in an α glycosidic linkage. Structure 25 was determined mainly by 1D and 2D NMR experiments, which allowed identification of the component sugars and the linkage between them using methods similar to those described for compound 23. Degradation analysis (methanolysis) was used to establish the composition of the ceramide part, which was a complex mixture of homologues, and as a further confirmation of the nature of the two saccharides.

26

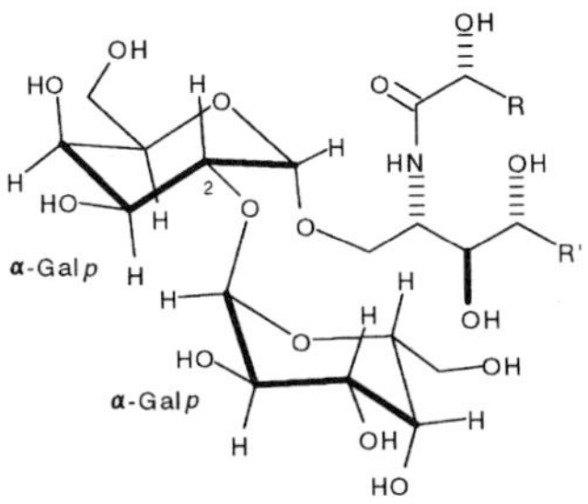

	(96)	(40)
Fatty acids: n-20:0	47.4%	
n-21:0	1.9%	
n-22:0	32.7%	18.6%
n-23:0	2.3%	15.6%
n-26:0	15.7%	65.6%
LCBs: n-16:0		21.7%
n-17:0		15.4%
iso-18:0	25.2%	13.7%
n-18:0	30.3%	7.8%
iso-19:0	17.2%	
n-19:0	18.8%	29.8%
n-21:0	15.5%	11.6%

26

Occurrence: *Agelas longissima* (sponge) (96), *Agelas dispar* (sponge) (40).

Physical Data: [α]$_D$= +21(DMSO); FABMS (peracetate), ^{1}H- and ^{13}C-NMR (peracetate, complete assignment).

About structure determination of digalactosylceramide 26, much the same can be said as for compound 23. It is worthy of note that saccharide chains of all three GSLs isolated from *A. longissima* (15, 24, and 26) are composed exclusively of galactose.

27a: Axiceramide A
27b: Axiceramide B

Occurrence: *Axinella* sp. (sponge) (59).

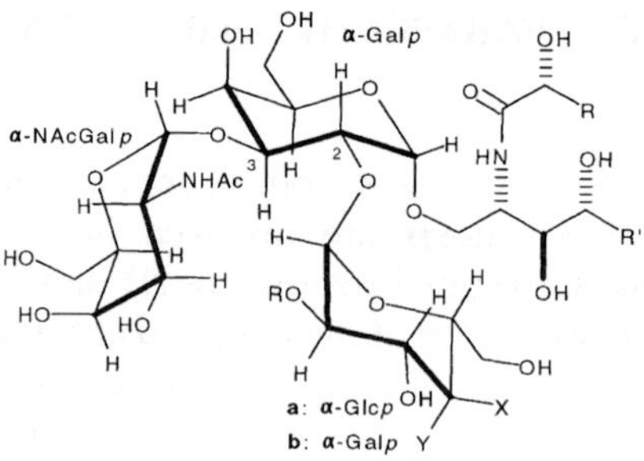

27

	a	b
Fatty acids: n-24:0	50.2%	50.2%
n-25:0	35.7%	35.7%
n-26:0	14.1%	14.1%
LCBs: n-16:0	5.4%	5.4%
n-17:0	11.2%	11.2%
iso-18:0	17.2%	17.2%
n-18:0	32.9%	32.9%
iso-19:0	9.4%	9.4%
n-19:0	14.4%	14.4%
n-21:0	9.5%	9.5%
	X = OH	X = H
	Y = H	Y = OH

Physical Data: $[\alpha]_D = +88$ (peracetate, CHCl$_3$); FABMS (peracetate), ^{1}H- and ^{13}C-NMR (peracetate, complete assignment) (59).

Axiceramide-A (**27a**) and -B (**27b**), isolated from a sponge of the genus *Axinella*, are the first examples of natural GSLs with three hexose units all engaged in α-glycosidic linkages. Compounds **27a** and **27b** were obtained as an inseparable mixture, but after acetylation their peracetyl derivatives could be separated. Both compounds were obtained as mixtures of homologues. Structures **27a** and **27b** were entirely established from FABMS spectra and ^{1}H- and ^{13}C-NMR experiments (including 2D NMR), except for the composition of the ceramide part which required methanolytic degradation. In addition, the absolute stereochemistry of the phytosphingosine obtained by methanolysis could be established. Superimposition of some key signals required most of NMR experiments to be performed in two different solvents, namely CDCl$_3$ and CD$_3$COCD$_3$.

28

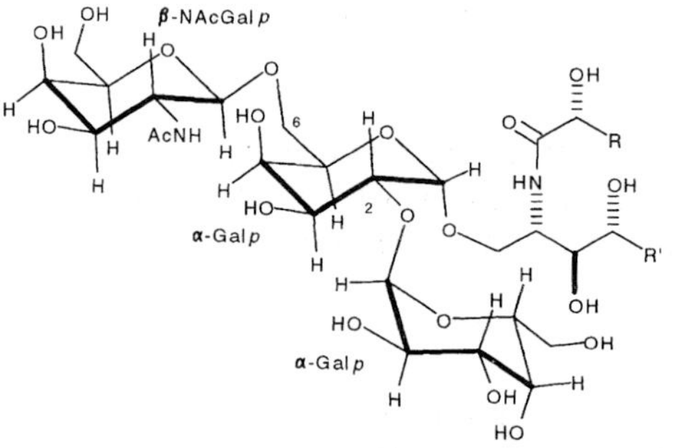

Fatty acids: n-22:0	24.4%
n-23:0	22.9%
n-24:0	53.5%
LCBs: n-16:0	20.7%
n-17:0	12.4%
iso-18:0	16.0%
n-18:0	6.7%
n-19:0	34.8%
n-21:0	9.4%

28

Occurrence: *Agelas dispar* (sponge) (40).

References, pp. 291–301

Physical Data: $[\alpha]_D = +39.6$ (DMSO); FABMS (peracetate), ^{1}H- and ^{13}C-NMR (complete assignment) (40).

Until now, triglycosylceramide **28** is the most complex GSL isolated from an *Agelas* sponge. Like its simpler analogues, it possesses an α-Gal as the first residue of the sugar chain, but for the first time in this kind of GSLs this galactose residue is glycosylated at position 6. In analogy with previous GSLs from *Agelas*, determination of structure **28** was based almost exclusively on NMR studies of the peracetyl derivative of **28**. In this case, an additional difficulty arose since in the ^{1}H-NMR spectrum of peracetylated **28** protons H-1 and H-2 of the terminal galactose were coincident, thus precluding determination of the anomeric configuration of this saccharide. This could only be accomplished by analysis of the NMR spectrum of the non-acetylated **28**.

29

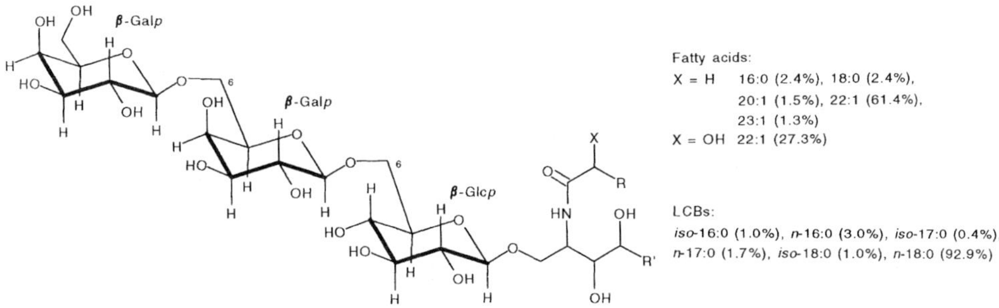

Fatty acids:
X = H   16:0 (2.4%), 18:0 (2.4%), 20:1 (1.5%), 22:1 (61.4%), 23:1 (1.3%)
X = OH 22:1 (27.3%)

LCBs:
iso-16:0 (1.0%), *n*-16:0 (3.0%), *iso*-17:0 (0.4%), *n*-17:0 (1.7%), *iso*-18:0 (1.0%), *n*-18:0 (92.9%)

29

Occurrence: *Hemicentrotus pulcherrimus* (sea urchin) (49).

Physical Data: FABMS and ^{1}H-NMR (anomeric protons) (49).

Triglycosylceramide **29** was isolated from the eggs of *H. pulcherrimus*, and its structure was clarified by negative-ion FABMS spectroscopy, sugar analysis, methylation analysis, and enzymatic hydrolysis. Anomeric configurations were established by CrO$_3$ oxidation, enzymatic hydrolysis, and ^{1}H-NMR spectroscopy. In addition, the β configuration of the linkage between the two galactose units was confirmed using the enzyme-linked immunosorbent assay (ELISA) with rabbit antisera anti-gala-6β and anti-gala-6α. Compound **29** is the first ceramide trihexoside from echinoderms, and has a novel carbohydrate structure. Its presence in *H. pulcherrimus* confirms the ability of sea urchins to produce a variety of unique sugar structures.

30

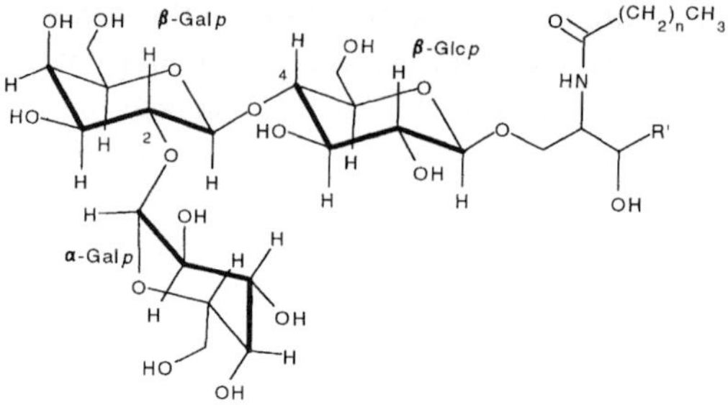

Fatty acids:
16:0 (61.1%), 17:0 (2.7%), 18:0 (19.1%), 20:0 (1.6%), 22:0 (2.1%)
LCBs:
n-17:0 (2.7%), n-18:0 (6.3%), n-18:1 (27.4%), br.19:0 (4.8%), br.19:1 (47.0%)

30

Occurrence: *Aplysia juliana* (mollusc) (*107*).

Physical Data: FABMS (*107*).

Triglycosylceramide **30** is the major neutral GSL component of *A. juliana*. The sugar sequence in the trisaccharide moiety was deduced by carbohydrate analysis, methylation analysis and examination of FABMS fragmentations. The ^{1}H-NMR spectrum of **30** displayed two β and one α anomeric protons; partial hydrolysis of **30** produced β-lactosylceramide and β-glucosylceramide, therefore the α-configuration was attributed to the terminal galactose. This triglycosylceramide is supposed to be one of the precursors of the phosphorus-containing glycolipids isolated from this and other species of the genus *Aplysia* (see below). Particularly, it could be the direct precursor of phosponoglycolipid, AJPnGL (**40**) isolated from the same organism which has almost the same structure with respect to the ceramide moiety (*119*).

31

Occurrence: *Aplysia juliana* (mollusc) (*107*).

Physical Data: FABMS (*107*).

The fucosylated triglycosylceramide **31**, also isolated from *A. juliana*, has almost the same structure as triglycosylceramide **30**, and

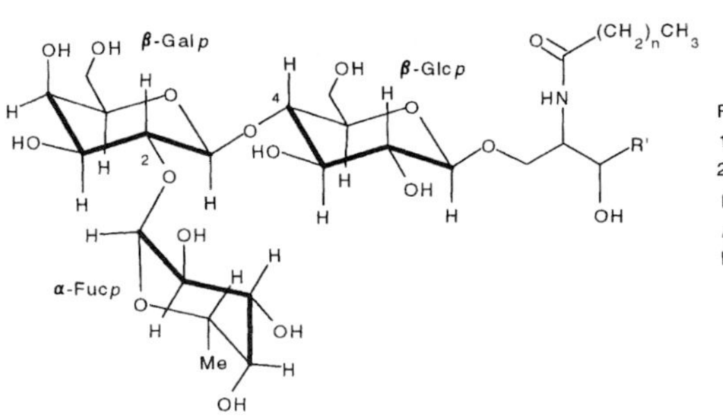

31

Fatty acids:
16:0 (59.7%), 17:0 (2.3%), 18:0 (12.3%),
20:0 (1.0%), 22:0 (2.7%)
LCBs:
n-17:0 (1.8%), n-18:0 (5.5%), n-18:1 (21.8%),
br.19:0 (5.5%), br.19:1 (57.5%)

its structure was determined similarly. It was reported for the first time as a metabolite in a marine organism in (*107*), but had already been isolated from the small intestine of the rat. Compound **31** could be the biogenetic precursor of those phosphonoglycolipids which possess a fucose branch, such as FGL-I (**48**) and FGL-IIb (**45**) from *Aplysia kurodai*.

32

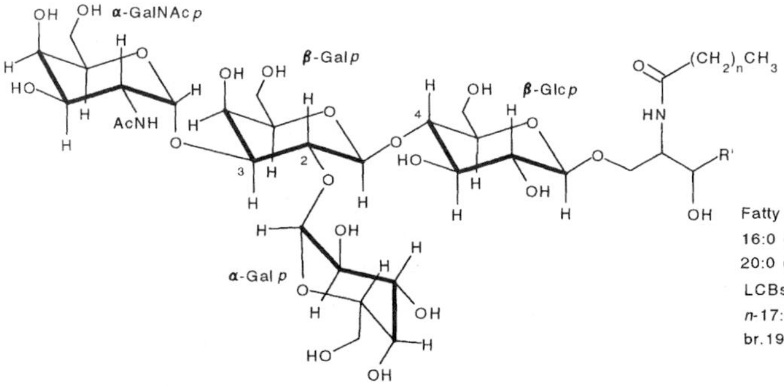

Fatty acids:
16:0 (57.5%), 17:0 (3.0%), 18:0 (20.6%),
20:0 (2.8%), 22:0 (3.2%)
LCBs:
n-17:0 (3.3%), n-18:0 (7.4%), n-18:1 (28.5%),
br.19:0 (4.2%), br.19:1 (45.7%)

32

Occurrence: *Aplysia juliana* (mollusc) (*107*).

Physical Data: FABMS (*107*).

Tetraglycosylceramide **32** is the most complex of the neutral GSLs isolated from *A. juliana*. In addition to the usual methylation analysis, structure elucidation of compound **32** was based on FABMS, which demonstrated the presence of a terminal acetylhexosamine, and partial hydrolysis, which yielded the triglycosylceramide **30**, lactosylceramide, and glucosylceramide, all already identified in *A. juliana*, and the cerebroside GalNAcα1-3Galβ1-4Glcα1-1Cer, characterized on the basis of FABMS, ^{1}H-NMR and methylation analysis. The oligosaccharide chain of **32** has the same structure as one common to most phosphoglycolipids found in *Aplysia kurodai*, to which it is supposed to be biogenetically related.

33

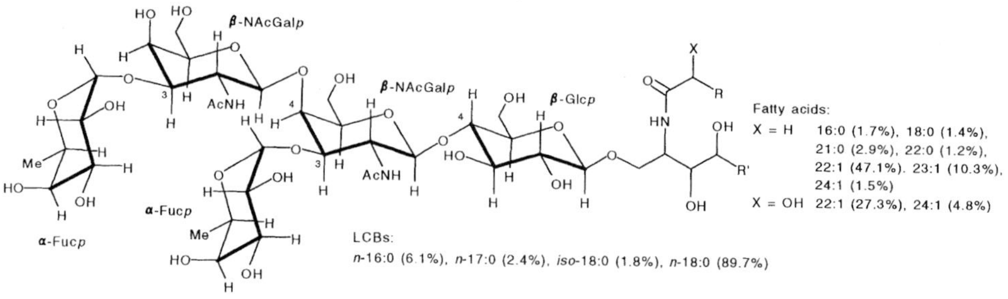

[Structure of compound 33 showing: α-Fucp, α-Fucp, β-NAcGalp, AcNH, β-NAcGalp, AcNH, β-Glcp units linked to ceramide]

Fatty acids:
X = H   16:0 (1.7%), 18:0 (1.4%), 21:0 (2.9%), 22:0 (1.2%), 22:1 (47.1%), 23:1 (10.3%), 24:1 (1.5%)
X = OH 22:1 (27.3%), 24:1 (4.8%)

LCBs: n-16:0 (6.1%), n-17:0 (2.4%), iso-18:0 (1.8%), n-18:0 (89.7%)

33

Occurrence: *Hemicentrotus pulcherrimus* (sea urchin) (*55, 120*).

Physical Data: FABMS (*120*), ^{1}H-NMR (complete assignment) (*55*).

The difucosylated cerebroside **33** was isolated from the eggs of *H. pulcherrimus* and is the first and until now the sole report of a fucosylated GSL from echinoderms. Although methanolysis and methylation analysis of **33** revealed only 1 mol of fucose per mol of GSL, the presence of two fucose units was demonstrated by FABMS and ^{1}H-NMR. This contradiction was ascribed to a relative instability of fucose residues to methanolysis and hydrolysis. ^{1}H-NMR was also used for determining anomeric configurations. Finally, a partial acid hydrolysis of **33** followed by methylation analysis was performed for elucidation of the linkages between sugars (*120*). In a related paper (*55*), the structure of the same GSL was redetermined and confirmed by using 2D NMR methods (DQF-COSY, HOHAHA, NOESY).

34

Occurrence: *Haliotis japonica* (mollusc) (*121*).

Physical Data: ^{1}H-NMR (anomeric protons) (*121*).

Structure of the difucosylated pentaglycosylceramide **34** was elucidated by sugar analysis and methylation analysis in combination with partial hydrolysis. The anomeric configurations were determined by ^{1}H-

References, pp. 291–301

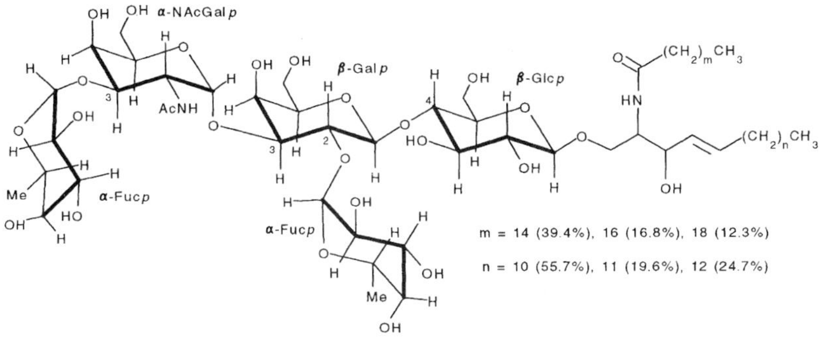

34

m = 14 (39.4%), 16 (16.8%), 18 (12.3%)
n = 10 (55.7%), 11 (19.6%), 12 (24.7%)

NMR. Even though the presence of fucose in GSLs from molluscs had already been reported (*17, 22*), compound **34** was the first GSL of this kind whose structure has been completely established.

5.2. Phosphorus-Containing Glycosphingolipids

35–36: AEP- and MAEP-galactosylceramide

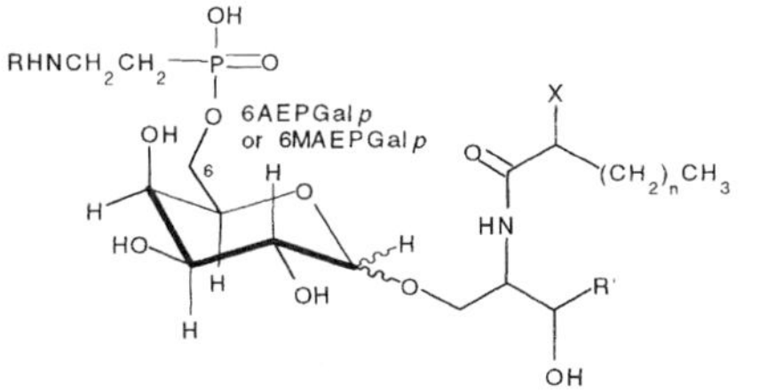

a: Major fatty acids:
  X = H   16:0 (32.6%), 17:0 (4.1%), 18:0 (3.0%)
  X = OH 16:0 (31.2%), br.17:0 (3.4%), 17:0 (6.1%), 18:0 (5.4%)

  Major dihydroxylated LCBs:
  16:1 (2.7%), 17:1 (5.7%), br.18:1 (7.0%), 18:1 (18.0%), 18:2 (14.7%), br.19:1 (3.1%), 22:2 (3.5%)
  Major trihydroxylated LCBs:
  18:0 (2.5%), 22:1 (11.2%)

b: Major fatty acids:
  X = H   16:0 (53.3%), 17:0 (8.1%), 18:0 (5.9%)
  X = OH 16:0 (14.6%), 17:0 (4.7%), 18:0 (3.7%)

  Major dihydroxylated LCBs:
  16:1 (5.3%), 17:1 (8.0%), br.18:1 (4.2%), 18:1 (14.6%), 18:2 (11.3%), br.19:1 (3.2%), 20:1 (2.5%), 22:2 (36.6%)
  Major trihydroxylated LCBs:
  22:1 (6.3%)

c: Major fatty acids:
  X = H   16:0 (49.7%), br.17:0 (3.7%), 17:0 (3.3%), 18:0 (6.3%), 18:1 (5.7%)
  X = OH 16:0 (12.2%), br.17:0 (3.1%), 18:0 (3.4%), 20:0 (2.4%)

  Major dihydroxylated LCBs:
  16:1 (9.0%), 17:1 (3.7%), *iso*-18:1 (9.2%), 18:1 (26.0%), 18:2 (12.4%), *iso*-19:1 (6.5%), *anteiso*-19:1 (18.3%), 22:2 (9.6%)
  Major trihydroxylated LCBs:
  18:0 (2.0%)

35 R = H   **36** R = Me

Occurrence: *Turbo cornutus* (mollusc) (*13, 122, 123*), *Monodonta labio* (mollusc) (*124*), *Chlorostoma argyrostoma turbinatum* (mollusc) (*111*).

N-Methyl-2-aminoethylphosphonylgalactosylceramide (MAEP-galactosylceramide) is the simplest example of a phosphoglycosphingolipid, a class of GSL characteristic of molluscs. MAEP-galactosylceramide (**36a**) was isolated for the first time from the viscera of *T. cornutus*. After a tentative characterization turned out to be incorrect (*13*), its structure was definitively determined (*122*) using chemical methods such as partial acid hydrolysis, alkaline hydrolysis, and periodate oxidation. Structures of the breakdown products were mainly identified by combined gas chromatography and mass spectrometry. In a subsequent paper (*123*), MAEP-galactosylceramide (**36b**) was also reported to be present in the muscle of *T. cornutus*, together with smaller amounts of the closely related 2-aminoethylphosphonylgalactosylceramide (AEP-galactosylceramide, **35b**). Compounds **35b** and **36b** were not separated, but were analyzed as a mixture with the same methods as for **36a**.

As regards other species of molluscs, AEP- and MAEP-galactosylceramide were also isolated from *M. labio* (**35c** and **36c**, respectively) (*124*), while *C. argyrostoma turbinatum* contained only MAEP-galactosylceramide (*111*). It is worthy of note that, to our knowledge, the anomeric configuration of the two phosphoglycosphingolipids has never been determined. It has been reported as β in a review paper (*25*), and indeed this is very probable, since this is the anomeric configuration of analogues of **35–36** such as **38** or **39**; however, we could not find any demonstration of this structural assignment.

37: MAEP-glucosylceramide

Fatty acids:
X = H   22:0 (3.0%), 22:1 (17.4%),
       24:0 (2.0%), 24:1 (10.5%)
X = OH 22:1 (37.7%), 24:1 (23.8%)

LCBs:
14:1 (17.4%), br.18:3 (24.3%), 18:3 (48.2%)

37

Occurrence: *Euphausia superba* (arthropod) (*39*).

Physical Data: FABMS, ^{1}H-NMR (anomeric proton).

N-Methyl-2-aminoethylphosphonylglucosylceramide (MAEP-glucosylceramide, **37**) is quite an exception among phosphonoglycolipids, since it was isolated from *E. superba*, an arthropod, and not from a mollusc like most phosphonoglycolipids isolated until now. Its structure

References, pp. 291–301

elucidation was based largely on the same methods as those described for compound **36**, except that in this case the anomeric configuration was determined (using ^{1}H-NMR).

38

a: m = 14, R = -CH=CH-CH$_2$-CH$_2$-CH=CH-(CH$_2$)$_8$CH$_3$
b: m = 14, R = -CH=CH-(CH$_2$)$_{12}$CH$_3$
c: m = 15, R = -CH=CH-(CH$_2$)$_{12}$CH$_3$
d: m = 16, R = -CH=CH-(CH$_2$)$_{12}$CH$_3$
e: m = 14, R = -(CH$_2$)$_{14}$CH$_3$

38

Occurrence: *Marphysa sanguinea* (annelid) (*125*), *Neanthes diversicolor* (annelid) (*126*).

Physical Data: **38a**: FABMS; ^{1}H- and ^{13}C-NMR (complete assignment) (*126*). **38b**: m.p. 173–188°C; $[\alpha]_D = +10.6$ (MeOH); FABMS; ^{1}H- and ^{13}C-NMR (complete assignment) (*125*). **38c**: m.p. 160–185°C; $[\alpha]_D = +9.5$ (MeOH); FABMS; ^{1}H- and ^{13}C-NMR (complete assignment) (*125*). **38d**: FABMS; ^{1}H- and ^{13}C-NMR (complete assignment) (*126*). **38e**: FABMS; ^{1}H- and ^{13}C-NMR (complete assignment) (*126*).

The galactosylceramides **38b** and **38c**, obtained in the pure state from the annelid *M. sanguinea* (*125*), are characterized by the presence of a phosphocholine group linked to the sugar chain, a feature reported here for the first time. Structures of compounds **38b** and **38c**, including the anomeric configurations, were suggested by FABMS and NMR and confirmed by HF hydrolysis of **38b**, which selectively cleaved phosphate ester bonds yielding galactosylceramide, and by acid methanolysis followed by acetylation, which generated a derivative of galactose identified as 1-*O*-methyl-2,3,4-tri-*O*-acetyl-α-D-galactopyranose-6-phosphocholine.

The same compounds **38b** and **38c**, together with three more homologues **38a**, **38d**, and **38e**, were isolated from another annelid, *N. diversicolor*. Separation of the five GSLs required application of the recycling HPLC technique. The position of the Δ^8 double bond in compound **38a** was established by conversion of the relevant sphingoid base into the dimethyl disulfide derivative by treatment with carbon disulfide and iodine, and subsequent EIMS analysis.

39

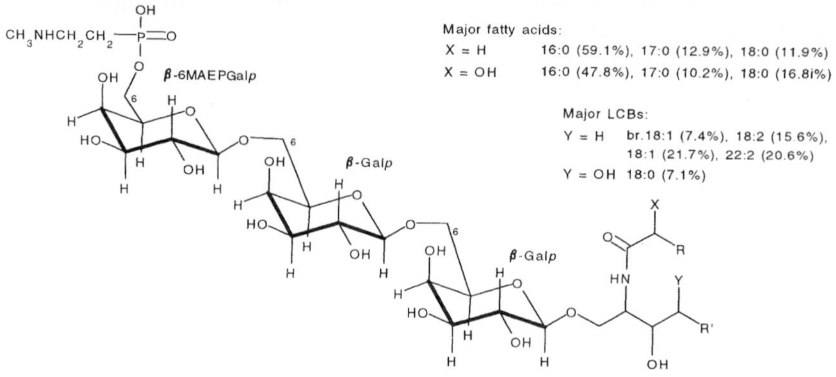

CH₃NHCH₂CH₂—P(OH)(=O)—O— β-6MAEPGal*p* ... β-Gal*p* ... β-Gal*p* —O—CH₂—CH(NHC(=O)CHXR)—CH(OH)—R'

Major fatty acids:
X = H   16:0 (59.1%), 17:0 (12.9%), 18:0 (11.9%)
X = OH   16:0 (47.8%), 17:0 (10.2%), 18:0 (16.8i%)

Major LCBs:
Y = H   br.18:1 (7.4%), 18:2 (15.6%), 18:1 (21.7%), 22:2 (20.6%)
Y = OH   18:0 (7.1%)

39

Occurrence: *Turbo cornutus* (mollusc) (*127*).

Physical Data: IR and FABMS (*127*).

The phosphonotriglycosylceramide **39**, isolated form the viscera of *T. cornutus*, can be considered as a higher homologue of the MAEP-galactosylceramide (**36**) from the same organism, or as the phosphonylated form of the trigalactosylceramide of the Gala-6 series (**21**), also isolated from *T. cornutus*. Treatment of **39** with HF produced **21**, thus determining the nature of its sugar moiety. Partial hydrolysis of **39** followed by trimethylsilylation and GC-MS analysis gave a penta-TMS-MAEP-hexose whose mass spectrum was coincident with that of an authentic sample of penta-TMS-6-*O*-MAEP-galactose. Structure **39** was confirmed using FABMS and methylation analysis.

40: AJPnGL

Occurrence: *Aplysia juliana* (mollusc) (*119*).

Physical Data: FABMS, ¹H-NMR (anomeric protons) (*119*).

The phosphonoglycolipid AJPnGL (**40**) was isolated from the eggs of *A. juliana*. Structure **40** was determined by methylation analysis, FABMS, ¹H-NMR, HF hydrolysis (dephosphonylation), and partial acid hydrolysis which gave an AEP-diglycosylceramide and an AEP-monoglycosylceramide, thus locating the aminoethylphosphonyl group on the inner sugar. The sugar backbone of compound **40** is rather

References, pp. 291–301

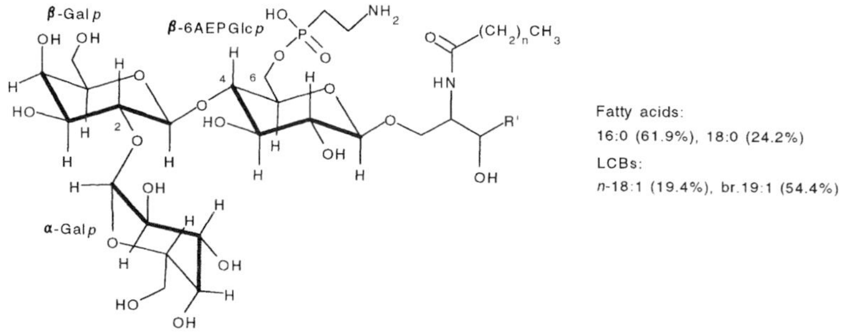

40

different from those of other phosphonoglycolipids of *Aplysia* species; a possible biosynthesis of **40** from the triglycosylceramide **31**, isolated from the same organism, has been proposed *(119)*.

41: SGL-I

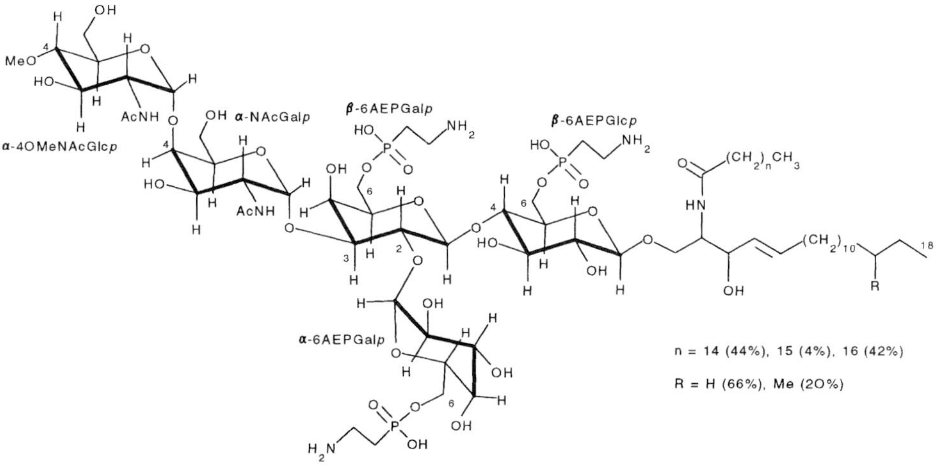

$n = 14$ (44%), 15 (4%), 16 (42%)

R = H (66%), Me (20%)

41

Occurrence: *Aplysia kurodai* (mollusc) *(128, 129)*.

Physical Data: FABMS and ^{1}H-NMR (anomeric protons) *(129)*.

The sea hare *Aplysia kurodai* produces a surprising variety of phosphorus-containing GSLs. Some of them were separated and isolated several years before their full structural elucidation became feasible. SGL-I (**41**) was isolated and partially characterized in 1980 *(128)*, but the complete structural determination appeared only in 1987 (*129*). The presence in SGL-I of 4-*O*-methylglucosamine, never detected before as a natural sugar, was demonstrated on the basis of the mass spectrum of its alditol acetate and of demethylation with BF_3, which gave glucosamine. The location of the three 2-aminoethylphosphonate groups was determined by comparing the results of methylation analyses of intact SGL-I and SGL-I pre-treated with HF in order to remove aminoethylphosphonyl groups. ^{1}H-NMR was used for determination of anomeric configuration. The anomeric signals were assigned to the relevant anomeric protons by comparison with other phosphonoglycolipids of the same family.

42: SGL-II

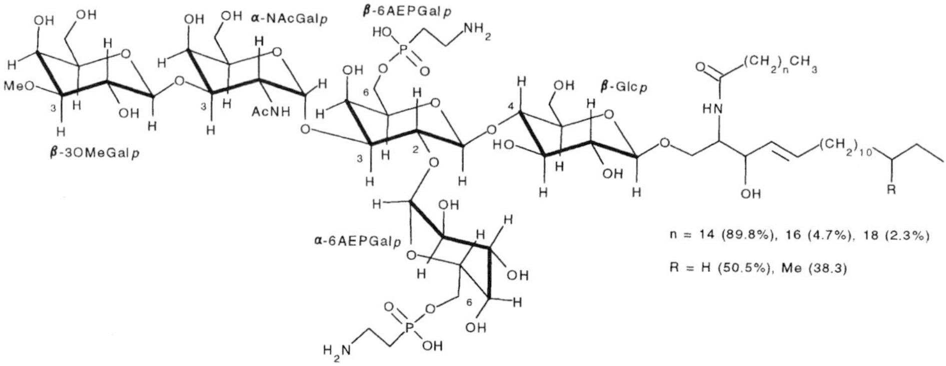

42

Occurrence: *Aplysia kurodai* (mollusc) (*128, 130, 131*).

Physical Data: FABMS and ^{1}H-NMR (anomeric protons) (*131*).

Of the over 30 phosphorus-containing glycolipids which are known to be produced by *A. kurodai* (*132*), the diphosphonoglycolipid SGL-II (**42**) was the first one whose structure was determined completely. The sugar backbone of SGL-II is the same as those of several other phosphonoglycolipids from *A. kurodai*, which differ only in the number and the kind of the groups, such as *O*-methyl or AEP, linked to the saccharide chain. SGL-II is a major component of the skin of *A. kurodai* and was rather easily purified compared with other glycolipids from *Aplysia*. After treatment with HF, the structure of the dephosphonylated glycolipid was established by partial hydrolysis and subsequent methylation analysis of the fragments. Comparison of the ^{1}H-NMR

References, pp. 291–301

spectrum of the HF-treated glycolipid with those of its fragments allowed assignment of the anomeric proton signals and consequently determination of the anomeric configurations, which were consistent with those deduced by CrO$_3$ oxidation. The presence of two AEP groups was demonstrated by colorimetric analysis. They were located as shown in **42** since controlled hydrolysis of the intact SGL-II produced only 6-*O*-(2-aminoethylphosphonyl)galactose, which was identified by comparing its GLC and GC-MS behavior with that previously reported (*123*).

SGL-II proved to be a hapten, and an antiserum raised against it reacted with SGL-II itself, SGL-I' (**43**), F-21 (**44**), and other minor glycolipids from *A. kurodai*. The same antiserum was used for immunohistochemical studies on *A. kurodai* (*133*).

43: SGL-I'

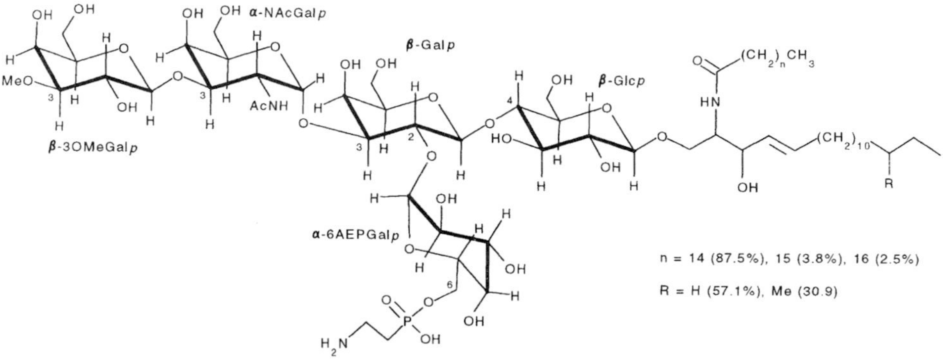

n = 14 (87.5%), 15 (3.8%), 16 (2.5%)

R = H (57.1%), Me (30.9)

43

Occurrence: *Aplysia kurodai* (mollusc) (*134*), *Dolabella auricolaria* (mollusc) (*135*).

Physical Data: ^{1}H-NMR (anomeric protons) (*134*).

The phosphonoglycolipid SGL-I' (**43**) is very similar to SGL-II (**42**), differing from the latter compound in the absence of one AEP group. Methods for structure determination were also similar. SGL-I' seems to be specifically concentrated in the skin of *A. kurodai* (*133*). The same GSL was subsequently found to be a component of another sea hare, *D. auricolaria* (*135*).

44: F-21

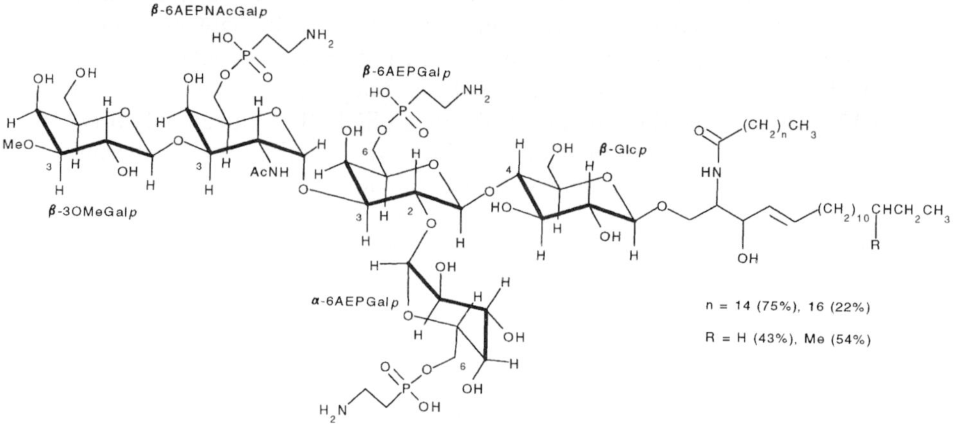

Occurrence: *Aplysia kurodai* (mollusc) (*136*).

Physical Data: FABMS and ¹H-NMR (selected sugar protons) (*136*).

The triphosphonoglycolipid F-21 (**44**) is specifically concentrated in the nervous system of *A. kurodai*. Its sugar backbone is the same as SGL-II (**42**); F-21 is also characterized by the presence of a 3-*O*-methylgalactose at the non-reducing end of the saccharide chain. Compared to SGL-II, F-21 contains one additional AEP group which is linked to the N-acetylgalactosamine. The presence of the three AEP groups was found to affect the chemical shifts of several sugar protons; this was interpreted as being due to interactions between these protons and the AEP groups in a possible three-dimensional conformation of F-21.

45: FGL-IIb

Occurrence: *Aplysia kurodai* (mollusc) (*137*).

Physical Data: FABMS and ¹H-NMR (anomeric protons) (*137*).

The phosphonoglycolipid FGL-IIb (**45**) is characterized by the presence of a pyruvic acid molecule linked as a ketal to *O*-3 and *O*-4 of the terminal galactose of the oligosaccharide chain. Even though pyruvic acid has been found in mycobacterial glycolipids, this is the first report of its presence in an animal GSL. Another noticeable structural feature of

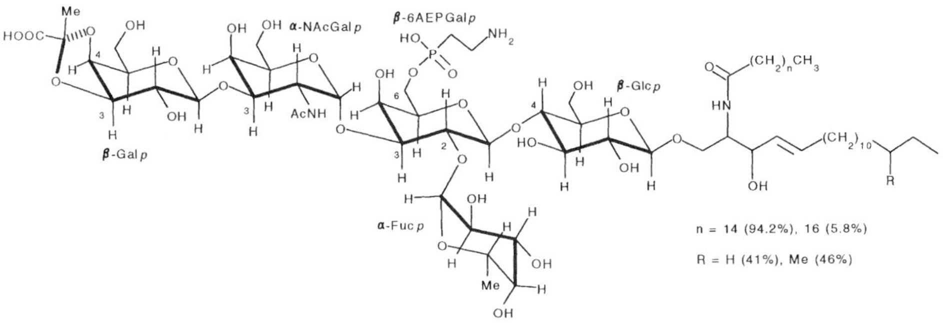

45

FGL-IIb is the presence of an α-fucosyl in place of the usual α-galactosyl as the side-chain sugar.

A problem in structure determination of FGL-IIb was that the ketal linkage of the pyruvic acid survived the acid methanolysis used for carbohydrate analysis, leading to a pyruvylated galactose which could not be identified by GLC. The glycolipid, however, could be depyruvylated without affecting glycosidic linkages with diluted aqueous HCl. The depyruvylated GSL was analyzed by the methods described above, while the pyruvic acid was determined as its 2,4-dinitrophenylhydrazone derivative. Location of the pyruvic acid residue was deduced by the presence of 2,6-di-*O*-methylgalactitol between the permethylated alditol acetates deriving from the intact FGL-IIb, which was replaced by 2,3,4,6-tetra-*O*-methylgalactitol when depyruvylated FGL-IIb was used. The stereochemistry of the ketal carbon of galactose, unassigned in (*137*), can now be assigned as *S* since the pyruvate methyl resonance is identical with that of compound **46** (see below).

Immunochemical studies showed that the antigenicity of FGL-IIb requires the presence of the free carboxyl group of pyruvate and that the GSL is concentrated in the nerve fibers of *A. kurodai* (*137*).

46: FGL-V

Occurrence: *Aplysia kurodai* (mollusc) (*138*).

Physical Data: FABMS and ^{1}H-NMR (selected protons) (*138*).

The phosphonoglycolipid FGL-V (**46**) is very similar to FGL-IIb (**45**), but carries a branched galactose instead of fucose and one more AEP group. Structure determination is also similar. In addition, the

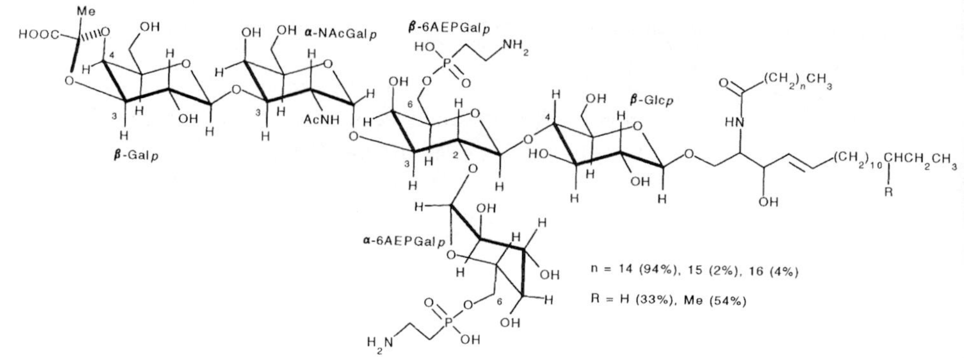

46

configuration of the ketal carbon of the pyruvate unit is assigned as *S* because of the ^{1}H-NMR chemical shift of the pyruvate methyl protons compared with literature data (*139*). FLG-V cross-reacted with the polyclonal antibody against FGL-IIb, in accordance with the fact that both FGL-V and FGL-IIb contain pyruvylated galactose at the non-reducing end of the saccharide chain, which appears to be essential for binding. Like FGL-IIb, FGL-V is localized in the nervous system, and a possible neurobiological role of these phosphonoglycolipids comparable to that of gangliosides in vertebrates is hypothesized.

47: FGL-IIa

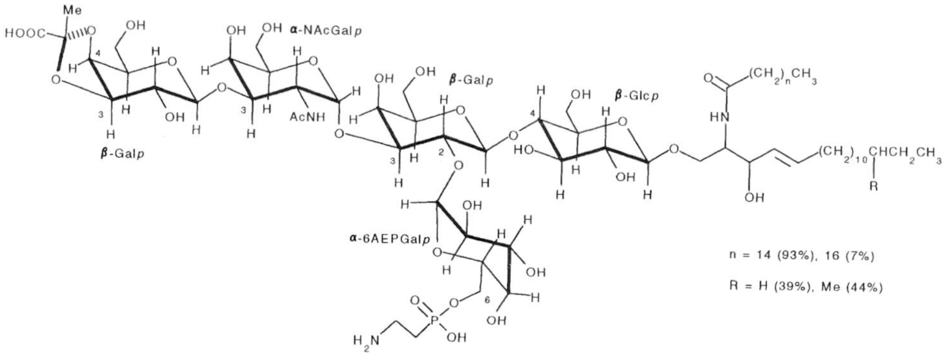

47

Occurrence: *Aplysia kurodai* (mollusc) (*138*).

References, pp. 291–301

Physical Data: FABMS and ^{1}H-NMR (selected protons) (138).

FGL-IIa (47) differs from FGL-V (46) only in the absence of the AEP group located on the internal galactose in FGL-V, so that the GSL obtained by dephosphonylation with HF of both compounds is the same. The immunochemical properties of FGL-IIa and its localization in nervous fibers are the same as for FGL-V.

48: FGL-I

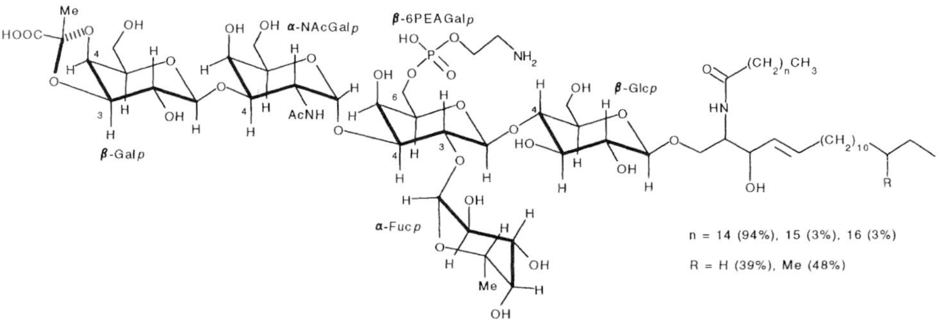

[Chemical structure of FGL-I showing β-Galp, α-NAcGalp, β-6PEAGalp, α-Fucp, β-Glcp residues with ceramide moiety; n = 14 (94%), 15 (3%), 16 (3%); R = H (39%), Me (48%)]

48

Occurrence: *Aplysia kurodai* (mollusc) (140).

Physical Data: FABMS and ^{1}H-NMR (anomeric protons) (140).

FGL-I (48), isolated from the nervous system of *A. kurodai*, is identical with FGL-IIb (45), except in bearing a phosphoethanolamine (PEA) residue instead of the AEP group present in FGL-IIb. The presence of the PEA group was demonstrated by colorimetric analysis which detected 1 mol of phosphate ester bond, but not a phosphorus-carbon bond, and by detection of ethanolamine after hydrolysis of the GSL. The linkage position of the PEA group was established by a comparison between the methylation analysis products obtained from intact and HF-treated FGL-I, and by isolation and MS identification of galactose-6-phosphate from the products of partial hydrolysis of FGL-I. The co-occurrence of aminoethylphosphonyl and aminoethylphosphoryl glycosphingolipids in the same animal is worthy of note.

49: F-9

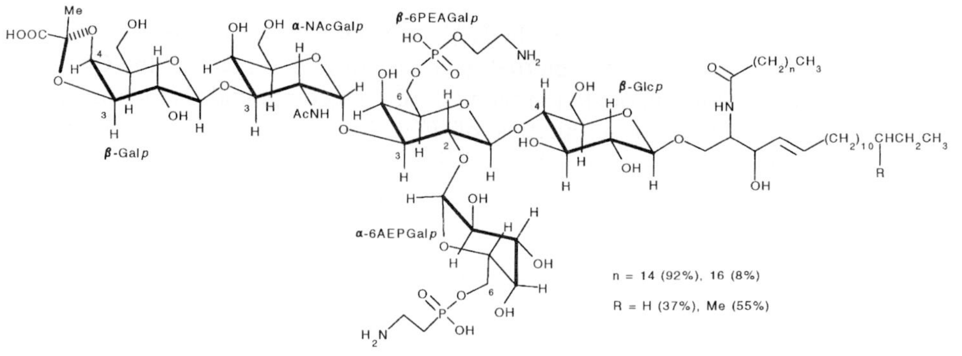

[Structure of compound 49 with labels: Me, HOOC, OH, α-NAcGalp, β-6PEAGalp, β-Galp, AcNH, α-6AEPGalp, β-Glcp, H₂N, (CH₂)ₙCH₃, HN, (CH₂)₁₀CHCH₂CH₃, R]

n = 14 (92%), 16 (8%)
R = H (37%), Me (55%)

49

Occurrence: *Aplysia kurodai* (mollusc) (*140*).

Physical Data: FABMS and ¹H-NMR (anomeric protons) (*140*).

The glycosphingolipid F-9 (**49**) isolated from the nervous system of *A. kurodai* differs from FGL-I (**48**) in the presence of a AEP-carrying α-galactosyl instead of a α-fucosyl residue. This GSL is unique in that it contains both aminoethylphosphonyl and phosphoethanolamino groups. Aminoacid analysis showed the presence of 1 mol each of phosphoethanolamine and aminoethylphosphonate. Comparative carbohydrate analysis of intact F-9 and HF-treated F-9 showed that both groups were linked to *O*-6 of a galactose; a fragment GSL from partial acid hydrolysis was identified as PEA-Gal-Glc-Cer by its FABMS spectrum, thus indicating that phosphoethanolamine is attached to the internal galactose, and AEP to the branched galactose.

50a–50e: PnGL-1

Occurrence: *Dolabella auricolaria* (mollusc) (*135*).

Physical Data: FABMS and ¹H-NMR (*135*).

The phosphonoglycolipid PnGL-1 (**50**), isolated as a mixture of homologues, was separated into five homogeneous compounds (**50a–50e**) by reversed-phase HPLC. Structure determination was based on carbohydrate analysis, methylation analysis, HF degradation, and extensive use of FABMS. It is worthy of note that structure elucidation of

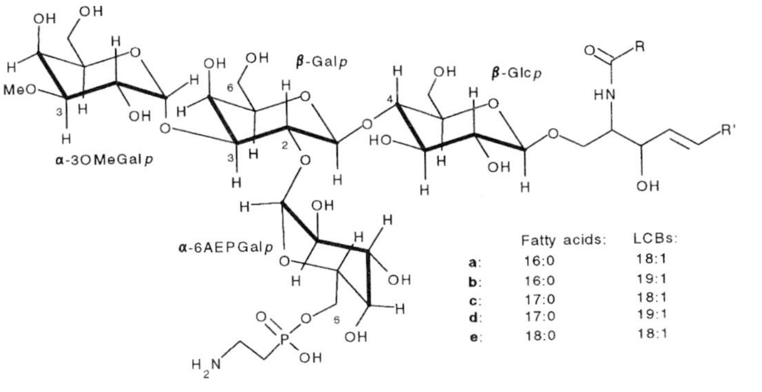

[Structure of compound **50** with substituent table:]

	Fatty acids:	LCBs:
a:	16:0	18:1
b:	16:0	19:1
c:	17:0	18:1
d:	17:0	19:1
e:	18:0	18:1

50

the ceramide part of the molecules, which were different in **50a–50e**, was based only on FABMS analysis. This technique consumes only a minute amount of sample compared with the usual methanolysis followed by GC-MS. On the other hand, it cannot distinguish between linear and branched alkyl chains. The authors also report the presence in the same organism of a neutral tetraglycosylceramide with the same structure of dephosphonylated **50**, which could be its biogenetic precursor; however, its isolation and characterization were not described.

51: EGL-I

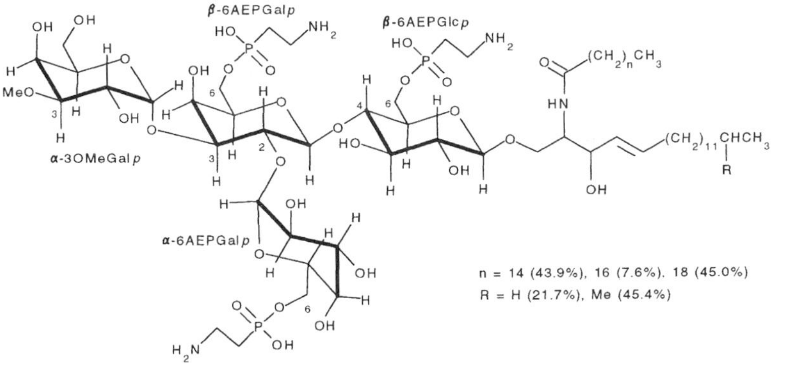

[Structure of compound **51** with:]

n = 14 (43.9%), 16 (7.6%), 18 (45.0%)
R = H (21.7%), Me (45.4%)

51

Occurrence: *Aplysia kurodai* (mollusc) *(141)*.

Physical Data: FABMS and ^{1}H-NMR (selected signals) *(141)*.

The triphosphonoglycolipid EGL-I was isolated from the eggs of *A. kurodai*; its structure is notably different from those of GSLs of adult specimens of *A. kurodai*. EGL-I is a tetraglycosylceramide instead of a pentaglycosylceramide, with the inner galactose glycosylated by 3-*O*-methyl-α-galactosyl in place of α-N-acetylgalactosamine. In contrast, the sugar backbone of EGL-I is the same as that of the phosphonoglycolipid PnGL-1 (**50**) isolated from *Dolabella auricolaria*.

As for immunochemical properties of EGL-I, antiserum against SGL-II (**42**), which specifically recognizes terminal 3-*O*-methylgalactose, did not react with EGL-I, showing specificity towards β-3-*O*-methylgalactoside; on the contrary, an antiserum raised against EGL-I reacted with SGL-II and with all 3-*O*-methylgalactose-containing glycolipids.

5.3. Gangliosides

52

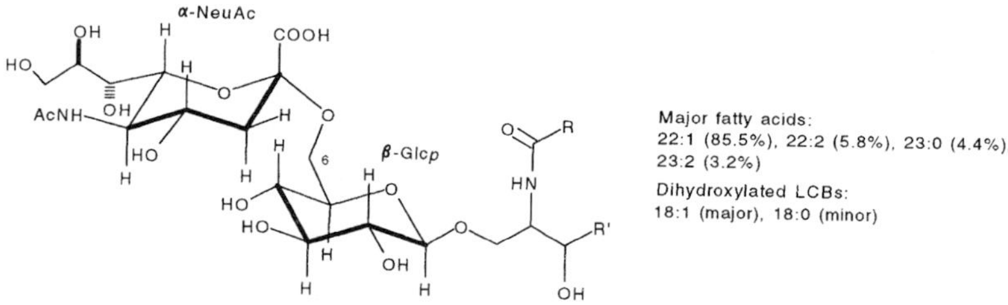

Major fatty acids:
22:1 (85.5%), 22:2 (5.8%), 23:0 (4.4%),
23:2 (3.2%)
Dihydroxylated LCBs:
18:1 (major), 18:0 (minor)

52

Occurrence: *Anthocidaris crassispina* (sea urchin) (*142*), *Tripneustes ventricosa* (sea urchin) (*143*).

Physical data: ^{1}H- and ^{13}C-NMR (complete assignment) (*143*).

Ganglioside **52** was the first ganglioside from a marine invertebrate whose structure was completely determined. It was isolated from spermatozoa of *A. crassispina* and its structure was determined by hydrolysis and methylation analysis. The sialic acid released by hydrolysis was identified by MS. The anomeric configurations were suggested by CrO$_3$ oxidation and enzymatic analysis (*142*). The presence of ganglioside **52** was restricted to spermatozoa of *A. crassispina*, whereas the eggs, the gonads and the somatic cells contained its hydroxylated analogue **53**; therefore, **52** was proposed as a differentiation marker of the spermatozoa. However, ganglioside **52** was subsequently isolated, together with **53**, from gonads of another sea urchin, *T. ventricosa* (*143*). In the same paper, the complete assignment of ^{1}H- and ^{13}C-NMR spectra of ganglioside **52** from *T. ventricosa* has been reported.

53: Ganglioside M5

	Major fatty acids:		Major LCBs:
a:	X = H (20%)	14:0 (4.9%), 16:0 (16.4%), 18:0 (49.4%), 20:0 (7.1%), 22:0 (4.0%), 26:0 (8.0%)	Trihydroxylated LCBs: 16:0 (11.9%), 17:0 (5.8%), 18:0 (68.3%), 19:0 (6.2%), 21:0 (5.5%)
	X = OH (80%)	18:0 (52.0%), 20:0 (12.0%), 22:0 (15.1%), 24:0 (8.2%)	
b:	X = H (35%)	14:1 (12.3%), 18:0 (6.0%), 22:0 (5.1%), 22:1 (24.9%), 23:1 (15.1%), 24:1 (17.4%)	Trihydroxylated LCBs: n-18:0 (95.4%)
	X = OH (65%)	22:1 (24.9%), 23:1 (13.8%), 24:1 (57.7%)	
c:	X = H (50%)	14:0 (6.3%), 15:0 (17.0%), 16:1 (7.1%), 16:0 (21.2%), 18:1 (13.1%), 18:0 (14.2%), 20:0 (6.0%), 22:0 (5.5%)	Dihydroxylated LCBs: 16:0 (9.4%), 17:0 (53.0%), 18:0 (19.0%), 19:0 (8.2%), 20:0 (10.4%)
	X = OH (50%)	16:0 (60.8%), 18:0 (28:6%)	

53

Occurrence: *Strongylocentrotus nudus* (sea urchin) (*144*), *Anthocidaris crassispina* (sea urchin) (*38*), *Hemicentrotus pulcherrimus* (sea urchin) (*145*), *Tripneustes ventricosa* (sea urchin) (*143*), *Cucumaria japonica* (holothurian) (*146*), *Ophtocoma echinata* (ophiuroid) (*147*), *Ophiomastix annulosa* (ophiuroid) (*147*), *Ophiura sarsi* (ophiuroid) (*148*).

Physical Data: FABMS and ^{1}H-NMR (*38*), ^{1}H- and ^{13}C-NMR (complete assignment) (*143*).

Ganglioside M5 [**53a**; the name M5 is from (*38*)] was first isolated from gonads of *S. nudus*. Its structure was established by total and partial acid hydrolysis, methanolysis and periodate oxidation. Enzymatic hydrolysis with neuraminidase and CrO_3 oxidation established the anomeric configurations. Ganglioside M5 (**53b**) was subsequently isolated from *T. ventricosa* (*143*) and *H. pulcherrimus* (*145*), and as the major ganglioside in the eggs of *A. crassispina*, where it is over 30 times more abundant than the second one, ganglioside T1 (**54**) (*38*). Ganglioside M5 from eggs of *A. crassispina* appears to be more hydroxylated than ganglioside **52**, isolated from spermatozoa of the same organism, in all of sialic acid, fatty acids, and long-chain bases.

The same ganglioside M5, but with a rather different ceramide composition, was subsequently isolated from other echinoderms, namely the holothurian *C. japonica* (**53c**) (*146*), and three species of *Ophiura*, *O. echinata*, *O. annulosa* (*147*), and *O. sarsi* (*148*). From the last *Ophiura*, ganglioside **53** was isolated as a mixture with **52**.

An antibody against M5 was used to determine the distribution of the ganglioside in the eggs using indirect immunofluorescence microscopy. The eggs of *H. pulcherrimus*, which also contain M5 as the major ganglioside, were used. This study showed a dramatic change of M5 distribution after fertilization (*145*).

A total synthesis of ganglioside M5 has been reported (*149*).

54: Ganglioside T1

Major fatty acids:
a: X = H (75%)   14:0 (5.5%), 15:0 (32.5%), 16:0 (9.2%), 18:0 (6.0%), 23:0 (6.8%), 24:0 (8.0%)
    X = OH (25%) 18:0 (11.1%), 22:0 (18.0%), 23:1 (6.4%), 23:0 (10.3%), 24:1 (8.8%), 24:0 (19.0%)
b: X = H (26%)   22:0 (5.8%), 22:1 (39.2%), 23:1 (21.5%), 24:1 (20.4%)
    X = OH (74%) 22:1 (28.0%), 23:1 (15.5%), 24:1 (53.1%)

Major LCBs:
16:0 (16.7%), 17:0 (8.8%), 18:0 (56.0%)

n-18:0 (96.5%)

54

Occurrence: *Echinocardium cordatum* (sea urchin) (*150*), *Echinarachnius parma* (sea urchin) (*151*), *Tripneustes ventricosa* (sea urchin) (*143*), *Anthocidaris crassispina* (sea urchin) (*38*), *Ophtocoma echinata* (ophiuroid) (*147*), *Ophiomastix annulosa* (ophiuroid) (*147*), *Ophiura sarsi* (ophiuroid) (*148*).

Physical Data: IR (*150*), FABMS and ^{1}H-NMR (*38*), ^{1}H- and ^{13}C-NMR (complete assignment) (*143*).

When ganglioside T1 [**54a**; the name T1 is from (*38*)] was isolated from the gonads of *E. cordatum*, it was the first example of a sialosphingolipid containing a sulfate group. The presence of a sulfated sialic acid

References, pp. 291–301

was demonstrated by mild acid hydrolysis of **54a** which yielded a sugar more acidic than genuine sialic acid, while the same experi-ment performed after solvolytic desulfation of **54a** with pyridine hydrochloride in dioxane gave *N*-glycolylneuraminic acid. The sulfate was located at C-8 since the sialic acid residue was resistant to periodate oxidation.

Ganglioside T1 was subsequently isolated from the sea urchins *E. parma* (*151*), *T. ventricosa* (*143*), and *A. crassispina* (**54b**) (*38*), and from three *Ophiura* species, *O. echinata*, *O. annulosa* (*147*), and *O. sarsi* (*148*). Ganglioside T1 from the last three species contained, in addition to trihydroxylated LCBs, significative amounts of dihydroxylated LCBs, which were the only sphingoid bases present in *O. sarsi*. From the same species, **54** was isolated admixed with minor amounts of its analogue **55**. In (*38*) the structure of ganglioside T1 was confirmed by FABMS and ^{1}H-NMR data, while in (*143*) its ^{1}H- and ^{13}C-NMR spectra were assigned.

55

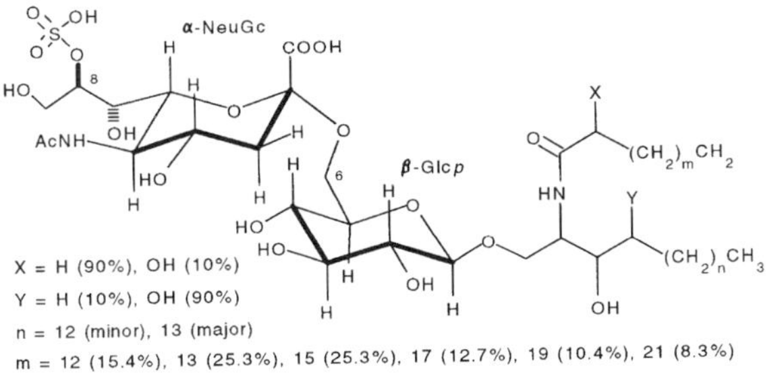

55

Occurrence: *Echinarachnius parma* (sea urchin) (*151*).

The ganglioside **55** isolated from the gonads of *E. parma* differs from **54** in that it contains *N*-acetylneuraminic acid instead of *N*-glycolylneuraminic acid. Its structure determination was based on degradation methods described for **54**. Until now, (*151*) is the sole report of isolation of pure **55** from a marine organism.

56

Occurrence: *Anthocidaris crassispina* (sea urchin) (*142*), *Tripneustes ventricosa* (sea urchin) (*143*).

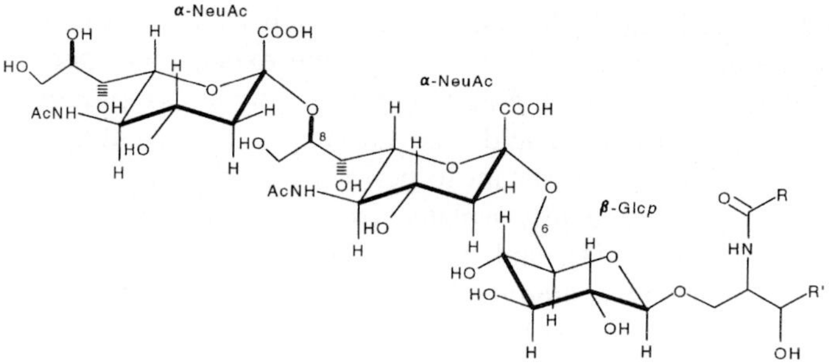

56

Major fatty acids:
22:1 (85.5%), 22:2 (5.8%),
23:0 (4.4%), 23:2 (3.2%)
Dihydroxylated LCBs:
18:1 (major), 18:0 (minor)

The disialoglycolipid **56** is the major glycolipid in the spermatozoa of *A. crassispina*. The sialic acid released by acid hydrolysis was identified as *N*-acetylneuraminic acid by EIMS of its silanized derivative. The linkage between the two sialic acid residues was determined as (2 →8) because about 50% of the sialic acid survived to periodate oxidation. The 6-*O* glycosylation of glucose was established by methylation analysis. Anomeric configurations were suggested by neuraminidase digestion and CrO_3 oxidation. Like monosialoglycolipid **52**, also isolated from spermatozoa of *A. crassispina*, compound **56** contains *N*-acetylneuraminic acid instead of *N*-glycolylneuraminic acid, commonly found in other tissues of sea urchins.

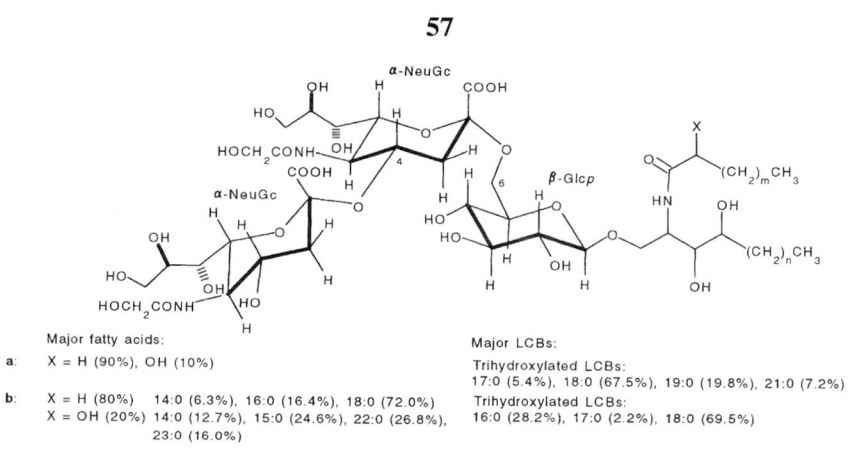

57

Major fatty acids:
a: X = H (90%), OH (10%)
b: X = H (80%)   14:0 (6.3%), 16:0 (16.4%), 18:0 (72.0%)
   X = OH (20%) 14:0 (12.7%), 15:0 (24.6%), 22:0 (26.8%),
                23:0 (16.0%)

Major LCBs:
Trihydroxylated LCBs:
17:0 (5.4%), 18:0 (67.5%), 19:0 (19.8%), 21:0 (7.2%)
Trihydroxylated LCBs:
16:0 (28.2%), 17:0 (2.2%), 18:0 (69.5%)

57

Occurrence: *Strongylocentrotus nudus* (sea urchin) (*144*), *Echinocardium cordatum* (sea urchin) (*152*).

Physical Data: IR (*152*).

References, pp. 291–301

The (2 → 4) bond between two *N*-glycolylneuraminic acid residues, unusual for sialic acids, is the most characteristic feature of the disialoganglioside **57**, isolated from the gonads of *S. nudus* (**57a**) (*144*) and *E. cordatum* (**57b**) (*152*). This linkage was determined by $NaIO_4$ oxidation which showed the presence of free hydroxyl groups at C-7, C-8, and C-9 of both sialic acid residues.

58

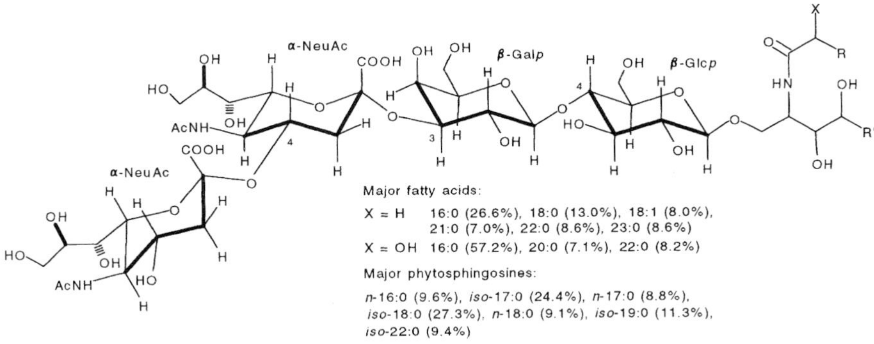
α-NeuAc
β-Galp
β-Glcp
α-NeuAc
Major fatty acids:
X = H 16:0 (26.6%), 18:0 (13.0%), 18:1 (8.0%), 21:0 (7.0%), 22:0 (8.6%), 23:0 (8.6%)
X = OH 16:0 (57.2%), 20:0 (7.1%), 22:0 (8.2%)
Major phytosphingosines:
n-16:0 (9.6%), *iso*-17:0 (24.4%), *n*-17:0 (8.8%), *iso*-18:0 (27.3%), *n*-18:0 (9.1%), *iso*-19:0 (11.3%), *iso*-22:0 (9.4%)

58

Occurrence: *Lethasterias fuska* (starfish) (*153*).

Disialoganglioside **58**, isolated from hepatopancreas of *L. fuska*, is the second example of a marine ganglioside with a (2 → 4) linkage between two sialic acid residues and the first one from a starfish. In this case, the linkage was demonstrated by mass spectral analysis of the sialic acid derivative, obtained by methanolysis of the permethylated ganglioside and subsequent acetylation. The mass spectrum was indicative of an acetyl group at position 4.

59

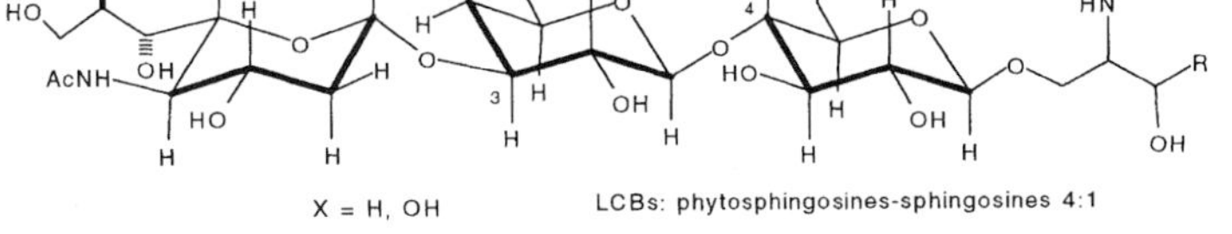
α-NeuAc
β-Galp
β-Glcp
X = H, OH
LCBs: phytosphingosines-sphingosines 4:1

59

Occurrence: *Lethasterias fuska* (starfish) (*153*).

Compound **59** was isolated as a minor ganglioside from hepatopancreas of *L. fuska*, and is the monosialo analogue of the more abundant **58**. Ganglioside **59** is also very similar to GAA-6 (**63**), which differs in containing a methylated sialic acid.

60

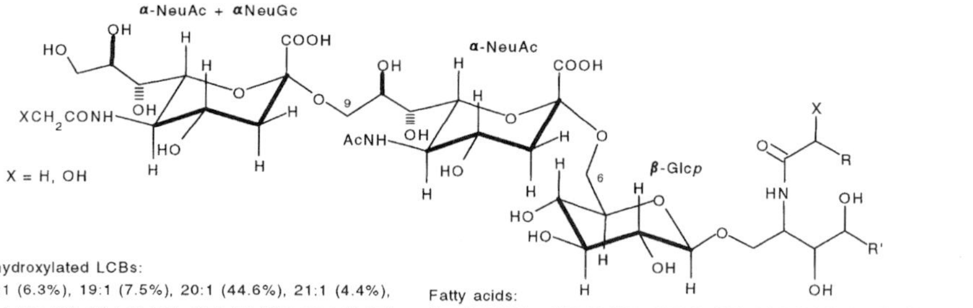

α-NeuAc + αNeuGc   α-NeuAc

XCH₂CONH—   AcNH—   β-Glcp

X = H, OH

Dihydroxylated LCBs:
18:1 (6.3%), 19:1 (7.5%), 20:1 (44.6%), 21:1 (4.4%),
22:1 (18.4%), 24:1 (4.4%), 26:1 (11.2%), 28:1 (3.2%)

Trihydroxylated LCBs:
17:0 (5.8%), 18:0 (8.3%), 19:0 (44.3%), 20:0 (4.7%),
21:0 (17.5%), 23:0 (4.2%), 25:0 (10.6%), 27:0 (10.6%)

Fatty acids:
X = H (50%)   14:0 (1.0%), 15:0 (1.2%), 16:0 (14.8%), 17:0 (2.4%), 18:0 (35.1%), 20:0 (4.3%), 21:0 (2.5%), 22:0 (18.6%), 23:0 (11.5%), 24:0 (8.6%)

X = OH (50%) 16:0 (1.6%), 18:0 (21.8%), 20:0 (3.7%), 21:0 (2.5%), 22:0 (32.3%), 23:0 (23.8%), 24:0 (14.3%)

60

Occurrence: *Ophtocoma echinata* (ophiuroid) (*147*).

Physical Data: ^{13}C-NMR (selected signals) (*147*).

In disialoglycolipid **60** from the *Ophiura* species *O. echinata*, the two *N*-acetylneuraminic acid residues are joined through a (2 → 9) bond. Methanolysis of methylated **60** generated 9-*O*-acetyl-4,7,8-tri-*O*-methyl-*N*-methyl-*N*-acetylneuraminic acid methyl ester methyl ketoside, whose structure was derived from its mass spectrum. The same analysis also demonstrated that **60** was not homogeneous, since the terminal sialic acid proved to be a mixture of *N*-acetyl- and *N*-glycolylneuraminic acid.

61: Ganglioside G-1

Occurrence: *Strongylocentrotus intermedius* (sea urchin) (*18, 154*).

Physical Data: EIMS (permethylated derivative) (*154*).

The saccharide chain of disialoganglioside G-1 (**61**) from eggs and embryos of *S. intermedius* is composed of alternating glucose and *N*-glycolylneuraminic acid residues, glucose being glycosylated at C-6 and sialic acid at C-8, as usually observed in gangliosides from sea urchins.

References, pp. 291–301

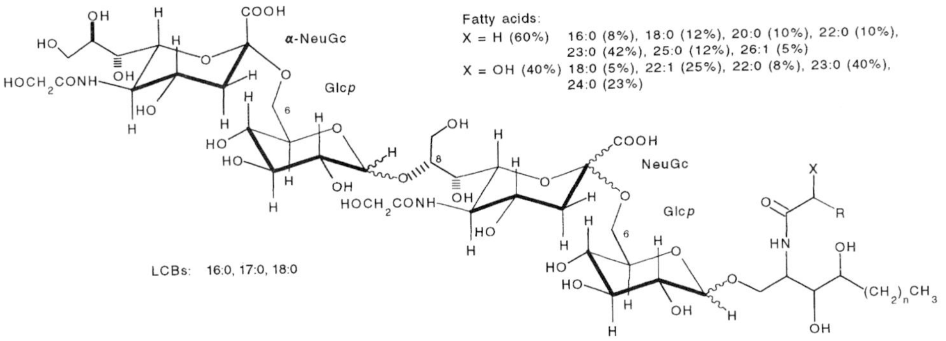

Marine Glycolipids

61

Compound **61** was first isolated in 1973 (*18*), but its structure determination appeared only in 1980 (*154*). Analysis of the products of periodate oxidation of **61** indicated the site of glycosylation of sialic acid, and methylation analysis those of glucoses. Only the anomeric configuration of the terminal *N*-glycolylneuraminic acid could be established by enzymatic digestion, while the remaining ones were left unassigned.

62: Ganglioside G-2

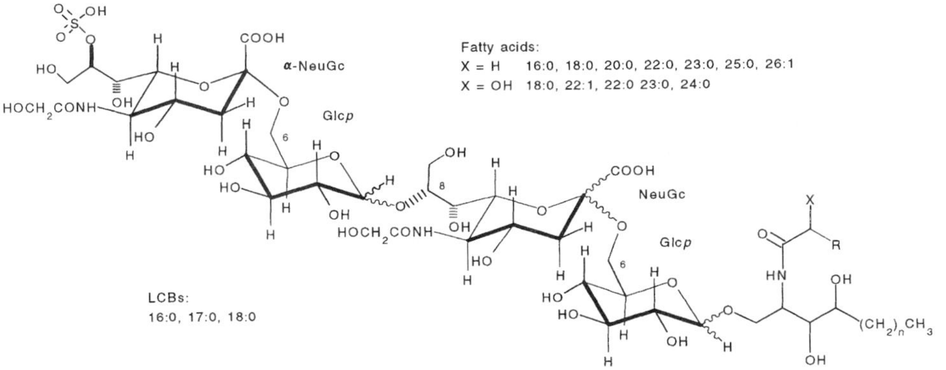

62

Occurrence: *Strongylocentrotus intermedius* (sea urchin) (*18, 154*).

Physical Data: IR (*154*).

Ganglioside G-2 (**62**), isolated from the same source as G-1 (**61**), is its sulfated analogue and is also the first example of a disialoganglioside carrying a sulfate group. Solvolytic desulfation of G-2 gave a ganglioside identical with G-1. In addition, both sialic acid residues were resistant to periodate oxidation and therefore were 8-*O*-substituted (by a glucose and a sulfate group, respectively).

Antisera against both G-1 and G-2 were prepared and were used for studying the surface localization of these gangliosides in embryos (*154*).

63: Ganglioside GAA-6

[Structure of ganglioside 63 showing α-NeuGc, β-Galp, β-Glcp residues with substituents and ceramide portion]

Major fatty acids:
a: R' = Me, R" = H   X = H (10%)   16:0 (19.3%), 17:0 (5.0%), 18:0 (62.6%)
   X = OH (90%) 16:0 (45.1%), 22:0 (12.1%), 23:0 (9.0%), 24:0 (9.1%)
b: R' = Me, R" = H or   X = OH   16:0, 18:0, 22:0, 23:0, 24:0 (major), 25:0
   R' = H, R" = Me

Major LCBs:
n-18:0 (8.8%), *iso*-19:0 (10.6%), *iso*-20:0 (10,9%)
n-22:1 (major)

63

Occurrence: *Aphelasterias japonica* (starfish) (*155*), *Asterias amurensis versicolor* (starfish) (*34*).

Physical Data: m.p. 155–160°C, IR, ^{1}H-NMR (selected signals), ^{13}C-NMR (complete assignment), FABMS (permethylated derivative) (*34*).

Ganglioside GAA-6 [the name GAA-6 is from (*34*)] was first isolated from hepatopancreas of *A. japonica* (**63a**) (*155*). The presence of a 8-*O*-methyl-*N*-glycolylneuraminic acid residue in **63a** was suggested by TLC of the sialic acid obtained from partial hydrolysis of the ganglioside and was confirmed by the mass spectrum of the sialic acid derivative from the trideuteriomethylated ganglioside. The anomeric configurations were determined by CrO$_3$ oxidation and neuraminidase digestion. Ganglioside GAA-6 was subsequently isolated from *Asterias amurensis versicolor* (**63b**) (*34*), and its structure re-determined using somewhat different methods. Hot water partial hydrolysis, which selectively cleaves the sialic acid-galactose linkage, gave lactosylceramide, identified by comparison with a synthetic sample. Analysis of the ^{13}C-NMR spectrum of the intact ganglioside showed that the sialic acid is present in **63b** as a mixture of 8-*O*-methyl- and 11-*O*-methyl-*N*-glycolylneuraminic acid. The α-configuration of sialic acid was deduced from the chemical shift of its equatorial H-3. In addition, the stereochemistry of the ceramide

References, pp. 291–301

part of the molecule was established by comparison with a synthetic phytosphingosine.

64: Ganglioside GAA-7

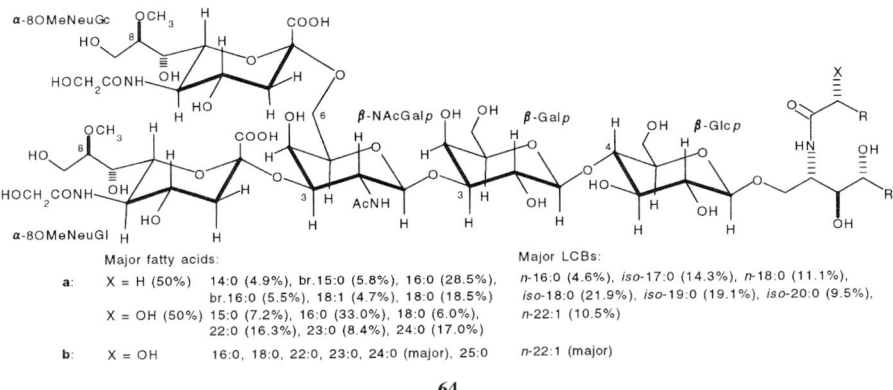

Major fatty acids:
a: X = H (50%)   14:0 (4.9%), br.15:0 (5.8%), 16:0 (28.5%),
                  br.16:0 (5.5%), 18:1 (4.7%), 18:0 (18.5%),
    X = OH (50%) 15:0 (7.2%), 16:0 (33.0%), 18:0 (6.0%),
                  22:0 (16.3%), 23:0 (8.4%), 24:0 (17.0%)
b: X = OH   16:0, 18:0, 22:0, 23:0, 24:0 (major), 25:0

Major LCBs:
n-16:0 (4.6%), iso-17:0 (14.3%), n-18:0 (11.1%),
iso-18:0 (21.9%), iso-19:0 (19.1%), iso-20:0 (9.5%),
n-22:1 (10.5%)

n-22:1 (major)

64

Occurrence: *Asterias amurensis* (starfish) *(156)*, *Asterias amurensis versicolor* (starfish) *(34)*, *Asterias rubens* (starfish) *(51, 157)*.

Physical data: m.p. 225–230°C, IR, ^{1}H-NMR (selected signals), ^{13}C-NMR (complete assignment), FABMS (permethylated derivative) *(34)*.

A sialic acid residue linked to NAcGal is rarely found in gangliosides; the presence of two sialic acids on the same galactosamine is reported for the first time in ganglioside GAA-7 [**64a**; the name GAA-7 is from *(34)*], isolated from hepatopancreas of *A. amurensis*. The carbohydrate chain structure was determined by partial hydrolysis and methylation analysis. The sialic acid was determined as 8-*O*-methyl-*N*-glycolylneuraminic acid from its mass spectrum after KBH$_4$ reduction and peracetylation. In order to confirm the presence of a methylated *N*-glycolylneuraminic, GAA-6 was subjected to BCl$_3$ demethylation; hydrolysis of the demethylated ganglioside gave *N*-glycolylneuraminic acid. The ganglioside was resistant to neuraminidase hydrolysis, probably because the sialic acid residues are sterically hindered, so that their anomeric configuration remained undetermined *(156)*. Later, the configuration of sialic acid was found to be α on the basis of the chemical shift of H-3 protons when GAA-7 was re-isolated from *A. amurensis versicolor* (**64b**) *(34)*. In the same paper, the stereochemistry of all the stereogenic centers of the ceramide was reported.

Ganglioside GAA-7 displays neuritogenic activity on a mouse neuroblastoma cell line (Neuro 2a) in a serum-free medium, while the activity is inhibited by the effect of the serum. It also exhibits a weak growth-inhibitory activity on the same cell line (*34*).

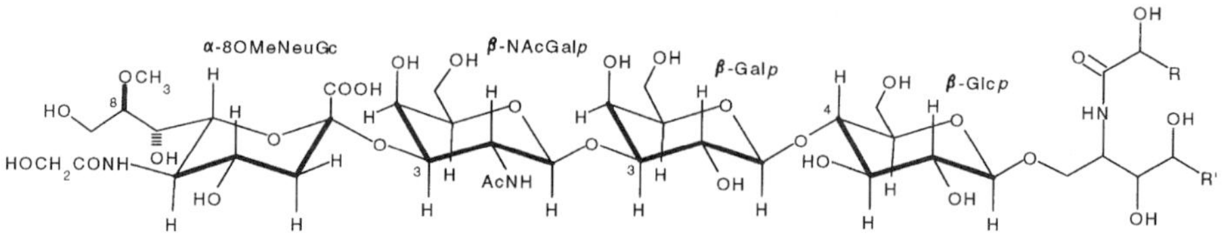

65

Occurrence: *Asterias rubens* (starfish) (*51*).

Physical Data: FABMS (*51*).

Ganglioside **65** was isolated as a minor component of a ganglioside mixture from *A. rubens* mainly composed of the disialoganglioside GAA-7 (**64**). Compound **65** can be considered the monosialo analogue of GAA-7. The structure determination was based on the fragmentation pattern in the FAB mass spectrum and on methylation analysis. The fatty acid and LCB composition was not determined nor was the stereochemistry of glycosidic linkages established. Anomeric configurations shown in structure **65** are hypothesized as analogous to those of GAA-7.

66

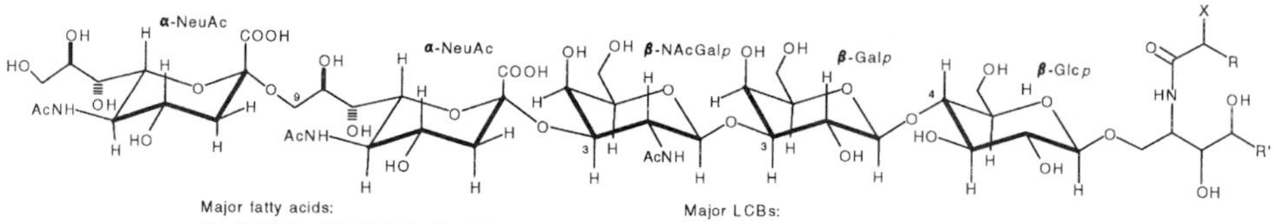

Major fatty acids:
X = H (20%) 14:0 (8.3%), 15:0 (5.6%), 16:0 (39.6%), 18:0 (23.5%), 22:0 (4.2%)
X = OH (80%) 14:0 (14.7%), 15:0 (22.0%), 16:0 (42.8%), 18:0 (5.6%)

Major LCBs:
n-16:0 (26.0%), n-17:0 (9.9%), iso-17:0 (27.2%), n-18:0 (11.3%), iso-18:0 (12.7%), n-19:0 (7.6%), iso-19:0 (3.1%)

Occurrence: *Evasteria retifera* (starfish) (*156*).

Physical data: EIMS (permethylated derivative) (*156*).

The linear nature of the saccharide chain of sialoganglioside **66** from *E. retifera* was clear from the mass spectra of the permethylated ganglioside, which showed fragment peaks for the terminal sialic acid, the terminal disialyl group composed of two *N*-acetylneuraminic acid residues, and a trisaccharide containing two sialic acid and one hexosamine residues. The (2 → 9) linkage between the two *N*-acetylneuraminic acid residues was demonstrated by methylation analysis which gave a 9-*O*-acetyl trimethylated *N*-acetylneuraminic acid, identified by its mass spectrum compared with literature data. Ref. (*156*) was the first report of a ganglioside with a (2 → 9) linkage between two sialic acid residues.

67

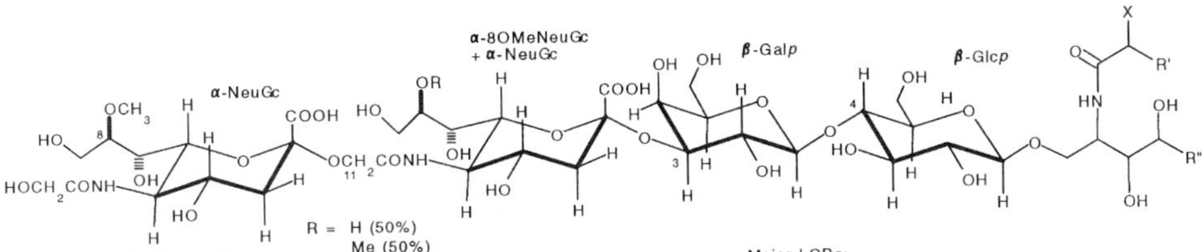

Major fatty acids:
X = H (minor)   16:0 (23.5%), 17:0 (4.5%), 18:0 (12.6%), (20:0 (4.9%), 22:0 (13.5%), 23:0 (5.7%), 24:0 (4.7%), 25:0 (5.7%)
X = OH (major) 14:0 (14.5%), 16:0 (30.0%), 22:0 (18:0%), 23:0 (4.7%), 24:0 (16.6%%)

Major LCBs:
n-16:0 (12.1%), *iso*-17:0 (30.0%), *n*-17:0 (6.6%), *iso*-18:0 (24.1%), *n*-18:0 (11.2%), *iso*-19:0 (9.7%), *n*-19:0 (6.3%)

67

Occurrence: *Aphelasterias japonica* (starfish) (*155*).

In the disialoganglioside **67** isolated from the hepatopancreas of *A. japonica* two *N*-glycolylneuraminic acid residues are joined together by a linkage involving the hydroxy group of the glycolic acid residue. After the nature and the sequence of the sugar residues were established by standard methods, this unique linkage was demonstrated by an extensive analysis of the EIMS spectrum of a sialic acid derivative (obtained by trideuteromethylation of the ganglioside, methanolysis, and acetylation) which was found to be acetylated at the glycolic acid hydroxy group. The same experiment showed that the terminal sialic acid was 8-*O*-methyl-*N*-glycolylneuraminic acid, while the internal sialic acid was a 1:1 mixture of 8-*O*-methyl- and non-methylated *N*-glycolylneuraminic acid.

68

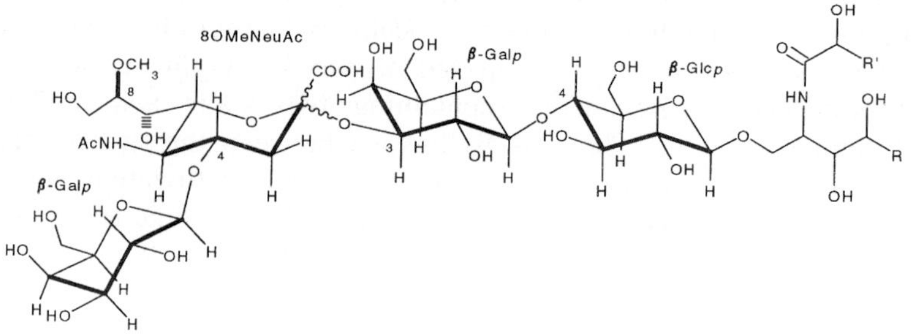

[Structure of ganglioside 68]

68

Occurrence: *Luidia quinaria bispinosa* (starfish) (*158*).

The structure of ganglioside **68** was determined mainly by standard chemical methods; in addition, analysis of the trideuteromethylated ganglioside (see above) was used in order to establish the position of the methyl group and the galactose unit on the sialic acid residue. As usual for gangliosides with an internal sialic acid neuraminidase digestion was ineffective, so that anomeric configuration of the sialic acid remained undetermined.

69: Ganglioside LG-1

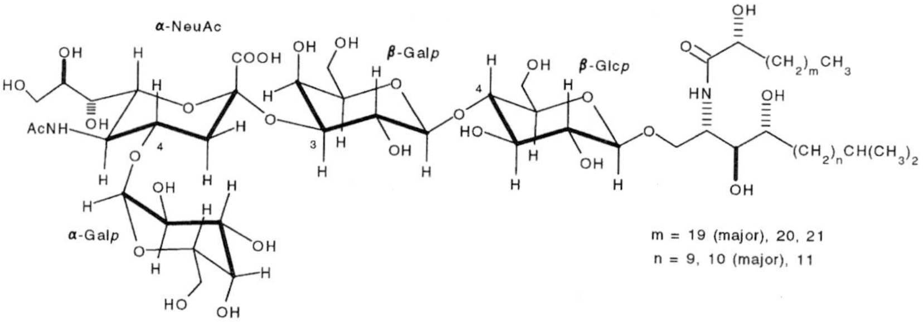

[Structure of ganglioside 69]

m = 19 (major), 20, 21
n = 9, 10 (major), 11

69

Occurrence: *Astropecten latespinosus* (starfish) (*30*).

Physical Data: m.p. 198–201°C, FABMS, IR, ^{1}H-NMR (selected signals), ^{13}C-NMR (complete assignment) (*30*).

Ganglioside LG-1 (**69**) was obtained from the water soluble lipid extract of *A. latespinosus*. Carbohydrate analysis showed the presence of 2 mol of Gal and 1 mol of Glc, while the presence of *N*-acetylneuraminic acid was deduced from its characteristic signals in the ^{13}C-NMR spectrum of **69**. The sugar sequence was suggested by fragmentation peaks in the FABMS spectrum, and linkages between sugars were elucidated using methylation analysis. The ^{1}H-NMR coupling constants of anomeric protons were used for determining the anomeric configurations of hexoses and the ^{13}C-NMR chemical shift value of C-2 for that of NeuAc.

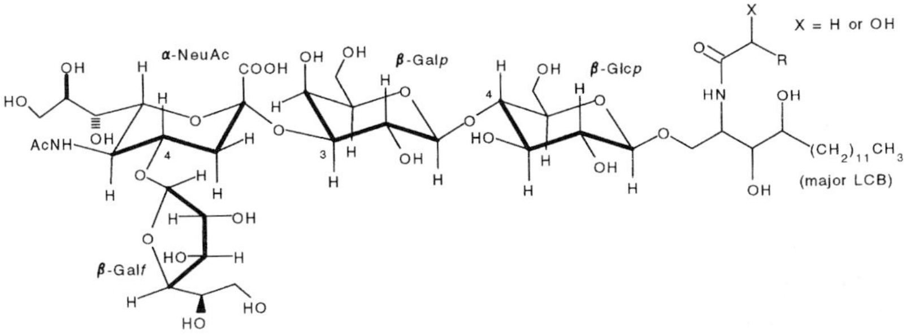

70

Occurrence: *Acanthaster planci* (starfish) (*159*).

Ganglioside **70** isolated from the whole body of the starfish *A. planci* differs from LG-1 (**69**) in that the terminal galactose is in the furanose instead of the pyranose form. The presence of galactopyranose is rare in a marine ganglioside and only two further examples are reported (**80** and **81**).

71: Ganglioside LG-2

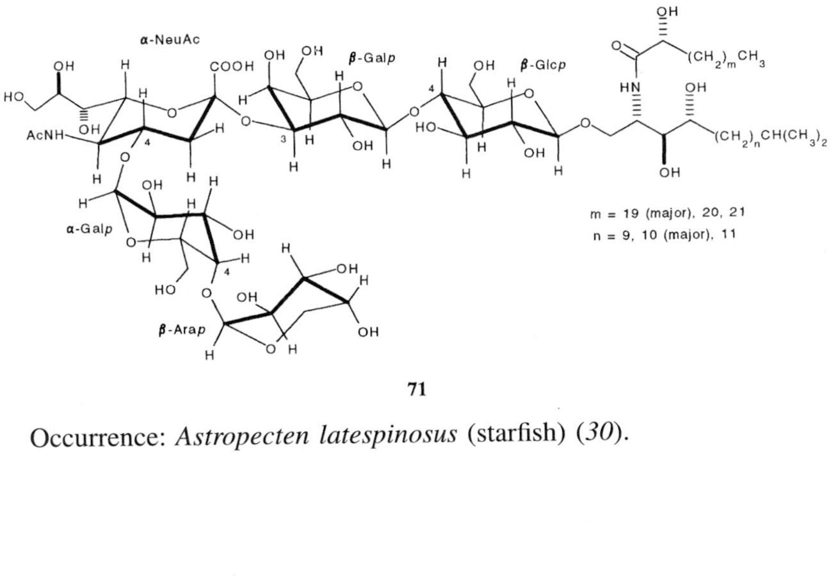

m = 19 (major), 20, 21
n = 9, 10 (major), 11

71

Occurrence: *Astropecten latespinosus* (starfish) (*30*).

Physical Data: m.p. 206–212°C, FABMS, IR, ^{1}H-NMR (selected signals), ^{13}C-NMR (complete assignment) (*30*).

Ganglioside LG-2 (**71**) from *A. latespinosus* is similar to LG-1 (**69**) isolated from the same source and differs from the latter only in the presence of an additional arabinose residue in the pyranose form. The nature of the sugars, the carbohydrate chain sequence, and the position of glycosidic linkages were determined by carbohydrate analysis and ^{13}C-NMR, FABMS, and Hakomori's method, respectively. Due to the presence in the ganglioside of two monoglycosylated galactose residues, the combined data led to two possible structures, Ara-(1 → 4)-Gal-(1 → 4)-NeuAc-(2 → 3)-Gal-(1 → 4)-Glc-Cer or Ara-(1 → 3)-Gal-(1 → 4)-NeuAc-(2 → 4)-Gal-(1 → 4)-Glc-Cer. The latter could be ruled out since the terminal trisaccharide fragment obtained by partial hydrolysis after methylation analysis gave an alditol acetate derived from 4-linked galactose.

Ganglioside LG-2 showed weak cytotoxicity against murine lymphoma L1210 cells (*30*).

72–73

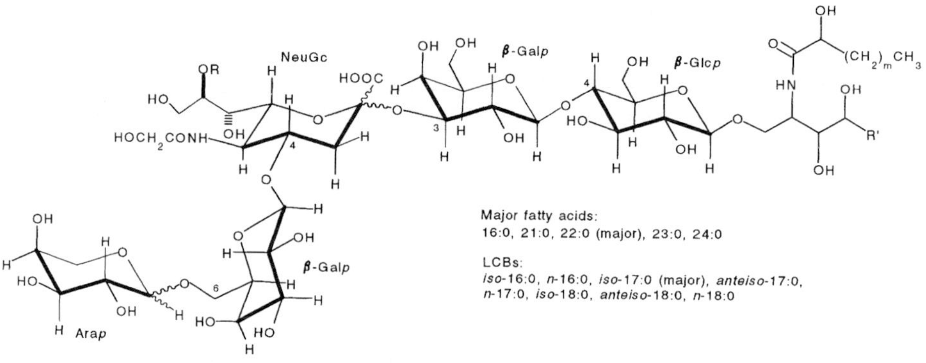

Major fatty acids:
16:0, 21:0, 22:0 (major), 23:0, 24:0

LCBs:
iso-16:0, *n*-16:0, *iso*-17:0 (major), *anteiso*-17:0, *n*-17:0, *iso*-18:0, *anteiso*-18:0, *n*-18:0

72 R = H   **73** R = Me

Occurrence: *Asterina* (= *Patiria*) *pectinifera* (starfish) (*160*).

The two arabinose-containing gangliosides **72** and **73** were isolated from *A. pectinifera*. Ganglioside **73** is the first ganglioside containing 8-*O*-methyl-*N*-glycolylneuraminic acid to be described in the literature. This methylated sialic acid has proved to be rather common in ganglioside from starfishes. The structure elucidation was based on methylation analysis of the ganglioside and of fragment oligosaccharides obtained by partial hydrolysis. As for the anomeric configurations, the β configuration assigned to the arabinose residue in **72** and **73** was based on its sensitivity to CrO_3 oxidation. Actually, sensitivity of aldopyranosides to CrO_3 oxidation is related to the equatorial position of the oxygen atom at C-1 (*52*), which for a β-arabinoside is not obvious; this point is not adequately discussed in the paper.

74

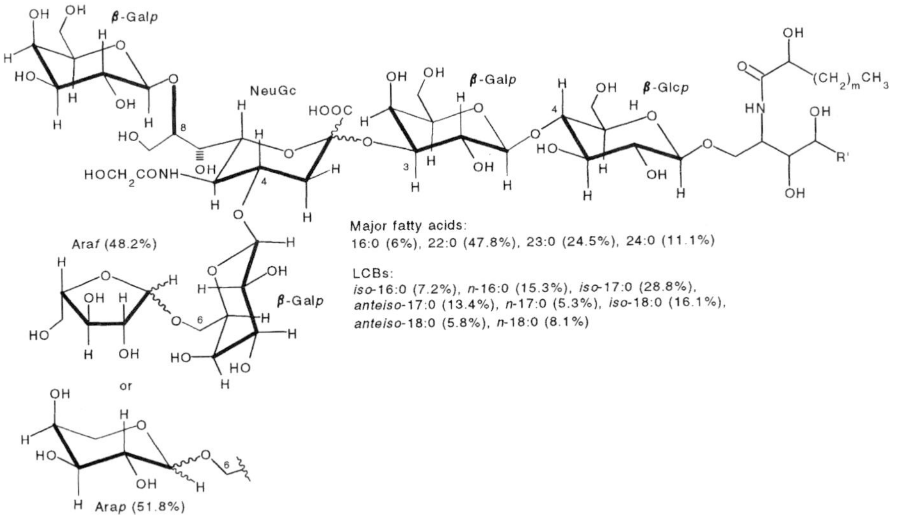
β-Galp
NeuGc
β-Galp
β-Glcp
HOCH2CONH
Araf (48.2%)
β-Galp
Arap (51.8%)
or
(CH2)mCH3
R'
Major fatty acids:
16:0 (6%), 22:0 (47.8%), 23:0 (24.5%), 24:0 (11.1%)
LCBs:
iso-16:0 (7.2%), n-16:0 (15.3%), iso-17:0 (28.8%),
anteiso-17:0 (13.4%), n-17:0 (5.3%), iso-18:0 (16.1%),
anteiso-18:0 (5.8%), n-18:0 (8.1%)

74

Occurrence: *Asterina* (= *Patiria*) *pectinifera* (starfish) (*161*).

Physical Data: IR (*161*).

Ganglioside **74** is the first reported arabinose-containing ganglioside; it is characterized by a branched carbohydrate structure at the sialic acid residue. Structure **74** was demonstrated by extensive application of methylation analysis on the intact ganglioside, on fragment gangliosides and oligosaccharides obtained by partial hydrolysis, and on an oligosaccharide obtained from digestion with β-galactosidase. This last fragment was decisive in locating the Ara(1 → 6)Gal(1 →) and Gal(1 →) fragments on the sialic acid. Methylation analysis also showed that the Ara residue was a mixture of its pyranose and furanose forms in a ratio of about 1 : 1.

75–76

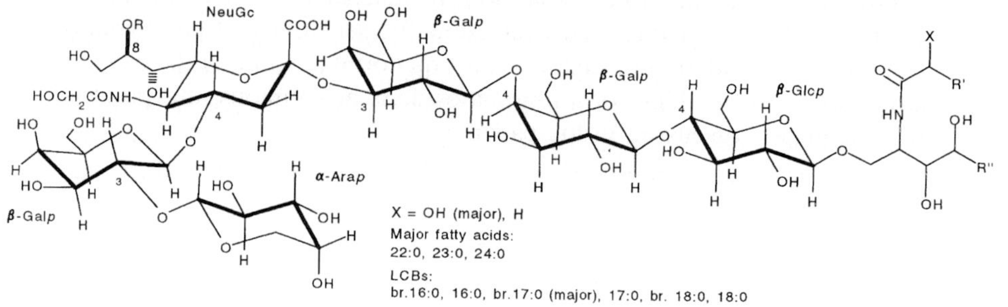

75 R = H   76 R = Me

X = OH (major), H
Major fatty acids:
22:0, 23:0, 24:0
LCBs:
br.16:0, 16:0, br.17:0 (major), 17:0, br. 18:0, 18:0

Occurrence: *Asterina* (= *Patiria*) *pectinifera* (starfish) (*162*).

Gangliosides **75** and **76** were isolated as a mixture from the body (except for gonads and hepatopancreas) of the starfish *A. pectinifera*. These gangliosides are unique with respect to all the other gangliosides from *A. pectinifera* because the sialic acid is linked to a triglycosylceramide [Gal-(1 → 4)-Gal-(1 → 4)-Glc-Cer] instead of a diglycosylceramide [Gal-(1 → 4)-Glc-Cer]. Gangliosides **75** and **76** differ for their sialic acids, which are *N*-glycolylneuraminic acid and 8-*O*-methyl-*N*-glycolylneuraminic acid, respectively.

77: Ganglioside GP-1a

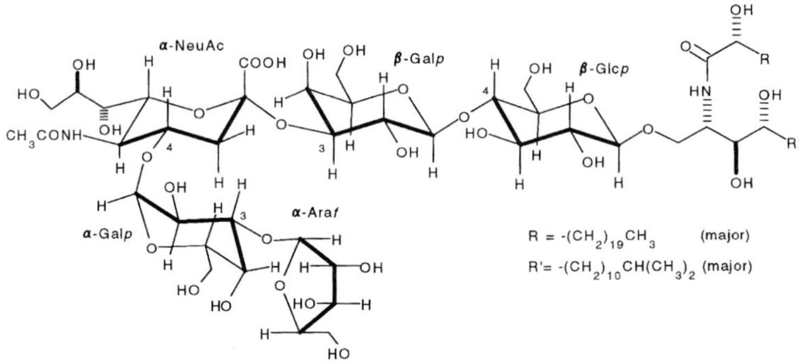

$R = -(CH_2)_{19}CH_3$ (major)
$R' = -(CH_2)_{10}CH(CH_3)_2$ (major)

77

Occurrence: *Asterina* (= *Patiria*) *pectinifera* (starfish) (*163*).

Physical Data: m.p. 215–217°C (Py/MeOH/H$_2$O), FABMS, IR, ^{1}H-NMR (selected signals), ^{13}C-NMR (complete assignment) (*163*).

References, pp. 291–301

The carbohydrate chain of ganglioside GP-1a (**77**) has the same carbohydrate sequence as ganglioside **72** also isolated from *A. pectinifera*, except for the presence of *N*-acetylneuraminic acid instead of *N*-glycolylneuraminic acid. In addition, in GP-1a the Gal residue linked to the sialic acid has the α-configuration and the arabinose is in the pyranose form. It is surprising that such a variety of gangliosides, all very similar but all different, is found in different specimens of the same species.

The carbohydrate sequence could be unequivocally established on the basis of the fragment ion peaks detected in the FABMS spectrum and by hot water hydrolysis of GP-1a which afforded a fragment ganglioside identified as ceramide lactoside. Methylation analysis elucidated the linkages between sugars, whereas ^{1}H- and ^{13}C-NMR data were used for clarifying the anomeric configurations. The stereochemistry of the ceramide part of the molecule was also determined.

78: Ganglioside GP-1b

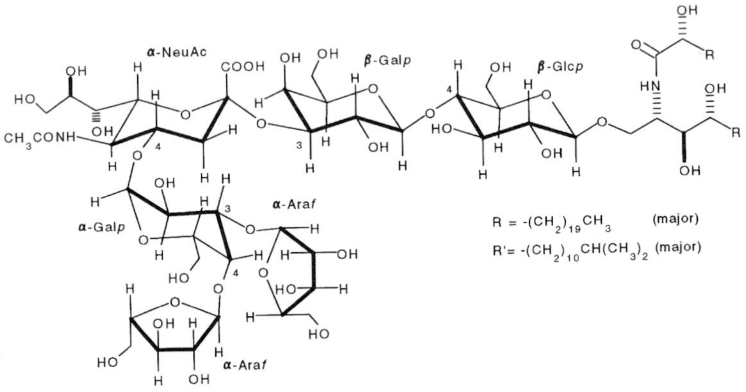
α-NeuAc
β-Galp
β-Glcp
α-Galp
α-Araf
α-Araf
COOH
CH3CONH
HN
R = -(CH2)19CH3 (major)
R'= -(CH2)10CH(CH3)2 (major)

78

Occurrence: *Asterina* (=*Patiria*) *pectinifera* (starfish) (*163*).

Physical data: m.p. 158–161°C, FABMS, IR, ^{1}H-NMR (selected signals), ^{13}C-NMR (complete assignment), CD (*163*).

Ganglioside GP-1b (**78**) contains an additional arabinose residue compared with GP-1a (**77**), linked to the same galactose. The structure determination was similar to that of GP-1a, involving methylation analysis of the intact ganglioside and of its fragments obtained by partial hydrolysis, and GC-MS and EIMS analysis of the obtained partially methylated monosaccharides. In addition, the configuration of the sialic acid was determined as α because of a negative Cotton effect at 221 nm in the CD spectrum of GP-1b.

79: Ganglioside GP-2 (asterinaganglioside-A)

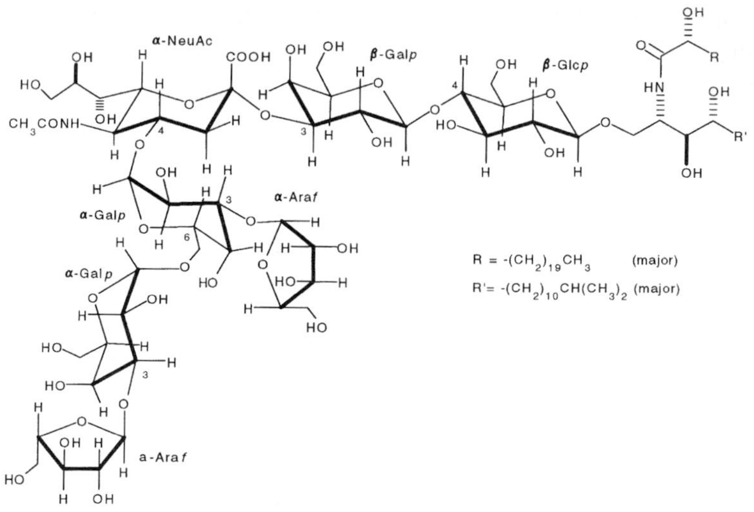

[Chemical structure of compound 79, showing α-NeuAc, β-Galp, β-Glcp, α-Galp, α-Araf, α-Galp, α-Araf units with:
R = -(CH$_2$)$_{19}$CH$_3$ (major)
R' = -(CH$_2$)$_{10}$CH(CH$_3$)$_2$ (major)]

79

Occurrence: *Asterina* (= *Patiria*) *pectinifera* (starfish) (*163, 164*).

Physical Data: m.p. 163–165°C, FABMS, IR, ^{1}H-NMR (selected signals), ^{13}C-NMR (complete assignment), CD (*163*).

Structure of ganglioside GP-2 (**79**) could not be completely elucidated using methylation analysis alone, even after partial hydrolysis, since two galactose units with a different glycosylation pattern were present in the terminal part of the molecule. Therefore, a NOESY spectrum was used to detect NOE effects between anomeric protons and neighbouring protons on the next sugar unit, which in turn were identified through a COSY spectrum. The anomeric configurations were established on the basis of the coupling constants of the anomeric protons in the ^{1}H-NMR spectrum, except for sialic acid where CD was used (*163*).

In an attempt to obtain gangliosides homogeneous in the ceramide part, ganglioside GP-2 was subjected to reversed-phase HPLC. Only a minor component of the mixture of homologues (containing a C$_{22}$ α-hydroxyacid and an unbranched C_{16} phytosphingosine) could be obtained in the pure state, and was named asterinaganglioside A. As for its biological activity, ganglioside GP-2 promotes survival of mammalian neuronal cells at a concentration of 1 µg/ml.

Possibly, ganglioside GP-2 is the same as a ganglioside previously isolated from the same starfish, *A. pectinifera* (*164*). The structure reported for this ganglioside differs from GP-2a only in the anomeric configuration of the galactose linked to the sialic acid, which is reported to be β instead of α. This assignment was based only on the fact that about 50% of the galactose is recovered after CrO_3 oxidation of the terminal tetrasaccharide obtained from the ganglioside, and it is known that this method may provide unreliable results (*165*). On the other hand, it would be surprising that two specimens of the same species elaborated two complex gangliosides, only differing in one of the anomeric configuration.

80: Ganglioside AG-2 (acanthagangliosides A–C)

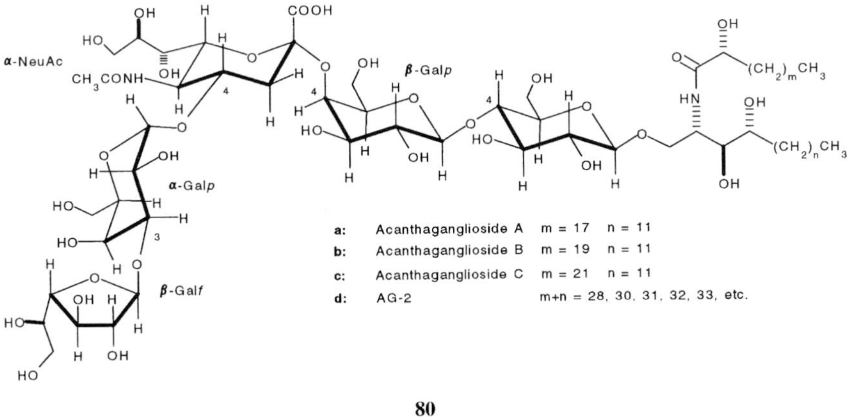

a: Acanthaganglioside A   m = 17   n = 11
b: Acanthaganglioside B   m = 19   n = 11
c: Acanthaganglioside C   m = 21   n = 11
d: AG-2   m+n = 28, 30, 31, 32, 33, etc.

80

Occurrence: *Acanthaster planci* (starfish) (*50*).

Physical Data: **80a**: m.p. 156–158°C (MeOH/H$_2$O), $[\alpha]_D = +27.9$ (MeOH), FABMS; **80b**: m.p. 155–157°C (MeOH/H$_2$O), $[\alpha]_D = +16.7$ (MeOH), FABMS; **80c**: m.p. 155–157°C (MeOH/H$_2$O), $[\alpha]_D = +19.4$ (MeOH), FABMS; **80d**: m.p. 153–156°C, FABMS, IR, ^{1}H-NMR (selected signals), ^{13}C-NMR (complete assignment), CD (*50*).

Even if ganglioside AG-2 (**80d**) has the same core carbohydrate sequence as those of many gangliosides from starfishes (NeuAc →

Gal → Glc → Cer), it differs from all of them in that the galactose is glycosylated by the sialic acid at the 4-OH, while in all gangliosides previously isolated from starfishes the sialic acid is linked at the 3-OH. In addition, this is also the first report of a ganglioside containing a galactofuranose residue. The structure determination was mainly based on analysis of the FABMS spectrum and on methylation analysis of the intact ganglioside and of the terminal trisaccharide, obtained by selective hydrolysis of the sialic acid glycosidic linkage using hot water/pyridine.

Like virtually every natural glycolipid, AG-2 is as a mixture of homologues. Its ceramide composition was not examined in detail. However, AG-2 was subjected to reversed-phase HPLC and three major homogeneous components were isolated, which were called acanthaganglioside A (**80a**), B (**80b**), and C (**80c**).

81: Ganglioside AG-3 (acanthagangliosides D–E)

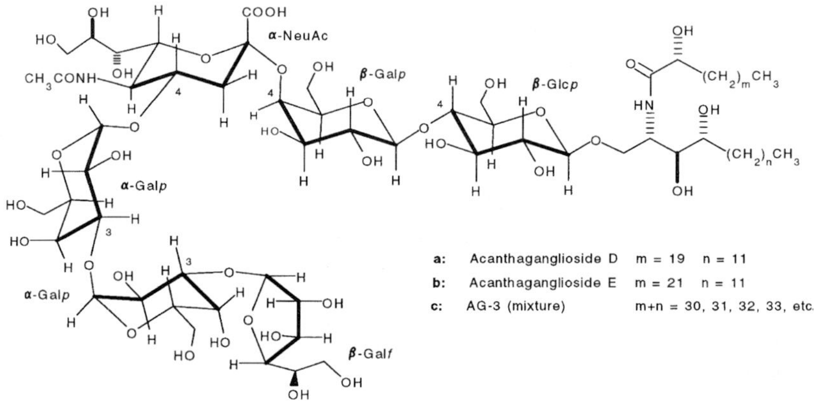

a: Acanthaganglioside D   m = 19   n = 11
b: Acanthaganglioside E   m = 21   n = 11
c: AG-3 (mixture)   m+n = 30, 31, 32, 33, etc.

81

Occurrence: *Acanthaster planci* (starfish) (*50*).

Physical data: **81a**: m.p. 161–163°C (MeOH/H$_2$O), $[\alpha]_D = +10.3$ (MeOH), FABMS; **81b**: m.p. 164–166°C (MeOH/H$_2$O), $[\alpha]_D = +8.1$ (MeOH), FABMS; **81c**: m.p. 164–166°C, FABMS, IR, ^{1}H-NMR (selected signals), ^{13}C-NMR (complete assignment), CD (*50*).

Ganglioside AG-3 (**81c**) shares the main structural features of AG-2 (**80**), differing from the latter ganglioside in the presence of an additional galactose residue between the sialic acid and the terminal galactofuranose. Its structure determination paralleled that of AG-2. Two components of AG-2, homogeneous in the ceramide part, could be obtained from AG-2 by reversed phase HPLC: acanthaganglioside D (**81a**) and E (**81b**).

82

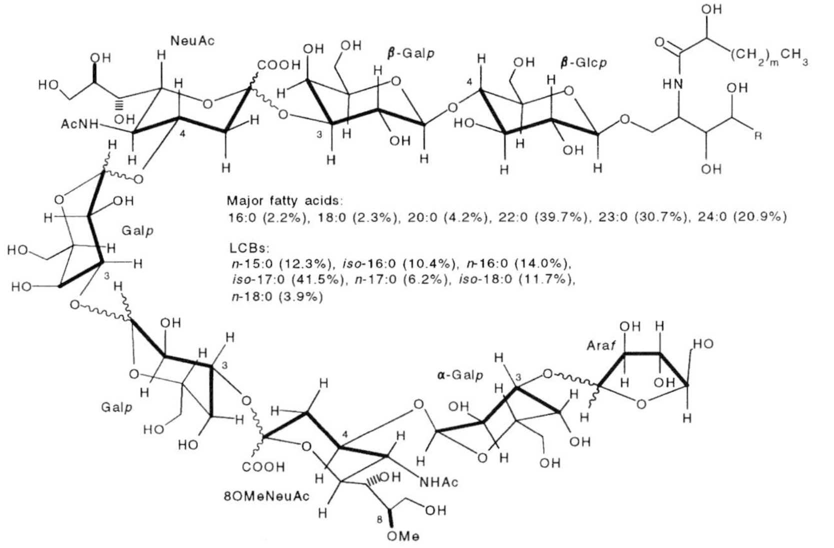
NeuAc
β-Galp
β-Glcp
Galp
Galp
α-Galp
Araf
8OMeNeuAc
Major fatty acids:
16:0 (2.2%), 18:0 (2.3%), 20:0 (4.2%), 22:0 (39.7%), 23:0 (30.7%), 24:0 (20.9%)
LCBs:
n-15:0 (12.3%), iso-16:0 (10.4%), n-16:0 (14.0%),
iso-17:0 (41.5%), n-17:0 (6.2%), iso-18:0 (11.7%),
n-18:0 (3.9%)

82

Occurrence: *Asterina* (= *Patiria*) *pectinifera* (starfish) (*166*).

Disialoglycolipid **82** is the most complex ganglioside isolated from a marine organism to date, possessing a carbohydrate chain composed of 8 saccharides. It is also the sole example of a ganglioside with two internal sialic acid residues. Interestingly, only one of the two *N*-acetylneuraminic acid residues is 8-*O*-methylated. Structure **82** was assigned to the ganglioside by combining information provided by carbohydrate analysis and methylation analysis of the ganglioside and of several fragment glycolipid and oligosaccharides obtained by mild acid hydrolysis. Many anomeric configurations remain undetermined.

6. Other Glycolipids

This section covers some marine glycolipids which are structurally composed of a lipid moiety of apparent non-mevalonate origin linked to the carbohydrate through a glycosidic bond. They cannot be classified either as glycoglycerolipids or glycosphingolipids. The random distribution of these uncommon metabolites and their limited number does not allow for any chemotaxonomic considerations.

83–85: Heterocyst glycolipids I and III

83	n = 19	R = H	R' = OH
84	n = 19	R = OH	R' = H
85	n = 19	R, R'	= O

Occurrence: *Anabaena cylindrica* (cyanobacterium) (*14, 23, 167*), *Anabaena torulosa* (*167*), *Nodularia harveyana* (cyanobacterium) (*168*).

Physical Data: **83**: $[\alpha]_D = +47.8$ (CHCl$_3$/MeOH) (*167*), FABMS, ^{1}H- and ^{13}C-NMR (complete assignment) (*168*); **84**: $[\alpha]_D = +40.0$ (CHCl$_3$/MeOH), FABMS, ^{1}H- and ^{13}C-NMR (complete assignment) (*168*); **85**: $[\alpha]_D = +38.8$ (CHCl$_3$/MeOH) (*167*), FABMS, ^{1}H- and ^{13}C-NMR (complete assignment) (*168*).

Heterocyst glycolipids are specifically found in specialized cyanobacterial cells known as heterocysts, which are capable of N$_2$ fixation. They were isolated for the first time from the marine cyanobacterium *A. cylindrica* (*14, 23*), but their structure have been recently revised when they were re-isolated from the same species (*167*) and from *N. harveyana* (*168*).

The structure of heterocyst glycolipid III (**83**) was successfully established in (*14*), but its stereochemistry was not. In a subsequent paper (*167*), the absolute configuration at C-3 was established as *R* through the CD exciton chirality method (*169*) using the tris-(*p*-bromobenzoate) of the aglycon triol, while that at C-25 was determined by the MOSHER method (*170*). Two syntheses of **83** have been reported (*171, 172*). In *N. harveyana* the heterocyst glycolipid **83** occurs together with its C-3 epimer **84**. Compounds **83** and **84** could be separated by HPLC as their peracetyl derivatives (*168*).

Heterocyst glycolipid I (**85**), isolated from both *A. cylindrica* and *N. harveyana*, is the oxidized form of **83–84**. Structure **85** was established from spectroscopic data and chemical correlation with **83–84** (*167, 168*),

References, pp. 291–301

showing that the previous characterization of **85** on the basis of MS data as a hydroxy fatty ester of glucose was incorrect (*23*). Finally, in (*23*) heterocyst glycolipids I and III from *A. cylindrica* were reported to contain about 10% of their α-galactosyl analogue, but this was not confirmed in the subsequent papers.

A biosynthetic study of **83–85** demonstrated that the long chain of the aglycon originates from a *de novo* biosynthesis and not by elongation of a preformed shorter molecule, and that there is no apparent interconversion between **83–84** and **85** (*173*).

86–87: Heterocyst glycolipids from *Cyanospira rippkae*

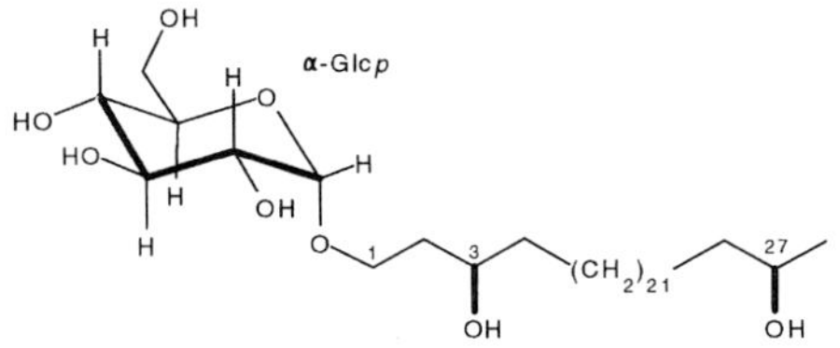

86

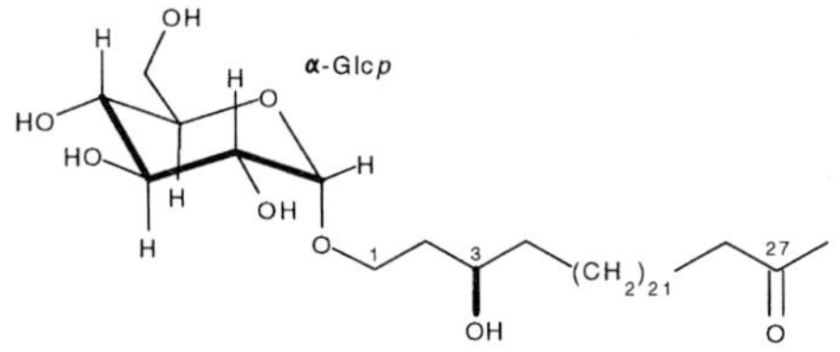

87

Occurrence: *Cyanospira rippkae* (*174*).

Physical Data: **86**: m.p. 118–120°; [α]$_D$ = +44.3 (CHCl$_3$/MeOH); FABMS; ^{1}H- and ^{13}C-NMR (complete assignment); **87**: FABMS and ^{1}H- and ^{13}C-NMR (complete assignment) (*174*).

Heterocysts glycolipids from *C. rippkae* are similar to those from *A. cylindrica* and *N. harveyana*. The major glycolipid **86** differs from **83** only in that the chain is C$_{28}$ instead of C$_{26}$, whereas the minor product **87** is a glycosylated ketodiol like **85**, but the keto function is at position ω-1 instead at position 3. The configuration of the stereogenic center was determined from the CD spectrum of the tris-(*p*-bromobenzoate) derivative of the triol aglycone, and confirmed using the MOSHER method. Glycolipids **86** and **87** were isolated not only from heterocysts, but also from akinetes, reproductive structures of *C. rippkae*. This confirms the close relationship between akinetes and heterocysts, previously suggested on the basis of comparative chemical analyses.

88–89: Heterocyst glycolipid IV and II

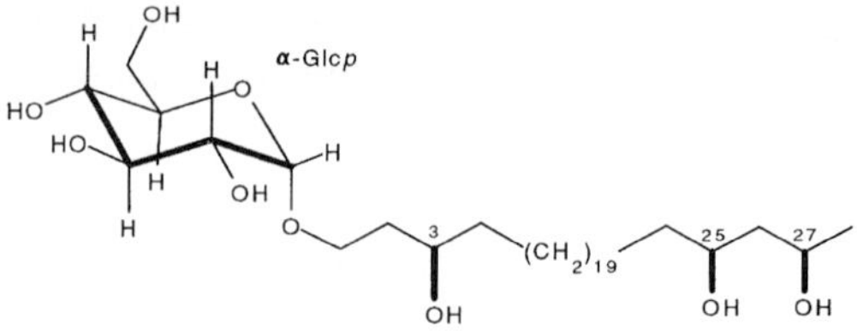

Occurrence: *Anabaena cylindrica* (cyanobacterium) (*14, 23, 167*), *Anabaena torulosa* (*167*).

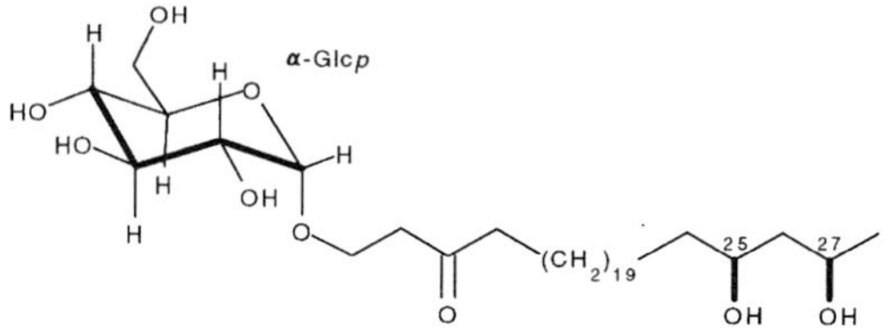

Physical Data: **88**: $[\alpha]_D = -48.9$ (CHCl$_3$/MeOH); FABMS, IR, ^{1}H- and ^{13}C-NMR (complete assignment) (*167*); **89**: $[\alpha]_D = +14.4$ (CHCl$_3$/MeOH); FABMS, IR, ^{1}H- and ^{13}C-NMR (complete assignment) (*167*).

The heterocyst glycolipids IV (**88**) and II (**89**) were isolated from *A. cylindrica* and *A. torulosa*, but not from *Nodularia harveyana*. They are similar to **83** and **85**, respectively, possessing a longer chain and one more hydroxy group (*167*). The absolute configuration at C-3 of compound **88** could be established by comparison with the NMR data of **83** because of the proximity of the chiral glucose residue. The relative configuration of C-25 and C-27 was determined as *syn* by analysis of the ^{13}C-NMR spectrum of the ^{13}C-enriched bis-isopropylidene derivative of the tetrol aglycone, while the absolute configuration at C-27 was obtained using the MOSHER method. The stereochemistry of **89** has been determined similarly. Structure **89** revises the previous characterization of heterocyst glycolipid II as a glycosylated dihydroxy fatty acid (*23*).

References, pp. 291–301

90: Rhizochalin

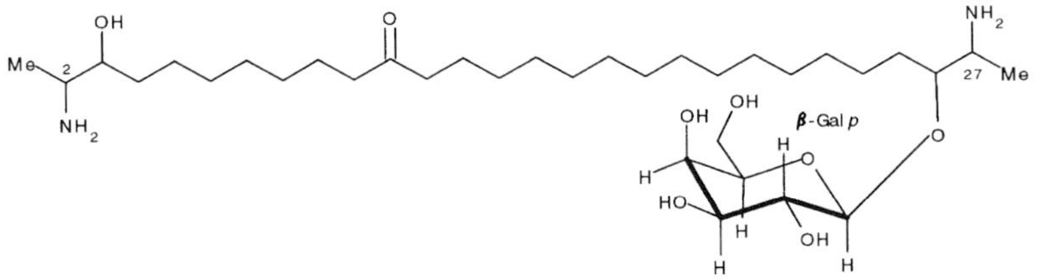
β-Gal p
27
2

90

Occurrence: *Rhizochalina incrustata* (sponge) (*175*).

Physical Data: m.p. 124–126°C (EtOH/EtOAc), $[\alpha]_D = -5$; EIMS, IR, ^{13}C-NMR, ^{1}H-NMR (peracetyl derivative) (*175*).

Rhizochalin (**90**) is an unprecedented glycolipid composed of a long-chain aminoalcohol glycosylated by a galactose unit. Its structure was established by spectroscopic (MS and NMR) methods, except for location of the keto group on the chain, which required perphthalic acid oxidation of the ketone function, hydrolysis of the esters thus obtained and identification of the fragments. The stereochemistry of the stereogenic centers on the alkyl chain remains unassigned.

91a–91e: Erylusamine A–E

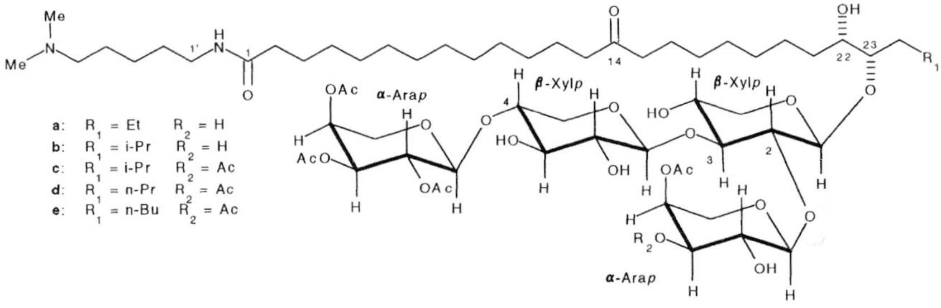
a: R1 = Et R2 = H
b: R1 = i-Pr R2 = H
c: R1 = i-Pr R2 = Ac
d: R1 = n-Pr R2 = Ac
e: R1 = n-Bu R2 = Ac
α-Arap
β-Xylp
β-Xylp
α-Arap

91

Occurrence: *Erylus placenta* (sponge) (*37,176*).

Physical Data: **91a**: $[\alpha]_D = -3.5$ (MeOH); IR, FABMS, ^{1}H- and ^{13}C-NMR (complete assignment); **91b**: $[\alpha]_D = -5.5$ (MeOH); IR, FABMS, ^{1}H- and ^{13}C-NMR (complete assignment); **91c**: $[\alpha]_D = -9.6$ (MeOH); IR, FABMS, ^{1}H- and ^{13}C-NMR (complete assignment); **91d**: $[\alpha]_D = -6.0$ (MeOH); IR, FABMS, ^{1}H- and ^{13}C-NMR (complete assignment); **91e**: $[\alpha]_D = -8.0$ (MeOH); IR, FABMS, ^{1}H- and ^{13}C-NMR (complete assignment) (*37*).

Erylusamine A-E (**91a–91e**) are unique glycolipids from the sponge *E. placenta*, characterized by a tetrasaccharide chain glycosidically linked to a ketodihydroxy fatty acid, which in turn is linked to a diamine through an amide bond. Structures **91a–91e** were determined by extensive use of spectroscopic methods. COSY and HOHAHA 2D NMR spectral data were used for determination of the nature of the four pentoses, while linkages between sugars were determined using a HMBC ^{1}H-^{13}C correlation. The keto group on the chain was located by analysis of the FAB mass spectrum of the aglycon part of the molecules. The absolute stereochemistry of the sugars was established by chiral GLC analysis of the acid hydrolysate. The relative configuration of C-22 and C-23 was determined through preparation of a cyclic acetonide at these positions and subsequent NMR analysis, while their absolute configuration was based on CD spectroscopy in the presence of Eu(fod)$_3$ (*177*). Erylusamine A-E are antagonists for interleukin-6 receptors, the most active being erylusamine E (**91e**) with IC$_{50}$ of 17 µg/ml.

92: Forbesin

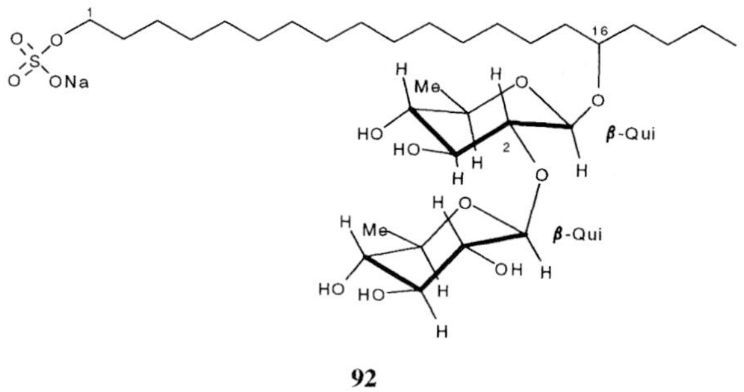

92

Occurrence: *Asterias forbesi* (starfish) (*178*), *Asterias vulgaris* (starfish) (*178*).

Physical Data: m.p. 168°C (dec.); FABMS. ^{1}H- and ^{13}C-NMR (complete assignment) (*178*).

Forbesin (**92**) is a unique diglycosylated long-chain diol sulfate from two starfishes of the genus *Asterias*. The nature of the two sugar units and the linkage between them were established by NMR studies on native forbesine and its peracetyl derivative. The position of the secondary hydroxy group was deduced from the mass spectrum of the aglycon. The presence of a sulfate ester at C-1 was confirmed by solvolytic desulfation. The ^{1}H-NMR spectrum of the desulfated glycolipid showed the expected upfield shift of the methylene protons at C-1.

7. Biological Activities

7.1. Immunological Activity

In general, GSLs are good haptens. In the presence of a good immunogen, they induce an immune response and the resulting antiserum is substantially specific toward their sugar part structure, thus providing an useful reagent for the localization of GSL molecules in cell membranes. Among marine organisms, immunochemical studies have been carried out on phosphoglycolipids of molluscs and gangliosides of echinoderms.

SATAKE and coworkers prepared polyclonal antibodies against three phosphonoglycosphingolipids, SGL-II (**42**) (*133*), FGL-IIb (**45**) (*137*) and EGL-I (**51**) (*141*), which are present in the skin, nerve fibers and eggs, respectively, of the sea hare *Aplysia kurodai*. The anti-SGL-II antiserum (*133*) reacted with SGL-II and other phosphonoglycosphingolipids of *A. kurodai* having β-3-*O*-methylgalactose at their non reducing end, such as SGL-I′ (**43**) and F-21 (**44**), but did not react with gangliosides from bovine brain or with globoside. Immunohistochemical studies revealed that SGL-II and GSLs immunologically related to it were present both in skin and in nervous tissues. The sugars recognized were 3-*O*-methylgalactose at the non-reducing end and galactose at the branched chain of the glycolipids (*133*). More recently (*141*), the anti-FGL-IIb antiserum was proved to be unreactive towards SGL-I (**41**) and EGL-I (**51**), both of which were isolated from *A. kurodai* and contained α-3-*O*-methylgalactose and α-4-*O*-methyl-*N*-acetylglucosamine, respectively, as the non-reducing end. Therefore, the antibody was judged to be specific to β-3-*O*-methylgalactosides. An attempt to raise antisera specific to α-3-*O*-methylgalactoside was also carried out, but an antiserum raised against EGL-1 reacted with both α- and β-3-*O*-methylgalactose-containing GSLs.

The anti-FGL-IIb antiserum (*137*) reacted with FGL-IIb (**45**) and other *Aplysia* GSLs such as FGL-I (**48**), FGL-IIa (**47**), FGL-V (**46**), and F-9 (**49**). All the antigenic GSLs were located specifically in the nerve bundles of *A. kurodai,* as indicated by immunohistochemical studies. The reactivity was suppressed by mild acid-methanolic treatment, and the lost reactivity was recovered by alkaline hydrolysis (*179*). These results suggested a key role of the carboxyl group of the pyruvic acid, present at the non reducing sugar end of all the immunoreactive GSLs.

In 1981 (*154*) antisera against two gangliosides, G-1 (**61**) and G-2 (**62**), present in the eggs and embryos of the sea urchin *Strongylocentrotus intermedius*, were prepared, and their specificity was revealed by immunoelectrophoresis and immunodiffusion. The surface localization of gangliosides in embryos incubated at different cell densities was studied by immunofluorescence microscopy.

In 1990 Kubo and Hoshi (*145*) raised an antiserum against M5 (**53**), the dominant ganglioside in the eggs of the sea urchin *Anthocidaris crassispina*, which was purified by affinity chromatography. Its specificity was verified by enzyme-linked immunosorbent assay and TLC immunostaining. Immunofluorescence microscopy with this antibody indicated the presence of M5 ganglioside in the eggs of another sea urchin, *Hemicentrotus pulcherrimus*, and evidenced dramatic variations in its intracellular distribution upon fertilization.

7.2. Pharmacological Activity

Natori and coworkers recently isolated five bioactive GSLs (**15a–15e**) named agelasphins from the extract of the marine sponge *Agelas mauritiana* (*35, 113*). Structurally, agelasphins are quite simple molecules, being composed of a phytosphingosine containing an amide linked α-hydroxy fatty acid at C-2 and an α-galactosyl residue at C-1. Compounds **15a–15e** differ from each other only in the length and/or branching of the sphingoid base. The alkyl chain of the α-hydroxyacyl residue is the same for compounds **15a–15d**, a linear chain of 26 carbon atoms, while in **15e** the chain possesses an additional methylene group.

All agelasphins exhibited a quite interesting antitumor activity; they markedly prolonged the survival period of B16 bearing mice, but did not prolonged the life span of P388-bearing mice. These findings, which were in contrast with the behavior of most chemotherapeutic agents, suggested that they were biological response modifiers, showing antitumor effects *via* activation of the immune system. This hypothesis was supported by a subsequent experiment which evidenced remarkable lymphocyte proliferation (LP) stimulatory effects of agelasphins on the allogenic mixed lymphocyte reaction (MRL) (*115*).

On account of the above antitumor properties, the most active agelasphin **15b** was synthesized (*114*). Various analogues of **15b** were also prepared to search for other candidates which possessed antitumor activity similar to that of **15b**, but could be synthesized more easily than

References, pp. 291–301

15b on a large scale (*116*). This investigation allowed a study of the relationships between the structures and the bioactivities of agelasphin analogues. Thus, taking into account the influence of the length of alkyl chains and the stereochemistry of the glycosidic linkage and the four chiral centers in the ceramide portion, as well as the role of the three ceramide secondary OH groups, a candidate, (2*S*,3*S*,4*R*)-1-*O*-(α-D-galactopyranosyl)-2-(*N*-hexacosanoylamino)-1,3,4-octadecanetriol (**93**), could be selected for clinical application (*180*).

93

In a more recent investigation (*181*) the *in vitro* and *in vivo* natural killer (NK) cell activity enhancing effects of α- and β-galactosylceramides and α- and β- glucosylceramides were examined. The results indicated that the α-types show stronger enhancing effects than β-types, with the α-galactosylceramides possessing the most potent activity. Analogous behaviour was observed when the above compounds were tested for their inhibitory effects on mice inoculated with B16 cells. These results suggested that the stereochemistry of the glycosidic linkage plays an important role in the antitumor activity of galactosylceramides.

Very recently, GSL analysis of four different *Agelas* species (*A. chlathrodes, A. longissima, A. dispar* and *A. conifera*) led to the isolation of five new GSLs [**23** (*40, 95*), **24** (*40, 56, 60, 95, 96*), **25** (*56*), **26** (*40, 96*), and **28** (*40*)] which are chemically related to agelasphins in that an α-galactosyl residue is linked to a ceramide composed of a phytosphingosine and an α-hydroxy fatty acid. The immunostimulation properties of these compounds were compared with those reported for agelasphins using the mixed lymphocyte reaction (MLR) assay (*40*). GSLs **23** and **24** exhibited a stimulatory effect on lymphocyte proliferation quite similar to that of agelasphins. By contrast, compounds **25**, **26**, and **28** did not exhibit any stimulatory activity. These results indicated that the immunostimulation activity of α-galactosylceramides is affected by a specific structural feature, namely glycosylation at position 2 of the inner sugar. In fact, compounds **23** and **24** possess a free 2-OH on the α-galactose directly linked to the ceramide moiety, while in compounds **25**, **26**, and **28** this position is glycosylated either by an α-galactosyl (**26** and **28**) or by an β-glucosyl (**25**) residue.

Erylusamines (**91a–91e**), glycolipids produced by the sponge *Erylus placenta*, exhibited potent antagonistic activity against an IL-6 receptor (*37, 176*). Interleukin-6 (IL-6) is a multifunctional cytokine which exerts several biological functions through binding with its specific receptor (*182*). Its activity is related to diseases such as inflammation, viral infection and cancer, so that inhibitors of IL-6 receptor are believed to be of potential therapeutic importance.

In the course of a screen for antitumor antiviral agents from natural sources, sulfoquinovosyldiacylglycerols (**7**) from the cyanobacteria *Lyngbya lagerheimii* and *Phormidium tenue* were found to be HIV-1 inhibitory compounds. At non-cytotoxic concentrations they were strikingly active against HIV-1 in cultured human lymphoblastoid cells (*80*). It is to be noted that sulfoquinovosyldiacylglycerols are structural components of chloroplast membranes and occur widely in higher plants, algae, and photosynthetic microorganisms, but had never been tested before for anti-HIV activity. On account of the potential role of these compounds as therapeutic agents the total synthesis of a cyanobacterial sulfoquinovosydiacyglycerol has been recently performed (*81*). The potency and formulability of the synthetic material were the same as were exhibited by material previously produced by fermentation.

As a result of their extensive studies on glycosides from Asteroidea, HIGUCHI and co-workers described some starfish gangliosides possessing biological activity. Ganglioside **77**, isolated from the starfish *Asterina pectinifera*, was shown to support the survival of cultural cortex cells, most of which were neuronal cells, of rat foetuses (*163*). Compound **64**, the main ganglioside of *Asterias amurensis versicolor*, exhibited neuritogenic and growth-inhibitory activities towards the mouse neuroblastoma cells. By examining the bioactivities of the other gangliosides present in the starfish, a key role of the terminal sialic acid residue present in **64** was hypothesized (*34*). Finally, ganglioside LG-2 (**71**), present in *Astropecten latespinosus*, exhibited weak in vitro antitumor activity against murine lymphoma L-1210 cells (*30*).

Crasserides (**9**), have been isolated from several marine sponges (*84*), where they are supposed to play the role of natural feeding deterrents (*58*). These unique compounds were found to exhibit 3-4-fold stimulation activity of nerve growth factor (NGF) synthesis in cultured

References, pp. 291–301

astroglials cells. It is to be noted that NGF-synthesis enhancers are considered as potential drugs for peripheral or central nerve disorders (*62*).

References

1. MINALE, L., R. RICCIO, and F. ZOLLO: Steroidal Oligoglycosides and Polyhydroxysteroids from Echinoderms. In: Progress in the Chemistry of Organic Natural Products (W. HERZ, G.W. KIRBY, R.E. MOORE W. STEGLICH, and CH. TAMM, eds.) Vol. 62, p. 75. Wien, New York: Springer. 1993.
2. GUNSTONE, F.D., J.L. HARWOOD, and F.D. PADLEY: The Lipid Handbook, p. 12. London: Chapman & Hall. 1994.
3. SWEELEY, C.C.: Sphingolipids. In: New Comprehensive Biochemistry (D.E. VANCE and J. VANCE, eds.) Vol. 20, Biochemistry of Lipids, Lipoproteins and Membranes, p. 327. Amsterdam: Elsevier. 1991.
4. RAJAGOPAL, M.V., and K. SOHONIE: Studies on the Sea Anemone *Gyrostoma* sp. Biochem. J., **65**, 34 (1957).
5. NAGAI, Y., and Y. ISONO: Occurrence of Animal Sulfolipid in the Gametes of Sea Urchins. Jpn. J. Exp. Med., **35**, 315 (1965).
6. ISONO, Y., and Y. NAGAI: Biochemistry of Glycolipids of Sea Urchin Gametes. 1. Separation and Characterization of New Type of Sulfolipid and Sialoglycolipid. Jpn. J. Exp. Med., **36**, 461 (1966).
7. ISONO, Y.: Changes of Glycolipids during Early Development of Sea Urchin Embryos. Jpn. J. Exp. Med., **37**, 87 (1967).
8. NICHOLS, B.W., and B.J.B. WOODS: New Glycolipid Specific to Nitrogen-Fixing Blue-Green Algae. Nature, **217**, 767 (1968).
9. WALSBY, A.E., and B.W. NICHOLS: Lipid Composition of Heterocysts. Nature, **221**, 673 (1969).
10. VASKOVSKY, V.E., E.Y. KOSTETSKY, V.I. SVETASKEV, I.G. ZHUKOVA, and G.P. SMIRNOVA: Glycolipids in Marine Invertebrates. Comp. Biochem. Physiol., **34**, 163 (1970).
11. KOEZUKA, I., M. KLOPPENBURG, and H. WIEGANDT: Characterization of Gangliosides from Fish Brain. Biochim. Biophys. Acta, **210**, 299 (1970).
12. KOMAI, Y., S. MITSUKAWA, and M. SATAKE: Glycolipids in Nervous Tissue of Invertebrates. J. Biochem., **70**, 367 (1971).
13. HAYASHI, A., and F. MATSUURA: Isolation of a New Sphingophosphonolipid Containing Galactose from the Viscera of *Turbo cornutus*. Biochim. Biophys. Acta, **248**, 133 (1971).
14. BRYCE, T.A., D. WELTI, A.E. WALSBY, and W.B. NICHOLS: Monohexoside Derivatives of Long-Chain Polyhydroxy Alcohols: a Novel Class of Glycolipid Specific to Heterocystous Algae. Phytochemistry, **11**, 295 (1972).
15. BJÖRKMAN, L.R., K.A. KARLSSON, and K. NILSSON: Existence of Cerebroside and Cholesterol Sulfate in the Tissues of the Sea Star *Asterias rubens*. Comp. Biochem. Physiol., **43B**, 409 (1972).
16. BJÖRKMAN, L.R., K.A. KARLSSON, I. PASCHER, and B.E. SAMUELSSON: Isolation of Large Amounts of Cerebroside and Cholesterol Sulphate in the Sea Star *Asterias rubens*. Biochim. Biophys. Acta, **270**, 260 (1972).
17. MATSUBARA, T., and A. HAYASHI: The Existence of Branched Structure in the Sugar Moiety of Oyster Sphingoglycolipid. J. Biochem., **74**, 853 (1973).

18. Kochetkov, N.K., I.G. Zhukova, G.P. Smirnova, and I.S. Glukhoded: Isolation and Characterization of a Sialoglycolipid from the Sea Urchin *Strongylocentrotus intermedius*. Biochim. Biophys. Acta, **326**, 74 (1973).
19. Komai, Y., S. Matsukawa, and M. Satake: Lipid Composition of the Nervous Tissue of the Invertebrates *Aplysia kurodai* (Gastropod) and *Cambarus clarki* (Arthropod). Biochim. Biophys. Acta, **316**, 271 (1973).
20. Kreps, E.M., N.F. Avrova, M.A. Chebotarêva, E.V. Chirkovskaya, V.I. Krasilnikova, E.E. Kruglova, M.V. Levitina, E.L. Obukhova, L.F. Pomazanskaya, N.I. Pravdina, and S.A. Zabelinskii: Phospholipid and Glycolipids in the Brain of Marine Fish. Comp. Biochem. Physiol., **52B**, 283 (1975).
21. Sugita, M., and T. Hori: New Types of Gangliosides with Sialic Acid Residues in the Inner Part of Their Carbohydrate Chains. J. Biochem., **80**, 637 (1976).
22. Matsubara, T., and A. Hayashi: Structural Studies on Glycolipid of Shellfish. II. Occurrence of 3-O-methylgalactosamine in Oyster Glycolipid. J. Biochem., **83**, 1195 (1978).
23. Lambein, F., and C.P. Wolk: Structural studies on the Glycolipid from the Envelope of the Heterocyst of *Anabaene cylindrica*. Biochemistry, **12**, 791 (1973).
24. Kochetkov, N.K., and G.P. Smirnova: Glycolipids of Marine Invertebrates. Adv. Carbohydr. Chem. Biochem., **44**, 387 (1986).
25. Hori, T., and M. Sugita: Sphingolipids in Lower Animals. Prog. Lipid Res., **32**, 25 (1993).
26. Kates, M.: Glycolipids of Higher Plants, Algae, Yeasts, and Fungi. In: Handbook of Lipid Research (M. Kates, ed.) Vol. 6, p. 235. New York, London: Plenum Press. 1990.
27. Higuchi, R., T. Natori, and T. Komori: Glycosphingolipids from the Starfish *Asterina pectinifera*. Isolation and Characterization of Acanthacerebroside B and Structure Elucidation of Related, Nearly Homogeneous Cerebrosides. Liebigs Ann. Chem., 51 (1990).
28. Higuchi, R., M. Inagaki, K. Togawa, T. Miyamoto, and T. Komori: Constituents of Holothuriodeae. IV. CE-2b, CE-2c and CE-2d, Three New Sphingosine-type Glucocerebrosides from the Sea Cucumber *Cucumaria echinata*. Liebigs Ann. Chem., 79 (1994).
29. Folch, J., M. Lees, and G.H. Sloane-Stanley: A Simple Method for the Isolation and Purification of Total Lipids from Animals Tissues. J. Biol. Chem., **226**, 497 (1957).
30. Higuchi, R., S. Matsumoto, M. Fujita, T. Komori, and T. Sasaki: Glycosphingolipids from the Starfish *Astropecten latespinosus*, 2. Structure of Two New Ganglioside Molecular Species and Biological Activity of the Ganglioside. Liebigs Ann. Chem., 545 (1995).
31. Oshima, Y., S.-H. Yamada, K. Matsunaga, T. Moriya, and Y. Ohizumi: A Monogalactosyl Diacylglycerol from a Cultured Marine Dinoflagellate, *Scrippsiella trochoidea*. J. Nat. Prod., **57**, 534 (1994).
32. Jin, W., K.L. Rinehart, and E.A. Jares-Erijman: Ophidiacerebrosides A-E, Five New Cytotoxic Glycosphingolipids from the Sea Star *Ophidiaster ophidiamus*. J. Org. Chem., **59**, 144 (1994).
33. Kobayashi, J., Y. Doi, and M. Ishibashi: Shimofuridin A, a Cytotoxic and Antimicrobial Nucleoside Derivative with an Acylfucopyranoside Unit from the Marine Tunicate *Aplidium multiplicatum*. J. Org. Chem., **59**, 255 (1994).
34. Higuchi, R., K. Inukai, J.X. Jhou, M. Honda, T. Komori, S. Tsuji, and Y. Nagai: GAA-6 and GAA-7, Two Ganglioside Molecular Species from the Starfish *Asterias amurensis versicolor*, Liebigs Ann. Chem., 359 (1993).

35. NATORI, T., M. MORITA, K. AKIMOTO, and Y. KOEZUKA: Agelasphin, Novel Antitumor and Immunostimulatory Cerebrosides from the Marine Sponge *Agelas mauritiana*. Tetrahedron, **50**, 2771 (1994).
36. HIGUCHI, R., M. INAGAKI, K. TOGAWA, T. MIYAMOTO, and T. KOMORI: Constituents of Holothuriodeae. V. Isolation and Structure of Cerebrosides from the Sea Cucumber *Pentacta australis*. Liebigs Ann. Chem., 653 (1994).
37. SATA, N., N. ASAI, S. MATSUNAGA, and N. FUSETANI: Erylusamines, Interleukin-6 Receptor Antagonists, from the Marine Sponge *Erylus placenta*. Tetrahedron, **50**, 1093 (1994).
38. KUBO, H., A. IRIE, F. INAGAKI, and M. HOSHI: Gangliosides from the Eggs of the Sea Urchin *Anthocidaris crassispina*. J. Biochem., **108**, 185 (1990).
39. ITONORI, S., K. KAMEMURA, K. NARUSHIMA, N. SONKU, O. ITASAKA, T. HORI, and M. SUGITA: Characterization of a New Phosphonocerebroside, N-Methyl-2-aminoethylphosphonylglucosylceramide, from the Antarctic Krill *Euphausia superba*. Biochim. Biophys. Acta, **1081**, 321 (1991).
40. COSTANTINO, V., E. FATTORUSSO, A. MANGONI, M. DI ROSA, A. IANARO, and P. MAFFIA: Glycolipids from Sponges. IV. Immunomodulating Glycosyl Ceramides from the Marine Sponge *Agelas dispar*. Tetrahedron, **52**, 1573 (1996).
41. SWEELEY, C.C., and R.V.P. TAO: Gas Chromatographic Estimation of Carbohydrates in Glycosphingolipids. Methods Carbohydr. Chem., **6**, 8 (1972).
42. SVENNERHOLM, L.: Quantitative Estimation of Sialic Acid. III. An Anion-Exchange-Resin Method. Acta Chem. Scand., **12**, 547 (1958).
43. HAKOMORI, S.: Rapid Permethylation of Glycolipids and Polysaccharides, Catalyzed by Methylsulfinyl Carbanion in Dimethyl Sulfoxide. J. Biochem., **55**, 205 (1964).
44. SANFORD, P.A., and H.E. CONRAD: The Structure of the *Aerobacter aerogenes* A3(S1) Polysaccharide. I. A Reexamination using Improved Procedures for Methylation Analysis. Biochemistry, **5**, 1508 (1966).
45. LI, Y.-T., and S.-C. LI: Glycosidases in Jack Bean Meal. I. Separation of Various Glycosidases by Isoelectric Focusing. J. Biol. Chem., **243**, 3994 (1968).
46. LI, S.-C., and Y.-T. LI: Glycosidases of Jack Bean Meal. II. Crystallization and Properties of β-N-Acetylhexosaminidase. J. Biol. Chem., **245**, 5153 (1970).
47. WEISSMAN, B., and D.F. HINRICHSEN: Mammalian α-Acetylagalactosaminase. Occurrence, Partial Purification, and Action on Linkages in Submaxillary Mucins. Biochemistry, **8**, 2034 (1969).
48. GATT, S., and M.M. RAPPORT: Isolation of β-Galactosidase and β-Glucosidase from Brain. Biochim. Biophys. Acta, **113**, 567 (1966).
49. KUBO, H., G.J. JIANG, A. IRIE, M. MORITA, T. MATSUBARA, and M. HOSHI: A Novel Ceramide Trihexoside from the Eggs of the Sea Urchin *Hemicentrotus pulcherrimus*. J. Biochem., **111**, 726 (1992).
50. KAWANO, Y., R. HIGUCHI, and T. KOMORI: Glycosphingolipids from the Starfish *Acanthaster planci*, 4. Isolation and Structure of Five New Gangliosides. Liebigs Ann. Chem., 43 (1990).
51. MURALIKRISHNA, G., G. REUTER, J. PETER-KATALINIC, H. EGGE, F.G. HANISH, H.C. SIEBERT, and R. SHAUER: Identification of a New Ganglioside from the Starfish *Asterias rubens*. Carbohydr. Res., **236**, 321 (1992).
52. HOFFMAN, J., B. LINDBERG, and S. SVENSSON: Determination of Anomeric Configuration of Sugar Residues in Acetylated Oligo- and Polysaccharides by Oxidation with Chromium Trioxide. Acta Chem. Scand., **26**, 661 (1972).
53. LAINE, R.A., and O. RENKONEN: Ceramide Di- and Trihexosides of Wheat Flour. Biochemistry, **13**, 2837 (1974).

54. LAINE, R.A., and O. RENKONEN: Analysis of Anomeric Configurations in Glyceroglycolipids and Glycosphingolipids by Chromium Trioxide Oxidation. J. Lipid Res., **16**, 102 (1975).
55. INAGAKI, F., S. TATE, H. KUBO, and M. HOSHI: A Novel Difucosylated Neutral Glycosphingolipid from the Eggs of the Sea Urchin *Hemicentrotus pulcherrimus*. II. Structural Determination by Two-Dimensional NMR. J. Biochem., **112**, 286 (1992).
56. COSTANTINO, V., E. FATTORUSSO, and A. MANGONI: Glycolipids from Sponges. III. Glycosyl Ceramide Composition of the Marine Sponge *Agelas conifera*. Liebigs Ann. Chem., 2133 (1995).
57. COSTANTINO, V., E. FATTORUSSO, A. MANGONI, M. AKNIN, A. FALL, A. SAMB, and J. MIRALLES: An Unusual Ether Glycolipid from the Senegalese Sponge *Trikentrion loeve* Carter. Tetrahedron, **49**, 2711 (1993).
58. COSTANTINO, V., E. FATTORUSSO, and A. MANGONI: Isolation of Five-Membered Cyclitol Glycolipids, Crasserides: Unique Glycerides from the Sponge *Pseudoceratina crassa*. J. Org Chem., **58**, 186 (1993).
59. COSTANTINO, V., E. FATTORUSSO, A. MANGONI, M. AKNIN, and E.M. GAYDOU: Axyceramide A and B, Two Novel Tri-α-glycosylceramides from the Marine Sponge *Axinella* sp. Liebigs Ann. Chem., 181 (1994).
60. CAFIERI, F., E. FATTORUSSO, Y. MAHAJNAH, and A. MANGONI: Longiside, a Novel Digalactosylceramide from the Caribbean Sponge *Agelas longissima*. Liebigs Ann. Chem., 1187 (1994).
61. KOBAYASHI, J., C. ZENG, and M. ISHIBASHI: Keruffaride, a New All-*cis*-Cyclopentanepentol-containing Metabolite from the Okinawan Marine Sponge *Luffariella* sp. J. Chem. Soc. Chem. Comm., 79 (1993).
62. ISHIBASHI, M., C.-M. ZENG, and J. KOBAYASHI: Keruffaride: Structure Revision and Isolation from Plural Genera of Okinawan Marine Sponges. J. Nat. Prod., **56**, 1856 (1993).
63. VAN HUMMEL, H.C.: Chemistry and Biosynthesis of Plant Galactolipids. Fortschr. Chem. Org. Naturst., **32**, 267 (1975).
64. DEMBITSKY, V. M., O.A. ROZENTSVET, and E.E. PECHENKINA: Glycolipids, Phospholipids and Fatty Acids of Brown Algae Species. Phytochemistry, **29**, 3417 (1990).
65. JONES, A.L., and J.L. HARWOOD: Comparative Aspects of Lipid Metabolisms in Marine Algae. Biochem. Soc. Trans., **15**, 482 (1987).
66. DEMBITSKY, V.M., E.E. PECHENKINA, and O.A. ROZENTSVET: Glycolipids and Fatty Acids of Some Seaweeds and Marine Grasses from the Black Sea. Phytochemistry, **30**, 2279 (1991).
67. DEMBITSKY, V.M., T. REZANKA, and O.A. ROZENTSVET: Lipid Composition of Three Macrophytes from the Caspian Sea. Phytochemistry, **33**, 1015 (1993).
68. KITAGAWA, I., K. HAYASHI, and M. KOBAYASHI: Heterosigma-glycolipids I and II, New Galactolipids from a Raphidophycean Dinoflagellate *Heterosigma* sp. Chem. Pharm. Bull., **37**, 849 (1989).
69. KOBAYASHI, M., K. HAYASHI, K. KAWAZOE, and I. KITAGAWA: Heterosigma Glycolipids I-IV, Four New Diacylglycerolipids from the Marine Dinoflagellate *Heterosigma akashiwo*. Chem. Pharm. Bull., **40**, 1404 (1992).
70. KIKUCHI, H., Y. KSUKITANI, T. MANDA, T. FUJII, H. NAKANISHI, M. KOBAYASHI, and I. KITAGAWA: Marine Natural Products. X. Pharmacologically Active Glycolipid from the Okinawan Marine Sponge *Phyllospongia foliascens* (Pallas). Chem. Pharm. Bull., **30**, 3544 (1982).
71. JIANG, Z.D., and W.H. GERWICK: Galactolipid from the Temperate Red Marine Alga *Gracilariopsis lemaneiformis*. Phytochemistry, **29**, 1433 (1990).

72. ARAO, T., and M. YAMADA: Positional Distribution of Fatty Acids in Galactolipids of Algae. Phytochemistry, **28**, 805 (1989).
73. MURAKAMI, N., H. SHIRAHSHI, J. SAKAKIBARA, and Y. TSUCHIDA: A Novel Gliceroglycolipid from the Nitrogen-fixing Cyanobacterium *Anabaena flos-aquae* F. *flos-aquae*. Chem. Pharm. Bull., **40**, 285 (1992).
74. JIANG, Z.D., and W.H. GERWICK: An Aldehyde-Containing Galactolipid from the Red Alga *Gracilariopsis lemaneiformis*. Lipids, **26**, 960 (1991).
75. FUSETANI, N., and Y. HASHIMOTO: Structures of Two Water Soluble Hemolysins Isolated from the Green Alga *Ulva pertusa*. Agric. Biol. Chem., **39**, 2021 (1975).
76. SON, B.W.: Glycolipids from *Gracilaria verrucosa*. Phytochemistry, **29**, 307 (1990).
77. SON, B.W.: Glycolipid from the Korean Marine Red Alga *Gracilaria verrucosa*. Bull. Korean Chem. Soc., **9**, 264 (1988).
78. KATSUOKA, M., C. OGURA, H. ETOH, K. SAKATA, and K. INA: Galactosyl- and Sulfoquinovosyldiacylglycerols Isolated from the Brown Algae, *Undaria pinnatifida* and *Costaria costata* as repellents of the blue mussel *Mytilus edulis*. Agric. Biol. Chem., **54**, 3043 (1990).
79. PETTIT, G.R., A.L. JONES, and L.H. HARWOOD: Lipids of the Marine Red Algae, *Chondrus crispus* and *Polysiphonia lanosa*. Phytochemistry, **28**, 399 (1989).
80. GUSTAFSON, K.R., J.H. CARDELLINA II, R.W. FULLER, O.S. WEISLOW, R.F. KISER, K.M. SNADER, G.M.L. PATTERSON, and M.R. BOYD: AIDS-Antiviral Sulfolipids from Cyanobacteria (Blue-Green Algae). J. Natl. Cancer Inst., **81**, 1254 (1989).
81. GORDON, D.M., and S.J. DANISHEFSKY: Synthesis of a Cyanobacterial Sulpholipid: Confirmation of its Structure, Stereochemistry, and Anti-HIV-1 Activity. J. Am. Chem. Soc., **114**, 659 (1992).
82. SON, B.W., Y.J. CHO, N.K. KIM, and H.D. CHOI: New Glyceroglycolipids from the Brown Alga *Sargassum thunbergii*. Bull. Korean Chem. Soc., **13**, 584 (1992).
83. KITAGAWA, I., Y. HAMAMOTO, and M. KOBAYASHI: Sulfonoglycolipid from the Sea Urchin *Antocidaris crassispina* A. Agassiz. Chem. Pharm. Bull., **27**, 1934 (1979).
84. COSTANTINO, V., E. FATTORUSSO, and A. MANGONI: The Stereochemistry of Crasserides. J. Nat. Prod., **57**, 1726 (1994).
85. HAKOMORI, S.: Chemistry of Glycosphingolipids. In: Handbook of Lipid Research, (J. N. KANFER and S. HAKOMORI, eds.) Vol. 3, p. 327. New York, London: Plenum Press. 1983.
86. NOJIRI, H., F. TAKAKU, Y. TERUI, Y. MIURA, and M. SAITO: Ganglioside GM3: an Acidic Membrane Component that Increase During Macrophage-like Cell Differentiation Can Induce Monocytic Differentiation of Human Myeloid and Monocytoid Leukemic Cell Lines HL-60 and U937. Proc. Natl. Acad. Sci. USA, **83**, 782 (1986).
87. HANAI, N., T. DOHI, G.A. NORES, and S. HAKOMORI: A Novel Ganglioside, De-*N*-acetyl-GM$_3$ (II^{3}NeuNH$_2$LacCer), Acting as a Strong Promoter fro Epidermal Growth Factor Receptor Kinase and as a Stimulator for Cell Growth. J. Biol. Chem., **263**, 6296 (1988).
88. ISHIDA, R., H. SHIRAHAMA, and T. MATSUMOTO: Coralipid, a New Glycosphingolipid from the Red Alga *Corallina pilulifera*. Chem. Lett., 9 (1993).
89. IRIE, A., H. KUBO, and M. HOSHI: Glucosylceramide Having a Novel Tri-Unsaturated Long-Chain Base from the Spermatozoa of the Starfish *Asterias amurensis*. J. Biochem., **107**, 578 (1990).
90. SCHMITZ, F.J., and F.J. McDONALD: Isolation and Identification of Cerebrosides from the Marine Sponge *Chondrilla nucula*. J. Lipid Res., **15**, 158 (1974).
91. HIRSCH, S., and J. KASHMAN: Structure of Ceramides and Cerebrosides, New Glycosphingolipids from Marine Organisms. Tetrahedron, **45**, 3873 (1989).

92. HAYASHI, A., Y. NISHIMURA, and T. MATSUBARA: Occurence of Ceramide Digalactoside as the Main Glycosphingolipid in the Marine Sponge *Halichondria japonica*. Biochim. Biophys. Acta. **1083**, 179 (1991).
93. ENDO, M., M. NAKAGAWA, Y. HAMAMOTO, and M. ISHIHAMA: Pharmacologically Active Substances from Southern Pacific Marine Invertebrates. Pure Appl. Chem., **58**, 387 (1986).
94. NAGLE, D.G., W.C. MCCLATCHEY, and W.H. GERWICK: New Glycosphingolipids from the Marine Sponge *Halichondria panicea*. J. Nat. Prod., **55**, 1013 (1992).
95. COSTANTINO, V., E. FATTORUSSO, and A. MANGONI: Glycolipids from Sponges, I. Glycosyl Ceramide Composition of the Marine Sponge *Agelas clathrodes*. Liebigs Ann. Chem., 1471 (1995).
96. CAFIERI, F., E. FATTORUSSO, A. MANGONI, and O. TAGLIALATELA-SCAFATI: Glycolipids from Sponges, II. Glycosyl Ceramide Composition of the Marine Sponge *Agelas longissima*. Liebigs Ann. Chem., 1477 (1995).
97. SUGITA, M.: Studies on Glycosphingolipids of the Starfish, *Asterina pectinifera*. I. The Isolation and Characterization of Ceramide Mono- and Di-Hexosides. J. Biochem., **82**, 1307 (1977).
98. KOMORI, T., Y. SANECHIKA, Y. ITO, J. MATSUO, T. NOHARA, and T. KAWASAKI: Strukturen eines neuen Cerebrosidgemischs und von Nucleosiden aus dem Seestern *Acantaster planci*. Liebigs Ann. Chem., 653 (1980).
99. KAWANO, Y., R. HIGUCHI, R. ISOBE, and T. KOMORI: Glycosphingolipids from the Starfish *Acanthaster planci*. Isolation and Structure of Six New Cerebrosides. Liebigs Ann. Chem., 19 (1988).
100. IRIE, A., H. KUBO, F. INAGAKI, and M. HOSHI: Ceramide Dihexosides from the Spermatozoa of the Starfish, *Asterias amurensis*, Consist of Gentobiosyl-, Cellobiosyl-, and Lactosylceramide. J. Biochem. **108**, 531 (1990).
101. HIGUCHI, R., J.X. JHOU, K. INUKAI, and T. KOMORI: Glycosphingolipids from the Starfish *Asterias amurensis versicolor*, 1. Isolation and Structure of Six New Cerebrosides, Asteriacerebrosides A-F, and two Known Cerebrosides, Astrocerebroside A and Acanthacerebroside C. Liebigs Ann. Chem., 745 (1991).
102. HIGUCHI, R., M. KAGOSHIMA, and T. KOMORI: Glicosphingolipids from the Starfish *Astropecten latespinosus*, I. Structure of Three New Cerebrosides, Astrocerebrosides A, B, and C, and of Related Nearly Homogeneous Cerebrosides. Liebigs Ann. Chem., 659 (1990).
103. KUBO, H., A. IRIE, F. INAGAKI, and M. HOSHI: Melibiosyl Ceramide as the Sole Ceramide Dihexoside from the Eggs of the Sea Urchin *Anthocidaris crassispina*. J. Biochem., **104**, 755 (1988).
104. BATRAKOV, S.G., V.B. MURATOV, O.G. SAKANDELIDZE, A.V. SULIMA, and B.V. ROSYNOV: Cerebrosides of the Far-East Sea Cucumber *Cucumaria japonica*. Bioorg. Khim., **9**, 539 (1983).
105. SHIMOMURA, K., S. HANJURA, P. F. KI, and Y. ISHIMOTO: An Unusual Glucocerebroside in the Crustacean Nervous System. Science, **220**, 1392 (1983).
106. KARLSSON, K.A., H. LEFFLER, and B.E. SAMUELSSON: Characterization of Cerebroside (Monoglycosylceramide) from the Sea Anemone *Metridium senile*. Identification of the Major Long-chain Base as an Unusual Dienic Base with a Methyl Branch at a Double Bond. Biochim. Biophys. Acta, **574**, 79 (1979).
107. YAMAGUCHI, Y., K. KONDA, and A. HAYASHI: Studies on the Chemical Structure of Neutral Glycosphinolipids in Eggs of the Sea Hare *Aplysia juliana*. Biochim. Biophys. Acta, **1165**, 110 (1992).

108. Sugiyama, S., M. Honda, and T. Komori: The Stereochemistry of the Four Diastereomers of the Phytosphingosine. Liebigs Ann. Chem., 1069 (1990).
109. Sugiyama, S., M. Honda, and T. Komori: Synthesis of Acanthacerebroside A from the Starfish *Acanthaster planci*. Liebigs Ann. Chem., 1063 (1990).
110. Matsubara, T., and A. Hayashi: Structural Studies on Glycolipid of Shellfish. III. Novel Glycolipids from *Turbo cornutus*. J. Biochem., **89**, 645 (1981).
111. Matsubara, T., and A. Hayashi: Structural Studies on Glycolipids of Shellfish. V. Gala-6 Series Glycosphingolipids of the Marine Snail *Chlorostoma argyrostoma turbinatum*. J. Biochem., **99**, 1401 (1986).
112. Li, H., S. Matsunaga, and N. Fusetani: Halicylindrosides, Antifungal and Cytotoxic Cerebrosides from the Marine Sponge *Halichondria cylindrata*. Tetrahedron, **51**, 2273 (1995).
113. Natori, T., Y. Koezuka, and T. Higa: Agelasphin, Novel α-Galactosylceramides from the Marine Sponge *Agelas mauritiana*. Tetrahedron Lett., **34**, 5591 (1993).
114. Akimoto, K., T. Natori, and M. Morita: Synthesis and Stereochemistry of Agelasphin-9b. Tetrahedron Lett., **35**, 5593 (1993).
115. Motoki, K., E. Kobayashi, T. Uchida, H. Fukushima, and Y. Koezuka: Antitumor Activity of α-, β-Monogalactosylceramides and Four Diastereomers of an α-Galactosylceramide. Bioorg. Med. Chem. Lett., **5**, 705 (1995).
116. Morita, M., T. Natori, K. Akimoto, T. Osawa, H. Fukushima, and Y. Koezuka: Syntheses of α-, β-Monoglycosylceramides and Four Diastereomers of an α-Galactosylceramide. Bioorg. Med. Chem. Lett., **5**, 699 (1995).
117. Kawano, Y., R. Higuchi, and T. Komori: Achantalactoside A and B, Two New Ceramide Lactosides from the Starfish *Acanthaster planci*. Liebigs Ann. Chem., 1181 (1988).
118. Sweeley, C.C., and B. Klionsky: Fabry's Disease. Classification as a Sphingolipidosis and Partial Characterization of a Novel Glycolipid. J. Biol. Chem., **238**, 3148 (1963).
119. Yamaguchi, Y., M. Otha, and A. Hayashi: Structural Elucidation of a Novel Phosphonoglycolipid in Eggs of a Sea Hare *Aplysia juliana*. Biochim. Biophys. Acta, **1165**, 160 (1992).
120. Kubo, H., G.J. Jiang, A. Irie, M. Suzuki, F. Inagaki, and M. Hoshi: A Novel Difucosylated Neutral Glycosphingolipid from the Eggs of the Sea Urchin *Hemicentrotus pulcherrimus*. I. Purification and Structural Determination of the Glycolipid. J. Biochem., **112**, 281 (1992).
121. Matsubara, T., and A. Hayashi: Structural Studies on Glycolipid of Shellfish. IV. A Novel Pentaglycosyl from Abalone *Haliotis japonica*. Biochim. Biophys. Acta, **711**, 551 (1982).
122. Matsuura, F.: Phosphonosphingolipid, a Novel Sphingolipid from the Viscera of *Turbo cornutus*. Chem. Phys. Lipids, **19**, 223 (1977).
123. Hayashi, A., and T. Matsuura: Characterization of Aminoalkylphosphonyl Cerebrosides in Muscle Tissues of *Turbo Cornutus*. Chem. Phys. Lipids, **22**, 9 (1978).
124. Matsuura, F.: The Identification of Aminoalkylphosphonyl Cerebrosides in the Marine Gastropod *Monodonta labio*. J. Biochem., **85**, 433 (1979).
125. Noda, N., R. Tanaka, K. Miyahara, and T. Kawasaki: Two Novel Galactosylceramides from the Marine Annelid *Marphysa sanguinea*. Tetrahedron Lett., **33**, 7527 (1992).
126. Noda, N., R. Tanaka, K. Miyahara, and T. Kawasaki: Isolation and Characterization of a Novel Type of Glycosphingolipid from *Neanthes diversicolor*. Biochim. Biophys. Acta, **1169**, 30 (1993).

127. Hayashi, A., and T. Matsubara: A new Homologue of Phosphonoglycosphingolipid, N-Methylaminoethylphosphonylgalactosylceramide. Biochim. Biophys. Acta, **1006**, 89 (1989).
128. Araki, S., Y. Komai, and M. Satake: A Novel Sphingophosphonoglycolipid Containing 3-O-Methylgalactose Isolated from the Skin of the Marine Gastropod *Aplysia kurodai*. J. Biochem., **87**, 503 (1980).
129. Araki, S., S. Abe, S. Odani, S. Ando, N. Fujii, and M. Satake: Structure of a Triphosphonopentaosylceramide Containing 4-O-Methyl-N-acetylglucosamine from the Skin of the Sea Hare *Aplysia kurodai*. J. Biol. Chem., **262**, 14141 (1987).
130. Araki, S., and M. Satake: Structure of a Novel Diphosphonoglycosphingolipid Isolated from the Skin of *Aplysia kurodai*. Biochem. Int., **10**, 603 (1985).
131. Araki, S., M. Satake, A. Ando, A. Hayashi, and N. Fujii: Characterization of a Diphosphonopentaosylceramide Containing 3-O-Methylgalactose from the Skin of *Aplysia kurodai* (Sea Hare). J. Biol. Chem., **261**, 5138 (1986).
132. Hori, T., O. Itasaka, H. Inoue, and K. Yamada: Structural Components of the Pyridine-Insoluble Sphingolipid from *Corbicula Sandai*, and the Distribution in Other Species. J. Biochem., **56**, 477 (1964).
133. Abe, S., Y. Watanabe, S. Araki, T. Kumanishi, and M. Satake: Immunochemical and Histochemical Studies on a Phosphonoglycosphingolipid, SGL-II, isolated from the Sea Gastropod *Aplysia kurodai*. J. Biochem., **104**, 220 (1988).
134. Araki, S., S. Abe, S. Ando, N. Fujii, and M. Satake: Isolation and Characterization of a Novel 2-Aminoethylphosphonyl Glycosphingolipid from the Skin of the Sea Hare *Aplysia kurodai*. J. Biochem., **101**, 145 (1987).
135. Matsubara, T., and A. Hayashi: Occurrence of Phosphonotetraglycosyl Ceramide in the Sea Hare *Dolabella auricolaria*. Biochim. Biophys. Acta, **1166**, 55 (1993).
136. Abe S., S. Araki, M. Satake, S. Fujiwara, K. Kon, and S. Ando: Structure of Triphosphonoglycosphongolipid containing N-Acetylgalactosamine-6-O-2-aminoethylphosphonate in the Nervous System of *Aplysia kurodai*. J. Biol. Chem., **266**, 9939 (1991).
137. Araki, S., S. Abe, S. Ando, K. Kon, N. Fujiwara, and M. Satake: Structure of Phosphonoglycosphingolipid Containing Pyruvylated Galactose in Nerve Fibres of *Aplysia kurodai*. J. Biol. Chem., **264**, 19922 (1989).
138. Araki, S., S. Abe, M. Satake, A. Hayashi, K. Kon, and S. Ando: Novel Phosphonoglycosphingolipids Containing Pyruvylated Galactose from the Nervous System of *Aplysia kurodai*. Eur. J. Biochem., **198**, 689 (1991).
139. Garegg, P.J., P.-E. Jansson, P. Lindberg, F. Lindh, J. Lönngren, I. Kvanrström, and W. Nimmich: Configuration of the Acetal Carbon Atom of Pyruvic Acid Acetals in Some Bacterial Polysaccharides. Carbohydr. Res., **78**, 127 (1980).
140. Araki, S., S. Abe, S. Yamada, M. Satake, N. Fujiwara, K. Kon, and S. Ando: Characterization of Two Novel Pyruvylated Glycosphingolipids Containing 2'-Aminoethylphosphoryl(→6)-galactose from the Nervous System of *Aplysia kurodai*. J. Biochem., **112**, 461 (1992).
141. Yamada, S., S. Araki, S. Abe, K. Kon, S. Ando, and M. Satake: Structural Analysis of a Novel Triphosphonoglycosphingolipid from the Egg of the Sea Hare *Aplysia kurodai*. J. Biochem., **117**, 794 (1995).
142. Hoshi, M., and Y. Nagai: Novel Sialosphingolipids from the Spermatozoa of the Sea Urchin *Anthocidaris crassispina*. Biochim. Biophys. Acta, **388**, 152 (1975).
143. Shashkov, A.S., G.P. Smirnova, N.V. Cekareva, and J. Dabrowski: Structural Study of Sialoglycolipids from the Sea Urchin *Tripneustes ventricosa* Gonads using ^{1}H- and ^{13}C-NMR Spectroscopy. Bioorg. Khim., **12**, 789 (1986).

144. KOCHETKOV, N.K., G.P. SMIRNOVA, and I.S. GLUKHODED: Structure of Sialolipids from the Gonads of the Sea Urchin *Strongylocentrotus nudus*. Bioorg. Khim., **4**, 1093 (1978).
145. KUBO, H., and M. HOSHI: Immunochemical Study of the Distribution of a Ganglioside in Sea Urchin Eggs. J. Biochem., **108**, 193 (1990).
146. CHEKAREVA, N.V., G.P. SMIRNOVA, and N.K. KOCHETKOV: Gangliosides of the Holothurian *Cucumaria japonica* Semper. Bioorg. Khim., **17**, 398 (1991).
147. CHEKAREVA, N.V., G.P. SMIRNOVA, and N.K. KOCHETKOV: Gangliosides from two Species of Ophiuria, *Ophtocoma echinata* and *Ophiomastix annulosa* Clark. Bioorg. Khim., **17**, 387 (1991).
148. SMIRNOVA, G.P., N.V. CHEKAREVA, and N.K. KOCHETKOV: Gangliosides of *Ophiura sarsi*. Bioorg. Khim., **12**, 507 (1986).
149. YAMAMOTO, T., T. TESHIMA, U. SAITOH, M. HOSHI, and T. SHIBA: Synthesis of Ganglioside M5 from Sea Urchin Eggs (*Anthocidaris crassispina*). Tetrahedron Lett., **35**, 2701 (1994).
150. KOCHETKOV, N.K., G.P. SMIRNOVA, N.V. CHEKAREVA: Isolation and Structural Studies on a Sulfated Sialosphingolipid from the Sea Urchin *Echinocardium cordatum*. Biochim. Biophys. Acta, **424**, 274 (1976).
151. SMIRNOVA, G.P., N.V. CHEKAREVA, and N.K. KOCHETKOV: Structure of Sialoglycolipids from the Gonad Tissue of the Sea Urchin *Echinarachnius parma*. Bioorg. Khim., **6**, 1667 (1980).
152. SMIRNOVA, G.P., N.V. CHEKAREVA, and N.K. KOCHETKOV: Structure of a Minor Sialoglycolipid from the Sea Urchin *Echinocardium cordatum*. Bioorg. Khim., **4**, 937 (1978).
153. SMIRNOVA, G.P., I.S. GLUKHODED, and N.K. KOCHETKOV: Gangliosides of the Starfish *Lethasterias fuska*. Bioorg. Khim., **12**, 679 (1986).
154. PROKAZOVA, N.V., A.T. MIKHAILOV, S.L. KOCHAROV, L.A. MALCHENKO, N.D. ZVEZDINA, G. BUZNIKOV, and L.D. BERGELSON: Unusual Gangliosides of Eggs and Embryos of the Sea Urchin *Strongylocentrotus intermedius*. Eur. J. Biochem., **115**, 671 (1981).
155. SMIRNOVA, G.P., N.K. KOCHETKOV, and V.L. SADOVSKAYA: Gangliosides from the Starfish *Aphelasterias japonica*, Evidence for a New Linkage Between Two N-Glycolylneuraminic Acid Residues through the Hydroxy Group of the Glycolyc Acid Residue. Biochim. Biophys. Acta, **920**, 47 (1987).
156. KOCHETKOV, N.K., G.P. SMIRNOVA, and I.S. GLUKHODED: Gangliosides with Sialic Acid Bound to *N*-Acetylgalactosamine from Hepatopancreas of the Starfish, *Evasterias retifera* and *Asterias amurensis*. Biochim. Biophys. Acta, **712**, 650 (1982).
157. SMIRNOVA, G.P., I.S. GLUKHODED, and N.K. KOCHETKOV: A Branched Disialoganglioside Containing N-Acetylgalactosamine from the Starfish *Asterias rubens*. Bioorg. Khim., **14**, 636 (1988).
158. SMIRNOVA, G.P., and N.K. KOCHETKOV: Gangliosides with Sialic Acid Located in the Inner Part of the Carbohydrate Chain Isolated from the Starfish *Luidia quinaria bispinosa*. Bioorg. Khim., **11**, 1650 (1985).
159. SMIRNOVA, G.P.: Gangliosides with a Sialic Acid Residue in the Inner Part of the Oligosaccharide Chain and with a Terminal Galactofuranose Residue from the Starfish *Achantaster planci*. Bioorg. Khim., **16**, 830 (1990).
160. SUGITA, M.: Studies on the Glycosphingolipids of the Starfish, *Asterina pectinifera*. III. Isolation and Structural Studies of Two Novel Gangliosides Containing Internal Sialic Acid Residues. J. Biochem., **86**, 765 (1979).
161. SUGITA, M.: Studies on Glycosphingolipids of the Starfish, *Asterina pectinifera*. II. Isolation and Characterization of a Novel Ganglioside with an Internal Sialic Acid Residue. J. Biochem., **86**, 289 (1979).

162. GLUKHODED, I.S., G.P. SMIRNOVA, and N.K. KOCHETKOV: Structures of Gangliosides from the Body of the Starfish *Patiria pectinifera*. Bioorg. Khim., **16**, 839 (1990).
163. HIGUCHI, R., K. INAGAKI, T. NATORI, T. KOMORI, and S. KAWAJIRI: Glycosphingolipids from the Starfish *Asterina pectinifera*. Structure of Three Ganglioside Molecular Species and a Homogeneous Ganglioside, and Biological Activity of the Ganglioside. Liebigs Ann. Chem., 1 (1991).
164. SMIRNOVA, G.P., and N.K. KOCHETKOV: A Novel Sialylglycolipid from Hepatopancreas of the Starfish *Patiria pectinifera*. Biochim. Biophys. Acta, **618**, 486 (1980).
165. HAKOMORI, S.: Chemistry of Glycosphingolipids. In: Handbook of Lipid Research, (KANFER, J.N. and S. HAKOMORI eds.) Vol. 3, p. 61. New York, London: Plenum Press. 1983.
166. KOCHETKOV, N.K., and G.P. SMIRNOVA: A Disialoglycolipid with Two Sialic Acid Residue Located in the Inner Part of the Oligosaccharide Chain from the Hepatopancreas of the Starfish *Patiria pectinifera*. Biochim. Biophys. Acta, **759**, 192 (1983).
167. SORIENTE, A., T. BISOGNO, A. GAMBACORTA, I. ROMANO, C. SILI, A. TRINCONE, and G. SODANO: Reinvestigation of Heterocyst Glycolipids from the Cyanobacterium *Anabaena cylindrica*. Phytochemistry, **38**, 641 (1995).
168. SORIENTE, A., G. SODANO, A. GAMBACORTA, and A. TRINCONE: Structure of the Heterocyst Glycolipids of Marine Cyanobacterium *Nodularia harveyana*. Tetrahedron, **48**, 5375 (1992).
169. HARADA, N., A. SAITO, H. ONO, J. GAWRONSKY, K. GAWRONSKA, T. SAGIOKA, H. UDA, and T. KURKI: A CD Method for Determination of Absolute Stereochemistry of Acyclic Glycols. 1. Application of the CD Exciton Chirality Method to Acyclic 1,3-Dibenzoate Systems. J. Am. Chem. Soc., **113**, 3842 (1991).
170. DALE, J.A., and H.S. MOSHER: Nuclear Magnetic Resonance Enantiomer Reagents. Configurational Correlation via Nuclear Magnetic Resonance Chemical Shifts of Diastereomeric Mandelate, *O*-Methylmandelate, and α-Metoxy-α-trifluoromethylphenylacetate (MTPA) Esters. J. Am. Chem. Soc., **95**, 512 (1973).
171. MORI, K., and Z.-H. QIAN: Synthesis of (3R, 25R)-3,25-Dihydroxyhexacosyl-α-D-glucopyranoside, the Heterocyst Glycolipid of the Marine Cyanobacterium *Nodularia harveyana*. Liebigs Ann. Chem., 35 (1994).
172. SORIENTE, A., A. LAUDISIO, M. GIORDANO, and G. SODANO: Enzymatic Desymmetrization of a Prochiral 1,3,5-Pentanetriol Derivative. Application to the Synthesis of a Cyanobacterial Heterocyst Glycolipid. Tetrahedron Asymm., **6**, 859 (1995).
173. GAMBACORTA, A., A. SORIENTE, A. TRINCONE, and G. SODANO: Biosynthesis of the Heterocyst Glycolipids in the Cyanobacterium *Anabaena cylindrica*. Phytochemistry, **39**, 771 (1995).
174. SORIENTE, A., A. GAMBACORTA, A. TRINCONE, C. SILI, M. VINCENZINI, and G. SODANO: Heterocyst Glycolipids of the Cyanobacterium *Cyanospira Rippkae*. Phytochemistry, **33**, 393 (1993).
175. MAKARIEWA, T.N., V.A. DENISENKO, V.A. STONIK, Y.M. MILGROM, and Y.V. RASHKES: Rhizochalin, a Novel Antimicrobial Secondary Metabolite from the Sponge *Rhizochalina incrustata*. Tetrahedron Lett., **30**, 6581 (1989).
176. FUSETANI, N., N. SATA, N. ASAI, and S. MATSUNAGA: Isolation and Structure Elucidation of Erylusamine B, a New Class of Marine Natural Products which Blocks an IL-6 Receptor, from the Marine Sponge *Erylus placenta*. Tetrahedron Lett., **34**, 4067 (1993).
177. PARTRIDGE, J.J., V. TOOME, and M.R. USKOKOVIC: A Stereoselective Synthesis of the 24(*R*),25-Dihydroxycholesterol Side Chain. J. Am. Chem. Soc., **98**, 3739 (1976).

178. FINDLAY, J.A., Z.-Q. HE, and L.A. CALHOUN: Forbesin, a Novel Sulphated Glycolipid from the Starfish *Asterias forbesi*. J. Nat. Prod., **53**, 1015 (1990).
179. WATANABE, J., S. ABE, S. ARAKI, T. KUMANISHI, and M. SATAKE: Characterization of Phosphonoglycosphingolipids Containing Pyruvate: Localization in *Aplysia* Nerve Bundles. J. Biochem., **106**, 972 (1989).
180. MORITA, M., K. MOTOKI, K. AKIMOTO, T. NATORI, T. SAKAI, E. SAWA, K. YAMAJI, Y. KOEZUKA, E. KOBAYASHI, and H. FUKUSHIMA: Structure-Activity Relationship of α-Galactosylceramides against B16-Bearing Mice. J. Med. Chem., **38**, 2176 (1995).
181. KOBAYASHI, E., K. MOTOKI, Y. YAMAGUCHI, T. UCHIDA, H. FUKUSHIMA, and Y. KOEZUKA: Enhancing Effect of α-, β-Monoglycosylceramides on Natural Killer Cell Activity. Bioorg. Med. Chem. Lett., **4**, 615 (1996).
182. KISHIMOTO, T.: Interleukin-6 and Its Receptors: a Paradigm for Cytokines. Science, **258**, 593 (1992).

(Received September 27, 1996)

Author Index

Page numbers printed in *italics* refer to References

Abaul, J.   *105*
Abdel Kader, M.S.   *113*
Abdel-Salam, N.A.   *113*
Abe, S.   *298, 301*
Abou-Elzahab, M.M.   *105*
Abyshev, A.Z.   *106*
Achenbach, S.H.   *116*
Acton, N.   133, 141, 149, 183, *203, 204, 207, 211*
Adam, W.   *105*
Afek, U.   *105*
Ager Jr., A.L.   *212, 214*
Aharoni, N.   *105*
Ahmad, M.S.   *208*
Ahmad, V.U.   *116*
Ahmed, A.F.   *109*
Ahond, A.   *105*
Ahsan, M.   *105*
Akimoto, K.   *293, 297, 301*
Aknin, M.   *294*
Al, Y.G.   *208*
Alam, M.S.   *111*
Albroscheit, G.   *117*
Alexander, J.A.   *108*
Ali, M.   *111*
Al-Meshal, I.A.   *206*
Al-Tel, T.H.   *105*
Al-Yahya, M.A.   *207*
Amaro-Luis, J.M.   *105*
Amer, M.E.   *113*
Ames, B.N.   *211*
Anders, J.C.   *205*
Andersen, S.L.   *209, 214*
Anderson, J.W.   *118*
Ando, S.   *298*
Annuziata, R.   *105*
Appendino, G.   *105*
Araki, S.   *298, 301*

Arantes, S.F.   *119*
Arao, T.   *295*
Archelas, A.   *119*
Arimoto, H.   *115*
Armstrong, J.A.   *108, 109, 115–117*
Arnold, K.   *212*
Asai, N.   *293, 300*
Asawamahasadka, W.   *208, 213*
Ashmore, R.W.   *205*
Astles, D.P.   *106*
Atta-ur-Rahman   *105*
Auprayoon, P.   *212*
Avery, B.A.   *210*
Avery, M.A.   133, 143, 147, 164, *166–168, 170–174, 204, 206, 209, 210*
Avrova, N.F.   *292*
Ayafor, J.F.   *114*
Ayalp, A.   *207*
Azuma, M.   *111, 118*

Baba, K.   *105, 119*
Bachi, M.D.   *213*
Baiwir, M.   *113*
Baker, J.K.   149, *204, 205, 207, 210, 211*
Balachandran, S.   *205*
Bal-Tembe, S.   *105*
Banerji, A.   *107*
Banerji, J.   *106*
Banthorpe, D.V.   *106*
Barnes, C.L.   *214*
Barros, S.M.G.   *113*
Barton, D.H.R.   *106*
Barua, A.K.   *106*
Basile, D.V.   *203*
Basnet, P.   *106*
Basu, K.   *106*
Batrakov, S.G.   *296*
Batsuren, D.   *114*

Behbud, A. *202*
Beier, R.C. *106*
Benakis, A. 195, *211*
Bergelson, L.D. *299*
Bernardi, A.C. *119*
Bernardo, R.R. *112*
Bhattacharjee, J. *106, 115*
Bhattacharyya, P. *106*
Bhedi, D.N. *105*
Bicchi, C. *106*
Bilia, A.R. *106*
Bill, H. 140, *205*
Binh, L.N. *212*
Bircher, J. *117*
Bisogno, T. *300*
Biswas, G.K. *106*
Björkman, L.R. *291*
Blasko, G. 137, *203*
Bodley, A.L. *213*
Bodo, B. *112*
Boeykens, M. *106, 107*
Bohlmann, F. *119*
Boiko, E.V. *107*
Bokesch, H.R. *106, 108, 109*
Bonk, J.D. *206, 209, 210*
Boralle, N. *119*
Borges, F. *106*
Borges, M.F.M. *106*
Bose, P. *106*
Botta, B. *107*
Botz, L. *109*
Boukouvalas, J. *205*
Bourgaud, F. *106*
Bourgeois, P. *105*
Boyd, M.R. *106, 108, 109, 111, 113, 295*
Bradol, J.H. *212*
Brewer, T.G. *205, 211*
Broegger Christensen, S. *112*
Brossi, A. 154, *204–206*
Brown, G.D. *106*
Bryce, T.A. *291*
Büchi, G. 129, 133, *203*
Buchs, P. *205*
Buckheit Jr., R.W. *111*
Bukreeva, T.V. *107*
Bunnag, D. *212*
Bunnelle, W.H. *214*
Bupp, J. *206*
Bupp, J.E. *210*
Burgin, H. *214*

Bustos, D.A. 162, *208, 209*
Butcher, G.A. *214*
Butler, A.R. 125, *202*
Buznikov, G. *299*
Bye, R. *113*

Cabrera, A. *119*
Cafieri, F. *294, 296*
Cai, F. *116*
Cairns, N. *106*
Calhoun, L.A. *301*
Camacho, M.R. *113*
Campos, M. *107*
Canfield, C.J. *211*
Caniato, R. *106*
Cappelletti, E.M. *106*
Cappiello, A. *106*
Cardellina, J.H. *106*
Cardellina II, J.H. *108, 109, 111, 113, 295*
Cardona, L. *107*
Carmeli, S. *105*
Carroll, F.I. *214*
Cavaleiro, C. *107*
Ceccherelli, P. *107*
Cecchini, C. *106*
Cekareva, N.V. *298*
Cerami, A. *212*
Cervera, E. *113*
Cha, J.K. *209*
Chan, K.L. *203*
Chang, C.-T. *107*
Charles, D.J. *203*
Chatterjee, A. *106, 107, 111, 113*
Chebotarëva, M.A. *292*
Cheetham, R.D. *203*
Chekareva, N.V. *299*
Chemesova, I.I. *107*
Chen, C.-T. *107*
Chen, F. *203*
Chen, H. *119*
Chen, H.R. *203*
Chen, I.-S. *107, 110*
Chen, M. *203, 207*
Chen, Q.-M. *211*
Chen, Q.-P. *115*
Chen, S.F. *207*
Chen, T.H. *202*
Chen, Y. *207*
Chen, Y.-X. *208*

Chen, Z.-X.   *107*
Chenera, B.   *107*
Cheng, C.H.   *204*
Chi, H.T.   149, *207*, *210*
Chirkovskaya, E.V.   *292*
Cho, Y.J.   *295*
Choi, H.D.   *295*
Chong, W.K.M.   *204*, *209*, *210*
Chongsuphajaisiddhi, T.   *212*
Chou, W.-S.   *202*, *207*
Christiansen, C.   *117*
Chulay, J.D.   *211*
Chung, H.   *205*
Chung, H.S.   *117*
Cisero, M.   *105*
Clardy, J.   *207*, *213*
Clark, A.   *213*
Clark, A.M.   *210*
Clark, I.A.   *214*
Clerivet, A.   *108*
Collado, I.G.   *109*, *119*
Connolly, J.D.   *114*
Conrad, H.E.   *220*, *293*
Conserva, L.M.   *115*
Cordell, G.A.   *113*, *203*
Corey, E.J.   *107*
Costantino, V.   *293–296*
Coussio, J.D.   *106*, *107*
Cowden, W.B.   *213*, *214*
Cragg, G.M.   *108*, *111*
Cravotto, G.   *105*
Croom, E.M.   *203*, *204*, *211*
Cruz, M.d.C.   *113*
Cubukcu, B.   *202*
Cuca Suarez, L.E.   *107*
Cui, S.   *202*
Cui, Y.-F.   *208*
Cumming, J.N.   *213*, *214*
Curini, M.   *107*
Currens, M.J.   *111*

Dabrowski, J.   *298*
D'Agostino, M.   *107*
Dai, L.   *206*
Dai, Y.   *116*
Dale, J.A.   *300*
Damadyan, B.   *202*
D'Amato, A.   *106*
Danishefsky, S.J.   *295*
Das, A.   *107*

Das, A.K.   *106*
Das, B.   *106*
Das, P.C.   *107*, *111*, *113*
Das G.F. da Silva, M.F.   *114*
Da Silva, A.J.R.   *112*
Debenedetti, S.L.   *106*, *107*
Declercq, J.P.   *107*
DeCoster, M.A.   *211*
Degáspari, L.R.O.   *114*
De Kimpe, N.   *106*, *107*
Delle Monache, F.   *107*
Delle Monache, G.   *107*
Del Rayo Comacho, M.   *117*
Dembitsky, V.M.   *294*
Denisenko, V.A.   *300*
Deshmukh, J.G.   *108*
Deshpande, A.R.   *108*
De Simone, F.   *107*
Desjardins, A.E.   *108*, *117*
Desjardins, R.E.   *211*
Desoky, E.K.   *108*
De Souza, N.J.   *105*
Detre, G.   *209*
DeVries, P.J.   *212*
Dhar, K.L.   *109*
Dhara, K.P.   *106*
Di Fazio, M.P.   *108*
Ding, S.F.   *206*
Dini, A.   *107*
Di Paolo, E.R.   *108*
Di Rosa, M.   *293*
Dobek, A.S.   *203*
Dohi, T.   *295*
Doi, Y.   *292*
Dolan, G.   *212*
Dondas, I.   *211*
Dong, H.-Q.   *213*
Dong, X.G.   *119*
Donnelly, D.M.X.   *106*
Döpke, W.   *108*, *110*
Dowd, P.F.   *112*, *118*
Downing, R.   *109*
Downum, K.R.   *118*
Dreyer, G.B.   *107*, *113*
Duan, G.L.   139, *204*
Dubois, M.-A.   *108*
Duc, D.D.   *212*
Duddeck, H.   *108*
Duh, C.-Y.   *107*, *108*

Eaton, J.W. *214*
Edlund, P.O. 142, *206*
Edwards, G. *206*
Egge, H. *293*
El-Domiaty, M.M. *206*
El-Feraly, F.S. 150, 183, *203, 204, 206–208, 211*
Elgamal, M.H.A. *108*
Ellis, D.S. *212*
Ellis, W.Y. *214*
El Marakby, S.A. *211*
El Modafar, C. *108*
El-Sherei, M.M. *203, 204*
El Sohly, H.A. *204*
El Sohly, H.N. 127, *203–205, 208–211*
El-Turbi, J.A. *108, 115*
Endo, M. *296*
Etoh, H. *295*

Fales, H.M. *205*
Fall, A. *294*
Famiglini, G. *106*
Fan, C.K. *202*
Fan, J.-F. *202, 207*
Fan, P. *210*
Fan, Z.C. *207*
Fang, Y. *205*
Farooqi, A.H.A. *202*
Fatima, N. *105*
Fattorusso, E. *293–296*
Feneau-Dupont, J. *107*
Fernald, M.L. *202*
Fernandes, J.B. *114*
Feroz, M. *105*
Ferreira, J.F.S. *203*
Filippini, R. *106*
Findlay, J.A. *301*
Finet, J.P. *106*
Finkelstein, J.A. *107*
Flamini, G. *106*
Fleckenstein, L. *212*
Fleuriet, A. *108*
Flippen-Anderson, J.L. *204, 205, 208*
Floch, J. 219, *292*
Follansbee, E. *203*
Fomum, Z.T. *114*
Fondekar, K.P.P. *115*
Foo, L.Y. *108*
Fraigui, O. *112*
Frattini, C. *106*

French, A.N. *213*
Frighetto, R.T.S. *113*
Fu, H.-N. *214*
Fu, L.C. *205, 206*
Fujii, J. *114*
Fujii, M. *109*
Fujii, N. *298*
Fujii, T. *294*
Fujita, E. *105*
Fujita, M. *292*
Fujita, T. *112*
Fujiwara, K. *110*
Fujiwara, N. *298*
Fujiwara, S. *298*
Fukai, T. *108*
Fukamiya, N. *111, 118*
Fukuda, T. *109*
Fukushima, H. *297, 301*
Fuller, R.W. *108, 109, 111, 295*
Fulzele, D.P. *203, 206*
Furstoss, R. *119*
Furukawa, H. *109–111, 118*
Fusetani, N. *293, 295, 297, 300*

Gai, Y.Z. *208*
Galán, R.H. *117*
Gallard, J.-F. *112*
Gambacorta, A. *300*
Gan, J. *209*
Gandhidasan, R. *117*
Gao, C. *119*
Gao, F. *210*
García, B. *107*
Garegg, P.J. *298*
Gatt, S. *293*
Gaudel, G. *117*
Gawronska, K. *300*
Gawronsky, J. *300*
Gaydou, E.M. *294*
Geiger, H. *111*
George, C.F. *205, 208*
Gerena, L. *213, 214*
Gerpe, L.D. *205*
Gerwick, W.H. *294–296*
Gilardi, R. *204*
Gindin, V.A. *106*
Giordano, M. *300*
Glocke, E. *214*
Gloer, J.B. *112, 118*
Glukhoded, I.S. *292, 299, 300*

Godbole, H.M.   *115*
Goins, D.K.   *210*
Gomez, M.A.   *106, 107*
Gonzalez, L.   *213, 214*
Gordon, D.M.   *295*
Gosser Jr., D.K.   *213*
Goswami, K.N.   *115*
Goto, K.   *114*
Gottlieb, H.E.   *119*
Gottlieb, O.R.   *119*
Grabowska, M.   *113*
Grate, S.J.   *211*
Gray, A.I.   *105, 108, 109, 115–117*
Gu, H.M.   *206, 212*
Gu, L.H.   *109*
Gu, Y.-X.   *208*
Gu, Z.   *109*
Guckert, A.   *106*
Guilhon, G.M.P.   *115*
Guiry, P.J.   *106*
Gunnewegh, E.A.   *109*
Gunstone, F.D.   *291*
Guo, G.   *206*
Guo, X.B.   *205, 206*
Guo, Y.-J.   *113*
Gupta, B.K.   *109*
Gupta, G.K.   *109*
Gupta, M.P.   *118*
Gupta, S.   *109*
Gustafson, K.R.   *108, 109, 111, 295*

Hachiya, K.   *207*
Hagemann, H.   *205*
Haginiwa, J.   *110*
Hakomori, S.   *220, 293, 295, 300*
Halim, A.F.   *109*
Halverson, J.M.   *205*
Hamamoto, Y.   *295, 296*
Hamburger, M.O.   *108*
Han, Y.S.   *119*
Hanai, N.   *295*
Hanish, F.G.   *293*
Hanjura, S.   *296*
Hanton, W.K.   *212*
Harada, N.   *300*
Harayama, T.   *109, 110*
Harinasuta, T.   *212*
Härmälä, P.   *109*
Haruna, M.   *118*
Harwood, J.L.   *291, 294*

Harwood, L.H.   *295*
Harwood, L.M.   *106*
Hasan, C.M.   *109*
Hashidoko, Y.   *199, 213*
Hashimoto, Y.   *295*
Hatano, K.   *118*
Hatano, T.   *109*
Hattori, M.   *109*
Hayashi, A.   *291, 292, 296–298*
Hayashi, K.   *294*
Hayashida, N.   *110*
Haynes, D.E.   *211*
Haynes, R.K.   *134, 163, 166, 185, 204, 209*
He, C.H.   *207*
He, D.   *205*
He, Z.-D.   *112*
He, Z.-Q.   *301*
Heble, M.R.   *206*
Heiffer, M.H.   *211*
Heinstein, P.   *203*
Herath, H.M.T.B.   *116*
Hernández, E.   *113*
Hernandez-Galan, R.   *109*
Hiegemann, M.   *108*
Higa, T.   *297*
Highet, R.J.   *211*
Higuchi, R.   *290, 292, 293, 296, 297, 300*
Higuchi, Y.   *114*
Hiltunen, R.   *109*
Hin, T.Y.Y.   *115*
Hinrichsen, D.F.   *293*
Hiremath, S.V.   *204*
Hirsch, S.   *295*
Ho, L.Y.   *205*
Hoch, J.M.   *203, 205*
Hoefnagel, A.J.   *109*
Hoffman, J.   *293*
Hofheinz, W.   *129, 204, 214*
Honda, M.   *292, 297*
Hong, M.   *111*
Hori, T.   *292, 293, 298*
Hoshi, M.   *288, 293–299*
Hossler, P.   *211*
Hostettmann, K.   *108, 113, 118*
Hu, H.   *210*
Hu, J.   *119*
Hu, Q.-S.   *213*
Hu, Y.   *176, 211*
Huang, B.-S.   *107*

Huang, D.Z. *204*, *213*
Huang, J.J. *204*
Huang, S.-C. *119*
Huang, Y.X. *205*, *206*
Hufford, C.D. 156, 157, *204*, *205*, *208*, *210*, *211*
Hughes, S.H. *111*
Hummelen, J.C. *207*
Huneck, S. *111*
Hunt, N.H. *213*
Huong, V.N. *108*
Husain, A. *202*, *203*
Hwang, H.-C. *111*

Ianaro, A. *293*
Idowu, O.R. 142, *206*
Iinuma, M. *114*
Ikemoto, T. *207*
Ikeshiro, Y. *109*, *110*
Ilarslan, R. *118*
Ilyas, M. *111*
Imakura, Y. 170, *207*, *210*
Imanari, M. *114*
Ina, K. *295*
Inagaki, F. *293*, *294*, *296*, *297*
Inagaki, K. *300*
Inagaki, M. *292*, *293*
Inoue, H. *298*
Inoue, K. *112*
Inoue, M. *109*, *111*, *118*
Inoue, T. *110*
Inukai, K. *292*, *296*
Iossifova, T. *114*
Irie, A. *293*, *295–297*
Isbell, T.A. *214*
Ishibashi, M. *292*, *294*
Ishida, R. *295*
Ishihama, M. *296*
Ishii, H. *107*, *109*, *110*
Ishikawa, M. *110*
Ishikawa, T. *107*, *110*
Ishimoto, Y. *296*
Ismail, H.B.M. *115*
Isobe, R. *296*
Isono, Y. *291*
Itasaka, O. *293*, *298*
Ito, C. *109–111*, *118*
Ito, K. *114*
Ito, Y. *296*
Itonori, S. *293*

Ittarat, I. *213*
Ivie, G.W. *106*
Iwashima, A. *114*

Jackson, Y.A. *111*
Jagdale, M.H. *108*
Jain, N. *111*
Jakupovic, J. *105*, *114*, *117*
Janick, J. *203*
Jansson, P.-E. *298*
Jaquet, C. *214*
Jares-Erijman, E.A. *292*
Jean, M.-Y. *111*
Jefford, C.W. 140, *205*
Jennings-White, C. *204*, *209*, *210*
Jeremic, D. 125, *202*
Jha, B.N. *109*
Jhou, J.X. *292*, *296*
Ji, R.-Y. *208*, *210*
Ji, Z. *202*
Jian, H.X. *205*, *206*
Jiang, G.J. *293*, *297*
Jiang, T. *111*
Jiang, Z.D. *294*, *295*
Jin, W. *292*
Jin, Y. *206*
Jinadasa, S. *203*
Johnson, T.L. *209*
Jokic, A. *202*
Jones, A.L. *294*, *295*
Jong, T.-T. *111*
Josephs, J.L. *115*
Joshi, P.C. *111*, *107*, *113*
Ju-ichi, M. *109–111*, *118*
Jung, D.J. *111*
Jung, M. *111*, 125, 153, 161–163, 165, *202*, *208*, *209*

Kadota, S. *106*
Kadushin, M.R. *109*
Kady, M.M. *112*
Kager, P.A. *211*, *212*
Kagoshima, M. *296*
Kajita, M. *110*
Kajiura, I. *109*
Kaltia, S. *109*
Kamchonwongpaisan, S. *202*, *211*, *213*
Kamemura, K. *293*
Kamil, M. *111*
Kaneko, Y. *109*, *110*

Kano, S. *212*
Kao, K.C. *202*
Kapil, R.S. *115*
Kappe, T. *112*
Karayannakos, P.E. *211*
Karbwang, J. *212*
Karle, J.M. *210, 211*
Karlsson, K.A. *291, 296*
Karnik, P.J. *208*
Kashman, J. *295*
Kashman, Y. *111*
Kates, M. *292*
Kato, A. *111*
Kato, H. *108*
Katsuno, K. *109*
Katsuoka, M. *295*
Kaul, V.K. *202*
Kawaguchi, H. *118*
Kawai, K. *114*
Kawai, S. *212*
Kawajiri, S. *300*
Kawano, Y. *293, 296, 297*
Kawasaki, C. *114*
Kawasaki, T. *296, 297*
Kawazoe, K. *294*
Kayser, O. *111*
Keane, P.J. *118*
Kelley, J.A. *205*
Kendrick, K. *203*
Kenmotsu, K. *110*
Kepler, J.A. *214*
Khalifa, S.I. *208, 211*
Khan, A.Q. *116*
Khan, I. *210*
Khanh, N.X. *212*
Ki, P.F. *296*
Kikuchi, H. *294*
Kikuchi, M. *113*
Kim, C.M. *112*
Kim, N.K. *295*
King, G.R. *209*
Kirtany, J.K. *115*
Kiser, R.F. *295*
Kishi, Y. *164, 209*
Kishimoto, T. *301*
Kitagawa, I. *294, 295*
Klayman, D.L. *125, 127, 202, 203, 205–209, 212*
Klinedinst, D. *213, 214*
Klionsky, B. *297*
Kloppenburg, M. *291*
Ko, F.-N. *107*
Kobayashi, E. *297, 301*
Kobayashi, H. *113*
Kobayashi, J. *110, 292, 294*
Kobayashi, J.-I. *110*
Kobayashi, M. *294, 295*
Kobayashi, T. *115*
Koch, M. *117*
Kocharov, S.L. *299*
Kochetkov, N.K. *292, 299, 300*
Koek, J.N. *207*
Koezuka, I. *291*
Koezuka, Y. *293, 297, 301*
Kojima, H. *114*
Kolodziej, H. *111*
Komai, Y. *291, 292, 298*
Komori, T. *292, 293, 296, 297, 300*
Kon, K. *298*
Konda, K. *296*
Kondo, Y. *111*
Kong, D.-Y. *109*
Kong, L.Y. *112*
Kongsaeree, P. *213*
Kopecky, K.R. *204*
Kostetsky, E.Y. *291*
Kostova, I. *114*
Kotake, K.-I. *110*
Kotsarelis, D. *211*
Kouam, J. *114*
Kowalski, P. *114*
Koyama, J. *210*
Kozawa, M. *105, 119*
Krasilnikova, V.I. *292*
Kreher, B. *112*
Kreps, E.M. *292*
Krishnaswamy, N.R. *117*
Kruglova, E.E. *292*
Krupadanam, G.L.D. *116*
Ksukitani, Y. *294*
Kubo, H. *288, 293–297, 299*
Kubota, Y. *111*
Kudakasseril, G.J. *127, 203*
Kulkarni, M.S. *115*
Kulkarni, S.A. *115*
Kumanishi, T. *298, 301*
Kumar, D. *112*
Kumar, M.A. *204*
Kumar, N.S. *116*
Kumar, V. *112, 116*

Kuoh, C.S. *114*
Kurki, T. *300*
Kurozumi, T. *118*
Kuster, R.M. *112*
Kuwajima, H. *112*
Kuypers, F. *213*
Kuznetsov, P.V. *117*
Kvanrström, I. *298*
Kwon, Y.S. *112*
Kyi, Z.Y. *210*

Laasko, J.A. *112*
Lahloub, M.F. *109*
Lai, J.-S. *119*
Laine, R.A. *293, 294*
Lakshmi, V. *115*
Lam, L. *203*
Lambein, F. *292*
Lamnaouer, D. *112*
Lamparczyk, H. *114*
Lankin, D.C. *203*
Lansbury, P.T. 136, 161, *204*
Larsen, I.K. *112*
Laschober, R. *112*
Laudisio, A. *300*
Laughlin, J.C. *203*
Leach, G. *105*
Leclerc, A. *116*
Lee, I.S. *204, 205, 210*
Lee, K.H. *210, 212*
Lee, M. *209*
Lee, Y.W. *214*
Lees, M. *292*
Leffler, H. *296*
Lehtonen, P. *109*
Leitão Filho, H.F. *113*
Lemmich, E. *112*
Lemmich, J. *112*
Levine, B.S. *211*
Levitina, M.V. *292*
Lewis, M.D. *209*
Li, G. *211*
Li, G.Q. *205, 206*
Li, G.Y. *207*
Li, H. *297*
Li, L. *202*
Li, L.N. *207*
Li, L.-Q. *208, 209*
Li, L.X. *206*
Li, R. *209*

Li, S.-C. *293*
Li, W. *205*
Li, X. *109, 112, 119, 208*
Li, X.-Y. *209*
Li, Y. *203, 207, 208*
Li, Y.-T. *293*
Li, Z.L. *212*
Lian, X.T. *207*
Liang, H.-Z. *209*
Liang, X.T. 139, 198, *205–207, 213*
Liersch, R. *203*
Lima, V. *213*
Lin, A.J. 155, 156, 160, 161, 185, 195, *203, 205, 206, 208, 209, 212*
Lin, H. *202*
Lin, J.K. *112*
Lin, L.-J. *113*
Lin, L.-Z. *113*
Lin, X.Y. *206*
Lin, Y.-C. *107*
Lin, Z. *116*
Linares, E. *113*
Lindberg, B. *293*
Lindberg, P. *298*
Lindh, F. *298*
Liu, C.X. *206*
Liu, J.-H. *113*
Liu, J.J. 139, *204*
Liu, J.-M. 128, *202, 207*
Liu, W.B. *119*
Liu, X.H. *119*
Llabres, G. *113*
Lönngren, J. *298*
Looareesuwan, S. *212*
Lotterer, E. *117*
Lou, X. *204*
Lou, Z. *119*
Lou, Z.-C. *115*
Lu, Y. *112*
Lugt, C.B. *202, 211*
Luider, T.M. *207*
Luis, F.R. *117, 119*
Luo, X.-D. 124, 125, 145, 155, 159, 188, *202, 205, 206*
Luxemburger, C. *212*

Macheix, J.J. *108*
Madhusudanan, K.P. *205*
Madruzza, G. *107*
Maffia, P. *293*

Magalhães, A.F. *113*
Magalhães, E.G. *113*
Maha, V.P. *202*
Mahajnah, Y. *294*
Mahier, T.J. *204*
Mahmoud, Z.F. *113*
Maia, J.G.S. *115*
Maillard, M. *118*
Maitland, D.J. *119*
Makariewa, T.N. *300*
Maki, S. *118*
Malchenko, L.A. *299*
Maldonado, E. *113*
Mali, R.S. *113*
Malik, A. *116*
Manandhar, K. *106*
Manas, A.R. *115*
Manda, T. *294*
Mandal, S. *107, 111, 113*
Mane, R.B. *108*
Manekar-Tilve, A. *113*
Mangani, F. *106*
Mangoni, A. *293–296*
Mannandhar, M.D. *106*
Marco, J.A. *113*
Marcotullio, M.C. *107*
Marion, D. *211*
Marsili, A. *106*
Marston, A. *113*
Martin, M.-T. *112*
Masciadri, R. *214*
Mase, I. *109, 110*
Massanet, G.M. *105, 109, 117, 119*
Masuda, T. *113*
Mata, R. *113, 117*
Matsubara, T. *291–293, 296–298*
Matsuda, N. *113*
Matsukawa, S. *292*
Matsumoto, S. *292*
Matsumoto, T. *295*
Matsunaga, K. *292*
Matsunaga, S. *293, 297, 300*
Matsuo, J. *296*
Matsuoka, M. *110*
Matsuura, F. *291, 297*
Matsuura, T. *297*
Maxwell, L.E. *213*
Mayuzumi, K. *115*
Mazur, J. *113*
McChesney, J.D. *205, 208, 209*

McClatchey, W.C. *296*
McDonald, F.J. *295*
McGarvey, D.J. *208*
McKee, T.C. *106, 108, 113*
McKeever, P. *211*
McMahon, J.B. *108, 111*
McNab, H. *113*
McPhail, A.T. *111, 207, 208, 210*
McPhail, D.R. *111, 207, 210*
Mehrotra, S. *209, 210*
Mele, S. *106*
Melendez, V. *141, 205*
Mendoza, S. *113*
Mendoza-Díaz, S. *117*
Meng, T. *119*
Menghini, A. *107*
Menon, R.B. *204*
Meou, A. *119*
Mericli, A.H. *118, 119, 202*
Mericli, F. *118, 119*
Merkus, F.W.H.M. *211*
Meshnick, S.R. *125, 194, 197, 202, 208, 211, 213*
Meyer, M. *114*
Michaelis, H.C. *117*
Mikhailov, A.T. *299*
Mil'grom, E.G. *114*
Milgrom, Yu.M. *300*
Milhous, W.K. *203, 208–210, 213, 214*
Miller, R. *209, 210*
Miller, R.E. *209, 211*
Min, Z.D. *112*
Minale, L. *291*
Minh, N.T. *108*
Miralles, J. *294*
Miski, M. *114*
Misra, L.N. *202*
Mitaku, S. *117*
Mitsukawa, S. *291*
Miura, Y. *295*
Miyahara, K. *297*
Miyaji, M. *114*
Miyamoto, T. *292, 293*
Miyauchi, N. *115*
Miyazaki, H. *110*
Mizuno, A. *114*
Mizuno, T. *109, 111*
Mizutani, S.J. *213*
Mohnhaupt, M. *205*
Molina, J. *204*

Morelli, I. *106*
Mori, K. *300*
Morita, M. *112, 293, 297, 301*
Moriya, T. *292*
Mors, W.B. *112*
Mosher, H.S. *300*
Motoki, K. *297, 301*
Mouncherou, S.M. *114*
Msonthi, J.D. *113*
Mukharya, D.K. *112*
Müller, A.H. *114*
Mungai, M. *117*
Muraguchi, M. *111*
Murakami, N. *295*
Muralikrishna, G. *293*
Muratov, V.B. *296*
Muroya, Y. *113*
Murray, R.D.H. *114*
Musallam, H.A. *214*
Myint, U.P.T. *208*

Na Bangchang, K. *212*
Nadinic, E.L. *106, 107*
Nadkarni, K.K. *115*
Nagai, M. *110*
Nagai, Y. *291, 292, 298*
Nagasawa, K. *114*
Nagle, D.G. *296*
Nagumo, S. *110*
Naing, U.T. *208*
Nair, M.S.R. *203*
Nakagawa, K. *118*
Nakagawa, M. *111, 296*
Nakajima, S. *114*
Nakanishi, H. *294*
Nakashima, T.T. *204*
Nakata, T. *111, 118*
Nakata, Y. *111, 118*
Nakatani, N. *113*
Namba, T. *106, 109*
Namsiripongpun, W. *212*
Nanduri, S. *117*
Nano, G.M. *105*
Narantuyaa, S. *114*
Narske, E.D. *112*
Narushima, K. *293*
Natori, S. *115*
Natori, T. *288, 292, 293, 297, 300, 301*
Natu, A.D. *108*
Nema, D. *114*

Neszmélyi, A. *112*
Ngadjui, B.T. *114*
Ni, M.Y. *202, 207*
Nicholls, K.M. *204*
Nichols, B.W. *291*
Nie, F. *116*
Nielsen, H.M. *112*
Niemeyer, H.M. *119*
Nilanonta, C. *213*
Nilhous, W. *205*
Nilsson, K. *291*
Nimmich, W. *298*
Nishike, S. *115*
Nishimura, Y. *296*
Nishino, A. *114*
Nishino, H. *114*
Nishioka, H. *109, 110*
Nishita, Y. *109*
Nishizawa, J. *108*
Niu, F. *116*
Niu, X.Y. *205*
Niwa, M. *110, 114*
Niyaz, N.M.M. *112*
Nkengfack, A.E. *114*
Noda, N. *297*
Nohara, T. *296*
Nojiri, H. *295*
Nomura, T. *108*
Nora, T. *109*
Nores, G.A. *295*
Nosten, F. *211, 212*
Novotny, J.F. *203*
Nowak, D. *117*
Nowak, D.M. *136, 161, 204*
Nozawa, K. *114*
Nozoe, S. *111*
Nwe, D.Y.Y. *208*
Nykolov, N. *114*

Obukhova, E.L. *292*
Ochocka, R.J. *114*
Odani, S. *298*
Oertli, E.H. *106*
Ognyanov, I. *156, 208*
Ogura, C. *295*
Oh, C.H. *208, 210, 213*
Ohashi, K. *115*
Ohi, K. *115*
Ohizumi, Y. *292*
Ohloff, G. *129, 203*

Ohta, S. *110*
Oka, T. *110*
Okada, Y. *114, 115*
Okano, M. *111, 118*
Okuda, T. *109*
Okumura, Y. *115*
Okuyama, T. *112, 114, 115, 118*
Oliveira, F.M. *115*
Omokhodion, S.I. *211*
Omura, M. *109–111, 118*
Ono, H. *300*
Ono, T. *111, 118*
Orabi, K.Y. *207*
Ortego, A. *113*
Osawa, T. *297*
Oshima, Y. *292*
Osoba, O.A. *115*
Otha, M. *297*
Oudman, D. *207*
Özen, H.C. *105*
Ozhatay, N. *202*

Padamanabhan, V.M. *115*
Padha, N. *115*
Padley, F.D. *291*
Paknikar, S.K. *106, 115*
Palacios, P.S. *107*
Paliwal, M.K. *117*
Palmer, C.J. *115*
Palmisano, G. *105*
Pan, H.-Z. *213*
Pan, J. *111*
Pan, J.U.P. *203*
Pan, X.-Q. *208*
Pandey, V.B. *117*
Pandhare, N.A. *113*
Pando, E. *105, 117*
Pannel, L.K. *205*
Paradkar, M.V. *108, 115*
Paranhos, A. *107*
Parente, J.P. *112*
Paris, M. *211*
Park, S.B. *213*
Partridge, J.J. *300*
Pascher, I. *291*
Patterson, G.M.L. *295*
Pechenkina, E.E. *294*
Pedro, J.R. *107*
Peggins, J.O. *205, 211*
Pei, Y.H. *112*

Pérez, J. *107*
Perler, J.P. *205*
Peter-Katalinic, J. *293*
Peters, W. *205, 206, 212*
Petras, J.M. *211*
Petrov, O. 156, *208*
Pettit, G.R. 153, *207, 295*
Phaipun, L. *212*
Philip, A. *214*
Philogene, E. *105*
Piatak, D.M. 153, *207*
Piedade, F.M. *112*
Pinheiro, R.M. *107*
Pinto, M.M.M. *106*
Pirani, J.R. *114*
Pistelli, L. *106*
Pizza, C. *107*
Plattner, R.D. *108, 117*
Plessas, Ch.T. *211*
Plessas, S.T. *211*
Ploypradith, P. *213*
Pomazanskaya, L.F. *292*
Popli, S.P. *205*
Porzel, A. *111*
Posner, G.H. 155, 197, *208, 210, 213, 214*
Potier, P. *105*
Poupat, C. *105*
Poutaraud, A. *106*
Pozetti, G.L. *113, 119*
Prakash, D. *115*
Pras, N. *202*
Pravdina, N.I. *292*
Proenca de Cunha, A. *107*
Prokazova, N.V. *299*
Pu, Y.-M. 143, 150, 155–157, *206–209, 213*

Qian, Z.-H. *300*
Qin, B. *115*
Qing, X.Y. *105*
Quader, M.A. *115*
Qualls, S. *214*
Quan-Long, P. *205*
Quirke, J.M.E. *118*

Rahmani, M. *115*
Raj, K. *115*
Rajagopal, M.V. *291*
Rajnikant, G.K.N. *115*

Rajzer, D. *114*
Raman, P.V. *117*
Ranz, A. *213*
Rao, G. *116*
Rao, M.N. *116*
Rapport, M.M. *293*
Rashid, M.A. *109, 116*
Rashkes, Y.V. *114, 300*
Rasool, N. *116*
Ravindranathan, T. *133, 204*
Reisch, J. *116*
Ren, Z.Y. *205*
Renkonen, O. *293, 294*
Reuter, G. *293*
Rezanka, T. *294*
Riccio, R. *291*
Ridder, W.E. *205*
Rimchala, W. *212*
Rinehart, K.L. *292*
Roberts, M.R. *115*
Robinson, B.L. *212*
Rodriguez-Luis, F. *105*
Rogers, C. *108*
Rogers, R.D. *211*
Roleira, F.M.F. *106*
Rollman, I.J. *203*
Romano, I. *300*
Rossier, J.C. *205*
Rosynov, B.V. *296*
Roth, R.J. *133, 149, 204, 207*
Rouessac, F. *116*
Rozentsvet, O.A. *294*
Ruangrungsi, N. *113*
Rustaiyan, A. *113*

Saad, H.-E.A. *109*
Sabri, S.S. *105*
Sadovskaya, V.L. *299*
Sagioka, T. *300*
Saha-Moeller, C.R. *105*
Saito, A. *300*
Saito, M. *295*
Saitoh, U. *299*
Sakai, T. *114, 301*
Sakakibara, J. *295*
Sakandelidze, O.G. *296*
Sakata, K. *295*
Sakurada, E. *110*
Salako, L.A. *211*
Sales, B.H.L.N. *113*

Salunkhe, M.M. *108*
Salvá, J. *109, 117*
Samb, A. *294*
Samuelsson, B.E. *291, 296*
Sandhu, P.K. *113*
Sanechika, Y. *296*
Sanford, P.A. *220, 293*
Sant'ana, A.E.G. *115*
Sanz, J.F. *113*
Sarker, S.D. *117*
Sasaki, T. *292*
Sata, N. *293, 300*
Satake, M. *287, 291, 292, 298, 301*
Sauter, M. *105*
Sawa, E. *301*
Schiebel, H.M. *205*
Schmid, G. *129, 204, 214*
Schmitz, F.J. *295*
Schopfer, C. *211*
Schulte, K.H. *129, 203*
Schulz, H. *117*
Schuster, B.G. *211*
Schuster, N. *117*
Schwartz, H. *205*
Scovill, J.P. *203*
Seki, H. *110*
Seong, B.W. *117*
Shah, Z. *105*
Shalaby, N.M.M. *108*
Shang, X. *146, 207*
Shapiro, T.A. *213, 214*
Sharifi, S. *117*
Sharma, R.P. *202*
Sharma, S. *202*
Shashkov, A.S. *298*
Shauer, R. *293*
She, Q.-L. *107*
Sheen, W.-S. *107*
Shen, C.C. *124, 125, 155, 159, 188, 202*
Shen, X.-W. *117*
Sheu, S.-J. *107*
Shi, L.-W. *115*
Shi, X.-C. *208*
Shiba, T. *299*
Shibata, S. *115, 118*
Shimomura, K. *296*
Shinde, S.L. *208*
Shirahama, H. *295*
Shirahshi, H. *295*
Shkarenda, V.V. *117*

Shu, H. *205*
Shu, Z. *205*
Shukla, A. *202*
Shukla, Y.N. *202*
Shwe, U.T. *208*
Siddiqui, I.R. *117*
Siebert, H.C. *293*
Sili, C. *300*
Silverton, J.V. *205*
Simon, J.E. *203*
Sindkhedkar, M.D. *113*
Singh, A. *202, 203*
Singh, J. *117*
Singh, R.P. *117*
Singh, S. *117*
Sinyh, A. *202*
Sipahimalani, A.T. 143, *203, 206*
Sipahimalani, M.R. *203*
Skalkeas, G. *211*
Skaltsounis, A.L. *117*
Slater, A.F.G. *212*
Sloane-Stanley, G.H. *292*
Smirnova, G.P. *291, 292, 298–300*
Smyth, M.H. *205*
Snader, K.M. *295*
Sneden, A.T. *108*
Sodano, G. *300*
Soejarto, D.D. *108, 109*
Sohonie, K. *291*
Soicke, H. *203*
Sokoloski, E.A. *205*
Son, B.W. *295*
Son, P.T. *108*
Sondengam, B.L. *114*
Song, Q.L. *206*
Song, Z.-W. *117*
Song, Z.Y. 194, *205, 206, 211*
Sonku, N. *293*
Sonmez, U. *118*
Soriano-García, M. *117*
Soriente, A. *300*
Sowumi, A. *211*
Spencer, G.F. *108, 117*
Srimannarayama, G. *116*
Sripathi, S.K. *117*
Srivastava, S.D. *118*
Srivastava, S.K. *114, 118*
Staba, E.J. *203*
Stefanovic, M. *202*
Stehr, C. *203*

Sticher, O. *109*
Still, W.C. 166, *209*
Stoeckli-Evans, H. *108, 118*
Stoev, G. *114*
Stohler, H. *214*
Stonik, V.A. *300*
Su, S. *119*
Sugino, H. *114*
Sugita, M. *292, 293, 296, 299*
Sugiyama, S. *297*
Suidema, J. *211*
Sukari, M.A. *115*
Sukontason, K. *212*
Sulima, A.V. *296*
Sulistyuwati, L. *118*
Sun, H. *116*
Sun, X.M. *205, 206*
Sun, Y.F. *119*
Suzuki, K. *115*
Suzuki, M. *212, 297*
Suzuki, Y. *110*
Svennerholm, L. *293*
Svensson, S. *293*
Svetaskev, V.I. *291*
Swain, L.A. *118*
Swearengen, J. *211*
Sweeley, C.C. *291, 293, 297*

Taglialatela-Scafati, O. *296*
Tagliapietra, S. *105*
Tahara, S. *213*
Takaishi, K. *112*
Takaku, F. *295*
Takashima, S. *114*
Takata, M. *114, 118*
Takayasu, J. *114*
Takemura, Y. *110, 111, 118*
Tamchompoo, B. *213*
Tan, N. *118, 119*
Tan, S. *110*
Tanahashi, S. *111*
Tanaka, E. *111*
Tanaka, K. *110, 111*
Tanaka, R. *297*
Tanaka, T. *114*
Tani, Y. *111*
Taniguchi, M. *105, 119*
Tao, R.V.P. *293*
Tao, S. *214*
Tate, S. *294*

Taufiq-Yap, Y.H.   *115*
Taylor, T.E.   *202*
Tempesta, M.S.   *114*
Ten, H.   *111*
Teng, C.-M.   *107, 111*
Teng, X.   *208*
Teo, C.K.H.   *203*
Teoh, K.   *203*
Te Paske, M.R.   *118*
Ter Kuile, F.   *212*
Terreaux, C.   *118*
Terui, Y.   *295*
Teshima, T.   *299*
Thakur, R.S.   *202*
Thanavibul, A.   *212*
Thebtaranonth, C.   198, *213*
Thebtaranonth, Y.   *213*
Theoharides, A.D.   143, *203, 205, 206*
Thiilborg, S.T.   *112*
Thomas, A.   *213*
Thombre, H.M.   *108*
Tian, Z.   *112*
Tillequin, F.   *114, 117*
Tin, T.   *212*
Tirillini, B.   *106*
Titulaer, H.A.C.   195, *211*
Togawa, K.   *292, 293*
Toma, L.   *105*
Tomita, Y.   *109, 110*
Toome, V.   *300*
Torok, D.S.   146, *207, 208, 211*
Tortella, F.C.   *211*
Tovey, G.   *212*
Toyonaga, T.   *110*
Trincone, A.   *300*
Tsai, I.-L.   *107* ,
Tsuchida, Y.   *295*
Tsuji, S.   *292*
Tsutsui, C.   *115*
Tu, Y.-Y.   125, *202, 207*
Tullner, H.U.   *203*
Tumsupapone, S.   *212*

Uchida, H.   *111, 118*
Uchida, T.   *297, 301*
Uda, H.   *300*
Udagawa, S.   *114*
Ulubelen, A.   *118, 119*
Urwyler, H.   *214*
Uskokovic, M.R.   *300*

Valio, I.F.M.   *113*
Van Bekkum, H.   *109*
Van Boxtel, C.J.   *212*
Van Hummel, H.C.   *294*
Vaskovsky, V.E.   *291*
Vassileva, E.   *114*
Venkatesan, S.   *210*
Vennerstrom, J.L.   *214*
Venugopalan, B.   155, *205, 208*
Vieira, P.C.   *114*
Vilegas, J.H.Y.   *113, 119*
Vilegas, W.   *113, 119*
Villena Iribe, R.   *117*
Vincenzini, M.   *300*
Vinciguerra, V.   *107*
Viravan, C.   *212*
Vishwakarma, R.A.   *203, 205*
Voerste, A.A.W.   *116*
Vonwiller, S.C.   134, 185, *204, 209*
Vouffo, T.W.   *114*
Vroman, J.A.   *209, 210*
Vuorela, H.   *109*

Wada, H.   *110*
Wagner, H.   *108, 112*
Waigh, R.D.   *119*
Walker, O.   *211*
Walsby, A.E.   *291*
Wang, C.L.   *119*
Wang, D.   *119, 213, 214*
Wang, D.S.   *208*
Wang, J.P.   *110*
Wang, L.   *117*
Wang, S.-K.   *108*
Wang, X.   *202*
Wang, Y.   *204, 205*
Wang, Y.-H.   *108*
Wang, Y.L.   *205, 206*
Ward, S.A.   *206*
Warhurst, D.C.   *206, 212*
Watanabe, H.   *115*
Watanabe, J.   *301*
Watanabe, Y.   *298*
Waterman, P.G.   *105, 108, 109, 115–117*
Weathers, P.J.   127, *203*
Wei, Z.X.   128, *203*
Weina, P.J.   *211*
Weislow, O.S.   *295*
Weissman, B.   *293*
Welti, D.   *291*

Author Index

Wesche, D.L. *211*
West, M.L. *107*
Westerlund, D.J. *206*
Wetsteyn, J.C.F.M. *211*
Whi, W. *209*
White, N.J. *212*
Wicklow, D.T. *112, 118*
Wickramaratne, D.B.M. *112, 116*
Wickramasinghe, A. *116*
Wiegandt, H. *291*
Wierer, M. *108*
Win, U.H. *208*
Winkle, S.A. *118*
Woerdenbag, H.J. *125, 202*
Wolk, C.P. *292*
Woo, W.S. *117*
Wood, J.K. *214*
Wood, K.V. *203*
Woods, B.J.B. *291*
Wu, B. *206*
Wu, B.-A. *208*
Wu, J. *210*
Wu, L. *119*
Wu, L.I. *107*
Wu, T.-S. *111, 112, 114, 119*
Wu, Y.-C. *108, 114*
Wu, Y.L. *134, 172, 202, 204, 207, 210*
Wu, Z. *206*
Wu, Z.H. *202, 207*
Wüest, H. *129, 203*
Wyandt, C. *210*
Wynberg, H.W. *207*

Xiao, Y.Q. *119*
Xie, G. *119*
Xie, G.H. *205, 206*
Xie, M. *206*
Xie, X. *213*
Ximen, L. *202*
Xu, C.-M. *213*
Xu, G. *205*
Xu, L.X. *206*
Xu, S.-X. *113*
Xu, X.-X. *133, 198, 204, 207, 213*

Yadava, V.S. *115*
Yagen, B. *150, 207, 208*
Yamada, K. *298*
Yamada, M. *295*
Yamada, S. *298*
Yamada, S.-H. *292*
Yamagish, T. *210*
Yamaguchi, K. *110*
Yamaguchi, Y. *296, 297, 301*
Yamaji, K. *301*
Yamamoto, T. *299*
Yamashita, S. *207*
Yan, S.Q. *109*
Yand, J.J. *207*
Yang, C.-R. *112*
Yang, L. *119*
Yang, Q. *209*
Yang, X. *109, 111*
Yang, X.B. *119*
Yang, Y.-Z. *213*
Yao, X.-S. *113*
Yarber, R.H. *205*
Yasuhara, T. *109*
Ye, B. *134, 172, 204, 210*
Yeh, H.J.C. *204–207, 211*
Yeng, M.Y. *206*
Yi, M.G. *206*
Yin, Z.-D. *117*
Yokoi, T. *210*
Yoneda, Y. *105*
Yook, C.S. *117*
Yu, D. *209*
Yu, L. *208*
Yu, P. *207, 208*
Yu, P.-L. *208*
Yuen, K.H. *203*
Yuste, A. *113*
Yuthavong, Y. *213*

Zabelinskii, S.A. *292*
Zaigan, D. *108*
Zaman, S.S. *125, 202*
Zange, K.D. *206*
Zarga, M.H.A. *105*
Zawadowski, T. *113*
Zdero, C. *119*
Zeng, C.-M. *294*
Zeng, G.F. *107*
Zeng, L. *108, 119*
Zeng, M.Y. *207*
Zeng, Y. *205*
Zerihun, B.M. *119*
Zhang, D. *109, 119*
Zhang, F. *116, 213*
Zhang, J.L. *207*

Zhang, L.   *207, 213*
Zhang, M.   *202*
Zhang, R.H.   *119*
Zhang, R.-Y.   *119*
Zhang, X.   *119*
Zhang, X.Q.   142, *206*
Zhang, Y.   *205*
Zhany, H.Z.   *206*
Zhao, K.C.   194, *206, 211*
Zhao, Y.   193, *211, 212*
Zheng, Q.   *112*
Zheng, Q.T.   *207*
Zheng, S.-Z.   *117*
Zheng, Y.P.   *208*
Zhong, F.   *203*
Zhong, J.J.   *119*
Zhong, Y.   *202*
Zhongshan, W.   *204*
Zhou, C.   *119*
Zhou, W.-S.   198, *204, 207, 213*
Zhou, Z.M.   141, *205, 206*
Zhu, H.S.   *205*
Zhu, J.   *204, 213*
Zhu, T.   *119*
Zhu, T.R.   *109, 112*
Zhukova, I.G.   *291, 292*
Ziffer, H.   143, 146, 150, 164, *205–209, 211, 213*
Zinsmeister, H.D.   *111*
Zollo, F.   *291*
Zou, A.-Q.   *206*
Zou, K.   *119*
Zubia, E.   *105, 119*
Zvezdina, N.D.   *299*

Subject Index

Acanthaganglioside A 279
Acanthaganglioside B 279, 280
Acanthaganglioside C 279, 280
Acanthaganglioside D 280, 281
Acanthaganglioside E 280, 281
Acanthalactoside-A 235
Acanthalactoside-B 235
Acanthaster planci 229, 234, 235, 273, 279, 280
Acetic anhydride 183
Acetone 156
Acetonitrile 127, 167
Acetoxyaurapten 6, 102
N-Acetylglucosamine 233
N-Acetylneuraminic acid 263, 264, 266, 271, 272, 277, 281
9-O-Acetyl-4,7,8-tri-O-methyl-N-methyl-N-acetylneuraminic acid 266
Acidic ethanol 148
Acrimarine-A 81, 102
Acrimarine-B 81, 102
Acrimarine-C 4, 81, 102
Acrimarine-D 13, 102
Acrimarine-E 12, 102
Acrimarine-F 12, 102
Acrimarine-G 11, 102
Acrimarine-H 11, 102
Acrimarine-I 12, 102
Acrimarine-J 12, 102
Acrimarine-K 11, 102
Acrimarine-L 4, 81, 102
Acrimarine-M 11, 102
Acrimarine-N 12, 102
Acrimarines 4
Aculeatin 4
(−)-Aculeatin 85, 102
6-O-Acyl-monogalactosyldiacylglycerol 225
Aegle marmelos 7, 13
Aeglin 7, 102

AEP-galactosylceramide 247, 248
Aesculetin 88, 102
Aflavarin 71, 102
Agelas clathrodes 230, 233, 234, 239, 289
Agelas conifera 230, 233, 234, 239, 240, 289
Agelas dispar 219, 239, 240, 242, 243, 289
Agelas longissima 233, 234, 239–241, 289
Agelas mauritiana 230, 233, 234, 240, 288
Agelas sp. 230, 234
Agelasphin-7a 233
Agelasphin-9a 233
Agelasphin-9b 233
Agelasphin-11 233
Agelasphin-13 233
Agelasphins 233, 234, 288
Agrobacterium 127
AJPnGL 244, 250
Albiflorin-1 9, 102
Albiflorin-2 18, 102
Albiflorin-3 18, 102
Algae 224, 226
Alicyclic allyl hydroperoxides 134
α-Alkylbenzylic esters 182
(+)-10-Alkyldeoxoartemisinin 163
10-Alkyldeoxoartemisinin derivatives 154, 186
9-Alkyl-9-desmethylartemisinin 170
Alloxanthoxyletol 33, 102
(+)-10β-Allyldeoxoartemisinin 164
3-Allyl-4-hydroxycoumarin 4
Allyltrimethylsilane 164
Alyxia reinwardti var. *lucida* 63
Amberlyst-15 167
Aminoethylphosphonate 258
6-O-(2-Aminoethylphosphonyl)galactose 253

Subject Index

2-Aminoethylphosphonylgalactosylceramide 248
2-Amino-1,3-octadecanediol 217
Ammi majus 15, 16
Ammonia 140
Amodiaquine 124
Amomum krervanth 198
Amphicerebroside B 232, 233
Amphicerebroside C 232, 233
Amphicerebroside D 232, 233
Amphicerebroside E 232, 233
Amphicerebroside F 232, 233
Amphimedon viridis 232
Anabaena cylindrica 282–284
Anabaena flos-aquae f. *flos-aquae* 225
Anabaena torulosa 282, 284
Angelica archangelica ssp. *archangelica* 21
Angelica dahurica 47
Angelica edulis 20, 21
Angelica flaccida 25, 26
Angelica gigas 71
Angelica pubescens 20
Angelica pubescens f. *biserrata* 9, 10
Angelicin 4, 83, 102
Angelidiol 20, 102
Angelitriol 9, 102
Angelol J 10, 102
Angelol K 10, 102
Angelol L 10, 102
Angustifolin 92, 102
Anhydrodihydroartemisinin 143, 149, 150, 155–157, 175, 185
Anhydronotoptoloxide 30, 102
Anisocoumarin H 6, 102
Anisocoumarin I 45, 102
Anisocoumarin J 45, 102
(+)-Anomalin 84, 102
Anopheles sp. 124
Antagonistic activity 290
Anthocidaris crassispina 226, 230, 236, 260–264, 288
Antifungal activity 85, 86, 232
Anti-gala-6α 243
Anti-gala-6β 243
Anti-HIV activity 226, 290
Antimalarial activity 124, 125, 127, 129, 148–150, 153, 155, 156, 158, 162, 164, 166, 167, 170, 172, 173, 192, 195–198, 200

Antipyretic activity 125
Antitumor activity 234, 288–290
Antiviral activity 290
Aotus trivirgatus 196
Aphelasterias japonica 268, 271
Aplysia juliana 230, 231, 234, 235, 244, 245, 250
Aplysia kurodai 245, 246, 251–260, 287
Aplysia sp. 251, 252
Aplysina cauliformis 227
Aplysina fistularis fulva 227
Artabotrys unciatus 198
Arteannuin 125
Arteannuin-B 136, 143
Arteether 139, 141, 143, 148, 150, 154, 158, 175, 177, 178, 187, 193–195
β-Arteether 138
9-*epi*-Arteether 150
Arteflene 199, 200
Artelinic acid 142, 160, 182, 195
Artemether 138, 139, 141, 143, 158, 160, 193, 194, 196
Artemisia absinthium 126
Artemisia annua 125–128, 133, 142, 183
Artemisia dracunculus 126
Artemisia laciniata 48, 50
Artemisia sacrorum 52
Artemisia tridentata 126
Artemisia vulgaris 126
Artemisia sp. 126
Artemisinic acid 127, 133, 134, 141, 143, 161–163, 165, 172, 175, 185
Artemisinin 124–134, 137–146, 148, 153–156, 158, 160, 161, 163–166, 168, 172, 173, 175, 178, 179, 186–188, 193–197, 200, 201
[^{14}C]-Artemisinin 143, 196
9-*epi*-Artemisinin 146
Artemisinin G 128
Artemisinin derivatives 148, 150, 155, 171, 178, 180, 187, 194, 196, 197
Artemisinine 125
Artemisitene 127, 134, 142, 183, 185
Artesunate 142, 193, 194, 196
^{3}H-Artesunate 143
Artesunic acid 160, 179
N-Aryl-10-azadihydroartemisinin derivatives 161
3-Aryl-oxygenated coumarins 68
3-Aryl-substituted coumarins 58

Subject Index

4-Aryl-substituted coumarins   61
Asacoumarin B   4, 79, 102
Aspergillus alliaceus   70
Aspergillus flavus   71
Aspergillus niger   175
Asterias amurensis   229, 234–236, 269
Asterias amurensis versicolor   230, 268, 269, 290
Asterias forbesi   286
Asterias rubens   269, 270
Asterias vulgaris   286
Asterinaganglioside-A   278, 279
Asterina pectinifera   229, 234, 235, 274–279, 281, 290
Asteroidea   290
Asterolasia squamiligera   56
Asterolasia trymalioides   52
Astropecten latespinosus   230, 272–274, 290
Atrichum undulatum   50, 54, 55
Axiceramide A   241, 242
Axiceramide B   241, 242
Axinella sp.   241, 242
Ayapin   88, 102
11-Azaartemisinin   147
11-Azaartemisinin derivatives   147

Badycoumarin A   25, 102
Badycoumarin B   26, 102
Bahia ambrosioides   65
Bakuchicin   4, 83, 102
Balsamiferone   92, 102
Bavacoumestan A   69, 102
Bavacoumestan B   69, 102
Beauveria sulfurescens   175, 176
Biemna sp.   227
Biological activity   170–172, 228, 279, 287, 290
Bisclausarin   73, 102
Biscoumarins   2, 3, 70
Bishassanidin   74, 102
6,9-Bisnorartemisinin   166
Bisnorponcitrin   73, 102
Bisosthenon   72, 102
Bisparasin   74, 102
Boenninghausenia albiflora   9, 18
Borohydride   131
Boronia algida   14
Boronia lanceolata   9, 26, 27
Boron trifluoride   153, 167

Boron trifluoride etherate   160, 164, 172
Bovine serum albumin   143
Bromine   155
Bromo derivatives   155
9-Bromodihydroartemisinins   185
N-Bromosuccinimide   155
Brosimum gaudichaudi   13
Bruceol   86, 102
Buntansin B   9, 102
Buntansin C   10, 102
Butyldimethylsilyl triflate   174
t-Butyldimethylsilyl triflate   175
t-Butylhydroperoxide   200
Butyllithium   136
t-Butylperoxide   197

Calanolide A   4, 60, 102
Calanolide B   4, 59, 102
Calanolide C   4
Calanolide D   4
Calanone   62, 102
Calophyllolide   95, 102
Calophyllum lanigerum   4, 59, 60
Calophyllum teysmannii   62
Camphorsulfonic acid   136, 138, 161
Capillary electrophoresis   3
Capillary gas chromatography   3
Carbaartemisinins   172, 173
Carduus tenuiflorus   41
cis-Casegravol   18, 102
Cellobiosylceramide   235, 236
Centrifugal partition chromatography   3
Cerebral malaria   124
Cerebrosides   229, 230, 232, 245, 246
Chamaejasmoside   71, 102
Chang-shan   124
Chloculol   17, 102
α-Chlorodimethyl ether   161
Chloroform   127, 139, 141, 218, 219
Chloromarmin   7, 102
m-Chloroperbenzoic acid   156
Chloroquine   124, 158–160, 179, 188, 192, 193, 195
Chlorostoma argyrostoma turbinatum   231, 238, 247, 248
Chondrilla nucula   230
Chondropsis sp.   231
Chromatin   196
Chromic oxide   156

Chromium trioxide   221
Citrumarin-A   76, 102
Citrumarin-B   75, 102
Citrumarin-C   76, 102
Citrumarin-D   76, 102
Citrus canariculata   72
Citrus funadoko   11–13, 31, 72
Citrus grandis   9, 10, 19
Citrus hassaku   6, 10, 23, 26, 31, 37–39, 57, 73–75, 77
Citrus kinokuni   37
Citrus paradisi   11, 12, 18, 19, 38, 56, 57, 74–76
Citrus sinensis   11, 74
Citrus sp.   3, 4
Citrus sulcata   9, 18
Citrus tamurana   37
Citrus tangerina   11, 12, 18, 38, 56, 57
Citrus unshiu   11
Citrus yuko   77
Citrusarin-A   39, 102
Citrusarin-B   39, 102
Claisen rearrangement   133
Claudimerin-A   77, 102
Clausena anisata   6, 45
Clausena indica   28
Clutia abyssinica   64, 66–68
Colchicum decaisnei   65
Collinin   89, 102
Columbianetin   82, 102
Compositae   126
Coptis trifolia   41
Corallina pilulifera   228, 231
Costatolide   94, 102
Cotton effect   278
Coumarin   79, 102
Coumarins   2–5, 63
Coumestans   2, 5, 69
Coumestrol   97, 102
Coutarea hexandra   62
Crasserides   223, 224, 227, 228, 290
Crellisin-B   6, 102
Cremanthodium ellisi   6
Cucumaria echinata   230
Cucumaria japonica   230, 261, 262
Cunninghamella elegans   175
Cyanobacteria   224, 226
Cyanospira rippkae   283, 284
Cyclobisuberodiene   98, 102
Cyclohexane   127

Cytochrome oxidase   197
Cytotoxic activity   232, 274

Dalbergia sissoides   66
Daphne arisanensis   46
Daphne mezereum   78
Daphneside   46, 102
Daucoidin A   21, 102
Daucoidin B   20, 102
Decalenone   131
Decumbensol   13, 102
Dehydroartemisinin   134
Dehydrogeijerin   80, 102
Demethylsuberosin   80, 102
Deoxoartemisinin   163–165, 172
(+)-Deoxoartemisinin   153, 161–163
10-Deoxoartemisinin   165
10-Deoxoartemisinin derivatives   164, 188
Deoxyarteether   154
Deoxyartemisinin   133, 139, 145, 146, 151, 152, 154
Deoxyazaartemisinin   147
Deoxybruceol   86, 102
5-Deoxyprotobruceol-II hydroperoxide regioisomer   27, 102
5-Deoxyprotobruceol-III hydroperoxide regioisomer   27, 102
5-Deoxyprotobruceol-I regioisomer   26, 102
Derris scandens   68
Derrusnin   97, 102
4,5-Desethanoartemisinin   170
9-Desmethylartemisinin   166
Diacetyldihydrofluorescein   142
Diazabicycloundecene   155
Diazomethane   132, 133
Dichloromethane   150
Dichroa febrifuga   124
Dicyclohexylcarbodiimide   159
Diels-Alder reaction   133, 166
Diethylaminosulfur trifluoride   156
Digalactosylceramide   237, 238, 241
Digalactosyldiacylglycerol   223, 225
Digalactosyldiacylglycerols   225, 226
Diglycosylceramides   239, 240
Dihydroarteannuin-B   136
Dihydroartemisinin   140, 141, 143, 145, 149, 150, 153, 158–161, 164, 175, 178, 182, 194, 196

[³H]-Dihydroartemisinin 194
Dihydroartemisinin-12-O-acetic acid 143
Dihydroartemisinin derivatives 155, 158, 160, 173, 181, 195
Dihydroartemisinin esters 190
Dihydroartemisinin ethers 182, 189
Dihydroartemisinins 138, 139
Dihydroartemisitene 150
Dihyrofuranocoumarins 5
Dihydropyranocoumarins 5
12,13-Dihydroxyicosapentaenoic acid 224, 226
5,8-Dimethoxycoumarin 89
p-Dimethyl-aminobenzaldehyde 141
Dimethylaminopyridine 159
4-Dimethylaminopyridine 159
2,6-Di-O-methylgalactitol 255
Diospyros kaki 71
Dioxinoacrimarine-A 18, 103
5,7-Dioxygenated coumarins 3, 29
6,7-Dioxygenated coumarins 3, 40
7,8-Dioxygenated coumarins 45
Diphosphonoglycolipids 252
Diplolophium buchananii 13
Disialogangliosides 265, 267, 270, 271
Disialoglycolipids 264, 266, 281
Dispiro-1,2,4,4-tetraoxanes 200
5,6-Disubstituted-7-oxygenated coumarins 28
6,8-Disubstituted-7-oxygenated coumarins 28
Dixylosylalkylglycerol 223
Dolabella auricolaria 253, 258, 260
Donatin 47, 103
Dorstenia brasiliensis 30
Dorstenia cayapiaa 30
Dorstenia contrajerva 30
Drummondita hassellii 48

Echinarachnius parma 262, 263
Echinocardium cordatum 262, 264, 265
Echinoderms 216, 229, 230, 243, 246, 262
Echinops niveus 61
Edgeworin 98, 103
Edgeworoside B 78, 103
Edgeworoside C 70, 103
Edgeworthia chrysantha 70, 78
Edulisin III 21, 103
Edulisin IV 20, 103

Edulisin V 21, 103
Edulisin VI 21, 103
EGL-I 259, 260, 287
(+)-Elisin 29, 103
Elsholtzia densa 47
Epiphyllocoumarin 32, 103
Epoxycollinin 45, 103
Eriobrucinol regioisomer-A 32, 103
Eriobrucinol regioisomer-B 37, 103
Eriostemon brucei 31, 32, 37
Eriostemon brucei var. *cinereus* 32, 33
Eriostemon cymbiformis 40
Eriostemon myoporoides 34,·35
Eriostemon spicatus 15, 16
Eriostemon tomentellus 11, 19
Erylusamine A 285, 286, 290
Erylusamine B 285, 286, 290
Erylusamine C 285, 286, 290
Erylusamine D 285, 286, 290
Erylusamine E 285, 286, 290
Erylus placenta 285, 286, 290
Erythrina sigmoidea 69
Esenbeckia grandiflora 56
N-(2'-Ethanal)-11-azaartemisinin 148, 179
Ethanol 127, 149, 150, 183
[³H]-Ethanol 143
Ethanolamine 257
1-Ethoxy-2-methylbutadiene 133
Ethuliacoumarin A 96, 103
Ethulia vernonioides 65
Ethyl acetate 127
Ethylene chloride 159
Ethyl ether 141
Ethyl iodide 158
Ethylnotopterol 30, 103
Euphausia superba 248
Evasteria retifera 270, 271

F-9 258, 287
F-21 253, 254, 287
Fatagarin 98, 103
Fatty acids 224, 231, 232, 234, 235, 237, 240, 241
Febrifugine 124
Fercoprenol 64, 103
Fercoprolone 63, 103
Ferric chloride 134
Ferula assafoetida 7
Ferula communis sp. *communis* 63, 64

Ferula communis var. *genuina* 63, 64, 74
Ferulenol 96, 103
Ferulenoloxyferulenol 74, 103
FGL-I 245, 257, 258, 287
FGL-IIa 256, 257, 287
FGL-IIb 245, 254–257, 287
FGL-V 255–257, 287
Fisher projection 216
Folch method 219
Forbesin 286
Forbesine 286
Frachinoside 40, 103
Fraxetin 91
Fraxinol 90, 103
Fraxinus chinensis 40
Freund's adjuvant 143
Frutinone A 64, 103
Frutinone B 66, 103
Frutinone C 64, 103
Furanocoumarins 5
Furobiclausarin 77, 103
Furobinordentatin 77, 103

Galabiosylceramide GL-3 236, 237
(2S,3S,4R)-1-O-(α-D-Galactopyranosyl)-2-(N-hexacosanoylamino)-1,3,4-octadecanetriol 289
Galactose 287
Galactose-6-phosphate 257
β-Galactosidase 275
Galactosylceramide 231
α-Galactosylceramides 289, 290
β-Galactosylceramides 231
Galactosylceramides 249, 289
Galactosyldiacylglycerol 224
Gala-6 series 237, 238, 250
Galbanic acid 4, 79
Gancaoin W 58, 103
Ganglioside AG-2 279–281
Ganglioside AG-3 280
Ganglioside G-1 266–268, 288
Ganglioside G-2 267, 268, 288
Ganglioside GAA-6 266, 268
Ganglioside GAA-7 269, 270, 290
Ganglioside GP-1a 276, 277, 290
Ganglioside GP-1b 277
Ganglioside GP-2 278, 279
Ganglioside LG-1 272–274
Ganglioside LG-2 273, 274, 290
Ganglioside M5 261, 262, 288

Ganglioside T1 261–263
Gangliosides 229, 260–263, 266, 270, 272–276
Geijerin 80, 103
Gentibiosylceramide 235, 236
Gerbera anandria 66, 70
Gigasol 71, 103
Gleinadiene 87, 103
Gleinene 86, 103
Glucosylceramide 229, 245
β-Glucosylceramide 230, 244
Glyasperin L 58, 103
Glycoglycerolipids 215, 216, 222–224, 227, 228
N-Glycolylneuraminic acid 263, 265–267, 269, 271, 276
Glycosides 3
Glycosphingolipids 215, 216, 219, 222, 223, 228–231, 233, 237–249, 251, 253–255, 257, 258, 260, 287–289
Glycosyl ceramides 219
Glycycoumarin 93, 103
Glycyrol 97, 103
Glycyrrhiza aspera 58
Glycyrrhiza inflata 61
Glycyrrhiza sp. 58
Glycyrrhiza uralensis 69
Gomortega keule 96
Graciliaropsis lemaneiformis 224, 226
trans-Grandmarin 37, 103
Gravelliferone 92, 103
Grignard reagents 163
Growth-inhibitory activity 270, 290

Hakomori method 220, 221, 235, 236, 274
Halichondria cylindrata 231
Halichondria japonica 230, 231, 236, 237
Halichondria panicea 231
Haliclona sp. 230
Halicylindroside A_1 231
Halicylindroside A_2 231
Halicylindroside A_3 231
Halicylindroside A_4 231
Halicylindroside B_1 232
Halicylindroside B_2 232
Halicylindroside B_3 232
Halicylindroside B_4 232
Halicylindroside B_5 232

Halicylindroside B_6 232
Haliotis japonica 246
Hansch analysis 173
Haplophyllum ptilostylum 36, 37
Haplophyllum thesioides 65
Hassmarin 75, 103
Hemicentrotus pulcherrimus 243, 244, 246, 261, 262, 288
Hemidesmin-1 52, 103
Hemidesmin-2 53, 103
Hemidesminine 41, 103
Hemidesmus indicus 41, 52, 53
Hemozoin 196
Heptaptera anisoptera 7, 8, 29
Heratomol 88, 103
Herniarin 79, 103
Hesperathusa crenulata 62
Heterocyst glycolipid I 282, 283
Heterocyst glycolipid II 284
Heterocyst glycolipid III 282, 283
Heterocyst glycolipid IV 284
Heterocyst glycolipids 283
Heterosigma akashiwo 224
Heterosigma-glycolipid I 224
Heterosigma-glycolipid II 224
Heterosigma-glycolipid III 224
(+)-Hexahydroisochroman-3-one 170
High-performance liquid chromatography 3
Hintonia latiflora 61
Hoehneliacoumarin 65, 103
(+)-Homodeoxoartemisinin 154
(+)-D-Homodeoxoartemisinin 162
Horeau's method 198
Hydrangea 124
α-Hydroxyacid 230, 231
Hydroxyacids 232
7α-Hydroxyarteether 176
7β-Hydroxyarteether 176
3α-Hydroxyartemisinic acid 175
3β-Hydroxyartemisinic acid 175
9α-Hydroxyartemisinin 157
9β-Hydroxyartemisinin 156, 157
9α-Hydroxy-10β-m-chlorobenzoate 156
9β-Hydroxy-10β-m-chlorobenzoate 156
7-Hydroxycoumarin 3
4α-Hydroxydeoxyartemisinin 197
9α-Hydroxydihydroartemisinin 151
9β-Hydroxydihydroartemisinin 151
10β-Hydroxydihydroartemisinin 156

14-Hydroxydihydroartemisinin 175
$4'\beta$-Hydroxyeriobrucinol 31, 103
[^{3}H]-Hypoxanthine 179

Immunological activity 287
Immunostimulatory activity 239, 289, 290
Imperatorin 90, 103
Inflacoumarin A 61, 103
Inositol 228
Interleukin-6 290
Isoarnottinin 15, 103
Isoartemisitene 155
Isobaisseoside 40, 103
Isobutane 140
Isoethuliacoumarin A 96, 103
Isoferprenin 64, 103
Isofraxidin 91, 103
Isoglycycoumarin 58, 103
Isoglycyrol 97, 103
Isokotanin A 70, 103
Isokotanin B 70, 103
Isokotanin C 70, 103
(+)-Isolimonene 133
Isopentenyl acetate 127
Isopulegol 129
(−)-Isopulegol 129, 130, 133
Isoscopoletin 88, 103
Ito multilayer coil separator-extractor 127

Jones oxidation 131
Jones reagent 157, 166

Keruffarides 224, 227, 228
(−)-*cis*-Khellactone 4, 22, 103
(+)-*trans*-Khellactone 4, 22, 103
(−)-*trans*-Khellactone 83, 103
Khelmarin-A 72, 103
Khelmarin-B 72, 103
Khelmarin-C 73, 103

Lactosylceramide 234, 235, 245, 268
β-Lactosylceramide 244
d-Laserpitin 22, 103
Ledum palustre 41
Leptodactylone 91, 103
Lethasterias fuska 265, 266
Leukemia cells P338 232
Lewis acids 150

Libanotis laticalycina 14
Licoarylcoumarin 58, 103
Licopyranocoumarin 93, 103
Ligularia persica 7, 8
Ligupersin A 8, 103
Ligupersin B 7, 103
Ligusticum daucoides 20, 21
Limettin 85, 103
Liquid ammonia 132
Lithium aluminum hydride 133, 153
Lithium diethylamine 146
Lithium methoxy(trimethylsilyl)-
 methylide 130
Ll-1 14, 103
Ll-2 14, 103
Lomatin 83, 103
Longiside 239, 240, 289, 290
Lonicera gracilipes var. *glandulosa* 41
Lotus creticus 69
Luffariella sp. 227, 228
Luidia quinaria bispinosa 272
Lymphoma L1210 cells 274
Lyngbya lagerheimii 226, 290

MAEP-galactosylceramide 247, 248
MAEP-glucosylceramide 248
Malaria 123, 124
Marine glycolipids 215, 217, 218, 221
Marmin 79, 103
(−)-(S)-*trans*-Marmin 6, 103
Marphysa sanguinea 249
Mass spectrometry 3
Mefloquine 124, 179, 196
Melanoma B16 234
Melibiosylceramide 236
Mercuric chloride 132
Methane 140
Methanol 139, 142, 153, 167, 218, 219, 223
Methanolic ammonia 146
5-Methoxy-8-prenyloxycoumarin 89
α-4-O-Methyl-N-acetylglucosamine 287
N-Methyl-2-aminoethylphosphonylgalac-
 tosylceramide 248
N-Methyl-2-aminoethylphosphorylglu-
 cosylceramide 248
Methylene blue 131
Methylene chloride 156
3-O-Methylgalactose 260
α-3-O-Methylgalactose 287

β-3-O-Methylgalactose 287
β-3-O-Methylgalactoside 260
4-O-Methylglucosamine 252
8-O-Methyl-N-glycolylneuraminic acid
 268, 269, 271, 274, 276
11-O-Methyl-N-glycolylneuraminic acid
 268
Methyl p-(hydroxymethyl)benzoate 160
Methyl lithium 165, 175
Methyl magnesium iodide 131
N-Methyl morpholine 157
Methyl orthoformate 132
3(R)-Methyl-6-phenylsulfinyl-cyclohexa-
 none 133
1-O-Methyl-2,3,4-tri-O-acetyl-α-D-galac-
 topyranose-6-phosphocholine 249
Metridium senile 230
Mevalonate 215
Micellar electrokinetic capillary chroma-
 tography 3
Michael addition 131
Microcybe multiflorus 75
Microcybin 75, 103
Micromelum minutum 20, 44
Microminutin 3
Microminutinin 20, 103
Minumicrolin 82, 103
Miroestrol 3
Moffat oxidation 157
Monodonta labio 247, 248
Monogalactosyldiacylglycerol 223, 224
Monoglycosylceramides 232
Mortierella remanniana 232
Mosher method 228, 282–284
Mucor mucedo 175
Murrangatin 82, 103
Murraya exotica 17
Murraya paniculata 15–17

Natural killer cell activity 289
Neanthes diversicolor 249
Neoacrimarine-A 57, 103
Neoacrimarine-B 56, 103
Neoacrimarine-C 23, 103
Neoacrimarine-D 37, 103
Neoacrimarine-E 38, 103
Neoacrimarines 4
Neofibularia nolitangere 227
Neoglycyrol 69, 103
Neuritogenic activity 270, 290

Nitrogen 142
Nivetin 61, 103
Nocardia corallina 175
Nodularia harveyana 282–284
NOESY experiment 138
NOESY interactions 149
Nordenletin 73, 103
Notopterygium forbesii 13, 31
Notopterygium incisum 30
Notoptolide 30, 103
Nuclear magnetic resonance 3

Oblongulide 94, 104
Obtusifol 4, 91, 104
Ophidiaster ophidiamus 230
Ophiomastix annulosa 261–263
Ophiura sarsi 261–263
Ophtocoma echinata 261–263, 266
Oreojasmin 98, 104
Oroselol 83, 104
Oroselone 83, 104
Osmium tetroxide 157
Ostruthin 80, 104
Oxaclausarin 57, 104
Oxanordentatin 38, 104
(2S,3R,6R)-2-(3-Oxobutyl)-3-methyl-6-
  [(R)-2-propanal]cyclohexanone 143
7-Oxygenated coumarins 6
Oxypeucedanin 85, 104
Oxypeucedanin hydrate 86, 104

Pachyrhizus tuberosus 56
Palustroside 41, 104
Paraformaldehyde 172
Paramignya monophylla 38
Patiria pectinifera 234, 274–279, 281
Pd-1a 83, 104
Pd-II 84, 104
Pelargonium sidoides 48, 52, 54
Penaeus aztecus 230
Penicillium chrysogenum 175
Pentacta australis 230
Pentaglycosylceramides 246
Penta-TMS-6-O-MAEP-galactose 250
Peroxytamarin 9, 104
Perphthalic acid 285
Petrocaulon alopecuroides 48
Petroleum ether 127, 141
Petromyces alliaceus 70, 71
Peucedanocoumarin I 25, 104

Peucedanocoumarin II 24, 104
Peucedanocoumarin III 24, 104
Peucedanum decumbens 13
Peucedanum japonicum 9, 22–25
Peucedanum praeruptorum 14, 22–26, 46
Peucedanum rubricaule 35
Peucedanum zhongdianensis 22
Peujaponiside 9, 104
Peujaponisin 25, 104
Peujaponisinol A 25, 104
Peujaponisinol B 23, 104
Pharmacological activity 288
Phebalium anceps 6, 13, 30
Phebalium bilobum 19
Phebalium coxii 19
Phebalium elatius ssp. *beckleri* 42–44
Phebalium filifolium 6, 46
Phebalium ralstonii 19
Phebalium phylicifolium 18
Philotheca citrina 39
Phormidium tenue 226, 290
Phosphoethanolamine 257, 258
Phosphoglycolipids 246, 287
Phosphoglycosphingolipids 248
Phosphonoglycolipids 244, 245, 248, 250–256, 258, 260
Phosphonoglycosphingolipids 287
Phosphonotriglycosylceramides 250
Photosynthetic organisms 224–226
Phyllocladus trichomanoides 32
Phyllocoumarin 32, 104
Phyllospongia foliascens 224, 226
Phytoalexin 79, 82, 88, 89
Phytosphingosine 217, 231
Phytosphingosines 230, 288
Pilocarpus goudotianus 29, 33, 47
Pilocarpus riedelianus 65, 67
Pituranthoside 6, 104
Pituranthos triradiatus 6
Plant coumarins 2
Plasmodium berghei 125, 142, 149, 158–160, 175, 179, 188–190, 192, 200
Plasmodium cynomolgi 160
Plasmodium falciparum 124, 148, 160, 166, 178, 179, 194–197, 199–201
Plasmodium malariae 124
Plasmodium ovale 124
Plasmodium sp. 124, 196, 199
Plasmodium vinckei 200

Plasmodium vivax 124
Plicadin 69, 104
PnGL-1 258–260
Polygala fruticosa 64, 66
Polyisoprenoidic glycolipids 215
Polytrichum formosum 50–52
Poncirus trifoliata 72, 75, 76
Potassium carbonate 145
Potassium selectride 176
(±)-Praeruptorin A 83, 104
(±)-Praeruptorin B 84, 104
Primaquine 124
Prionanthoside 40, 104
Propargyl alcohol 155
Protobruceol-I 33, 104
Protobruceol-II 33, 104
Protobruceol-II hydroperoxide 33, 104
Protobruceol-III 33, 104
Protobruceol-III hydroperoxide 33, 104
Protobruceol-IV 32, 104
Prunus prostrata 49
Pseudobruceol-I 31, 104
Pseudobruceol-II 31, 104
Pseudocalanolide C 4, 59, 104
Pseudocalanolide D 4, 59, 104
Pseudoceratina crassa 227, 228
Psoralea corylifolia 69
Psoralea plicata 69
Psoralen 81, 104
Pterocarpus santalinus 40
Pterocaulon lanatum 49
Pterocaulon purpurascens 52, 54
Ptilin 36, 104
Ptilostin 37, 104
Ptilostol 36, 104
Pueraria mirifica 56
Pulegone 133
Pummeloquinone 4, 19, 104
Purpurasol 52, 104
Purpurenol 54, 104
Pyracantha coccinea 44
Pyracanthin A 44, 104
Pyracanthin B 44, 104
Pyranocoumarins 5
Pyridine 159, 183
Pyridine hydrochloride 263
Pyrimethamine 158, 179
Pyruvic acid 254, 255

Qianhucoumarin A 22, 104
Qianhucoumarin B 23, 104
Qianhucoumarin C 22, 104
Qianhucoumarin D 23, 104
Qianhucoumarin E 26, 104
Qianhucoumarin F 14, 104
Qianhucoumarin G 46, 104
Qinghao 123–125
Qinghaosu 124, 125
Quinine 124, 179, 196

Ramosinin 92, 104
Rhizochalin 285
Rhizochalina incrustata 285
Rhizopogon sp. 175
Rosa rugosa 199
Rubricauloside 35, 104
Ruta graveolens 46, 47
Rutamarin 92, 104
Rutaretin 89, 104
Ruthenium periodate 134
Ruthenium trichloride 172

Sargassum thunbergii 226
Schinicoumarin 67, 104
Schinifolin 89, 104
Schinilenol 45, 104
Schininallylol 45, 104
Schinindiol 45, 104
Schizonticidal activity 199
Scoparone 88, 104
Scopoletin 88, 104
(+)-4,5-Secoartemisinin 168
(+)-8a,9-Secoartemisinin 167
(+)-D-Secoartemisinin 167
Seselin 84, 104
Seselinal 86, 104
Seselinol 26, 104
Seselinol isovalerate 26
Seshadrin 95, 104
Sesibiricin 87, 104
Setaria italica 63
Setarin 4, 63, 104
SGL-I 251, 252, 287
SGL-I' 253, 287
SGL-II 252, 253, 254, 260, 287
Sialic acid 229, 262–269, 271, 272, 274–281
Sialogangliosides 271
Sialoglycolipids 220

Subject Index

Sialosphingolipids 262
Silica gel 127, 147, 150–152, 156
Silver oxide 158
Simsia cronquistii 49
Skimmia japonica 46
Sodium 132
Sodium artesunate 141, 160, 196, 197
Sodium bicarbonate 194
Sodium borohydride 138, 153, 165
Sodium carbonate 156
Sodium methoxide 146
Sodium naphthalenide 161
Sodium periodate 172
Sodium thiophenolate 133
Solvent extraction 3
Sphinganine 217
Sphingosine 217
Sphingosines 223, 230–232
Sphondin 88, 104
Stellera chamaejasme 71
Steroidal glycosides 215
Streptomyces lavendulae 175
Strongylocentrotus intermedius 266, 267, 288
Strongylocentrotus nudus 261, 264, 265
Suberenol 80, 104
Suberosin 80, 104
9-Substituted artemisinin derivatives 183
N-Substituted 11-azaartemisinins 178
3-Substituted coumarins 56
4-Substituted coumarins 59
6-Substituted-7-oxygenated coumarins 9
8-Substituted-7-oxygenated coumarins 15
Succinic anhydride 160
Sulfadoxine 158, 179
Sulfonic acid 226
Sulfoquinovosyldiacylglycerol 223, 226, 290
Sulfuric acid 141, 147, 156

Tenuidin 80, 104
Terpenic glycosides 215
Tetragalactosylceramide 238
Tetraglycosylceramides 245, 259, 260
2,3,4,6-Tetra-*O*-methylgalactitol 255
5,6,7,8-Tetraoxygenated coumarins 3, 54
Tetraphis pellucida 50, 51, 54

Thesiolen 65, 104
Toddacoumalone 36, 104
Toddacoumaquinone 4, 36, 104
Toddalenol 29, 104
Toddalenone 35, 104
Toddalia asiatica 4, 29, 35, 36, 75
Toddalolactone 4, 85, 104
Toddalosin 75, 104
Toddanol 85, 104
Toluene 146
p-Toluenesulfonic acid 150, 154
Tortuoside 82, 104
Tributyl tin hydride 155
Triethylamine 159
Triethylsilicon trioxide 175
Triethylsilyl hydrotrioxide 174
Trigalactosylceramide 237, 238, 250
Triglycosylceramides 243–245, 251
Trikentrion loeve 227
3,6,9-Trimethyl-9,10b-epidioxyperhydropyrano[4.3.2-jk]benzoxepin-2-one 129
2,3,4-Tri-*O*-methylgalactoside 238
3-Trimethylsilyl-butenone 131
Trimethylsilylether 167
1,2,4-Trioxanes 174, 197, 201
5,6,7-Trioxygenated coumarins 48
5,7,8-Trioxygenated coumarins 50
6,7,8-Trioxygenated coumarins 52
Triphasia trifolia 16
Triphenylphosphine hydrobromide 150
Triphosphonoglycolipids 254, 260
Tripneustes ventricosa 260–263
Triscoumarins 2, 3, 78
4,6,8-Trisubstituted-5,7-dioxygenated coumarins 4
Triumbelletin 78, 104
Triumbellin 78, 104
Tungsten hexachloride 136
Turbo cornutus 231, 238, 247, 248, 250
Two-dimensional planar chromatography 3

Ultrasound-assisted furanocoumarin extraction 3
Ulva pertusa 226
Umbelliferone 79, 104

Vanillin 141
Verongula gigantea 227

Vinyl silanes   129
Viola prionantha   40
(−)-Visnadin   24, 104

Wittig reaction   162, 170, 175

Xanthotoxin   89, 104
Xestospongia sp.   227

Yingzhaosu A   198
Yingzhaosu B   198
Yingzhaosu C   198
Yingzhaosu D   198
Yuehgesin-A   4, 17, 104
Yuehgesin-B   16, 104
Yuehgesin-C   16, 104

Zanthoxylum schinifolium   6, 45, 67

SpringerChemistry

Fortschritte der Chemie organischer Naturstoffe

Progress in the Chemistry of Organic Natural Products

Founded by L. Zechmeister
Edited by W. Herz, G. W. Kirby, R. E. Moore, W. Steglich, and C. Tamm

Volume 71

1997. 11 partly coloured figures. IX, 358 pages.
Cloth DM 320,–, öS 2240,–
Subscription price:
Cloth DM 288,– , öS 2016,–
ISBN 3-211-82850-8

Contents:
G. Gäde: The Explosion of Structural Information on Insect Neuropeptides.
S.B. Christensen, A. Andersen, and U.W. Smitt: Sesquiterpenoids from Thapsia Species and Medicinal Chemistry of the Thapsigargins.
D. Deepak, S. Srivastav, and A. Khare: Pregnane Glycosides.

Volume 70

1997. 86 partly coloured figures. VII, 307 pages.
Cloth DM 290,–, öS 2030,–
Subscription price:
Cloth DM 261,–, öS 1827,–
ISBN 3-211-82825-7

Contents:
G.R. Pettit: The Dolastatins.
A. Cavé, B. Figadère, A. Laurens, and D. Cortes: Acetogenins from Annonaceae.

SpringerWienNewYork

Sachsenplatz 4-6, P.O.Box 89, A-1201 Wien, Fax +43-1-330 24 26,
e-mail: order@springer.at, Internet: http://www.springer.at
New York, NY 10010, 175 Fifth Avenue • Heidelberger Platz 3, D-14197 Berlin
Tokyo 113, 3-13, Hongo 3-chome, Bunkyo-ku

SpringerChemistry

Fortschritte der Chemie organischer Naturstoffe

Progress in the Chemistry of Organic Natural Products

Founded by L. Zechmeister
Edited by W. Herz, G. W. Kirby, R. E. Moore, W. Steglich, and C. Tamm

Volume 69

1996. 17 figures. IX, 268 pages.
Cloth DM 250,–, öS 1750,–
Subscription price:
Cloth DM 225,–, öS 1575,–
ISBN 3-211-82824-9

Contents:
J.F. Grove: Non-Macrocyclic Trichothecenes, Part 2.
D. Deepak, S. Srivastava, N.K. Khare, A. Khare: Cardiac Glycosides.
E. Haslam: Aspects of the Enzymology of the Shikimate Pathway.

Volume 68

1996. VIII, 498 pages.
Cloth DM 330,–, öS 2310,–
Subscription price:
Cloth DM 297,–, öS 2079,–
ISBN 3-211-82702-1

Contents:
G.W. Gribble: Naturally Occurring Organohalogen Compounds – A Comprehensive Survey.

SpringerWienNewYork

Sachsenplatz 4-6, P.O.Box 89, A-1201 Wien, Fax +43-1-330 24 26,
e-mail: order@springer.at, Internet: http://www.springer.at
New York, NY 10010, 175 Fifth Avenue • Heidelberger Platz 3, D-14197 Berlin
Tokyo 113, 3-13, Hongo 3-chome, Bunkyo-ku